T0241742

Contributions from Biology Education Research

Founding Editors
Marida Ergazaki
Kostas Kampourakis

Series Editors
Marcus Hammann, Zentrum für Didaktik der Biologie, University of Münster, Münster, Nordrhein-Westfalen, Germany
Anat Yarden, Department of Science Teaching, Weizman Institute of Science, Rehovot, Israel

Editorial Board Members
Jörg Zabel, Leipzig University, Leipzig, Germany
Constantinos Korfiatis, University of Cyprus, Nicosia, Cyprus
Maria Pilar Jimenez Aleixandre, University of Santiago de Compostela, Santiago de Compostela, Spain, Ute Harms, Kiel University, Kiel, Germany
Michael Reiss, University College London, London, UK
Niklas Gericke, Karlstad University, Karlstad, Sweden
Charbel Nino El-Hani, Federal University of Bahia, Salvador-Bahia, Brazil
Vaille Dawson, The University of Western Australia, Perth, Australia
Ross Nehm, Stony Brook University, Stony Brook, NY, USA
William McComas, University of Arkansas, Fayetteville, USA
Cynthia Passmore, University of California Davis, Davis, CA, USA
Marcus Grace ⓘ, University of Southampton, Southampton, UK
Marie Christine Knippels ⓘ, Utrecht University, Utrecht, The Netherlands

Contributions from Biology Education Research (CBER) is the international book series of the European Researchers in Didactics of Biology (ERIDOB). The series includes edited collections of state-of-the-art papers presented at the ERIDOB international conferences, and monographs or edited collections of chapters by leading international scholars of the domain. The aim of the series is to shed light on global issues and trends in the teaching and learning of biology by gathering cutting edge research findings, theoretical views, and implications or concrete suggestions for everyday school practice regarding biology. The books may serve as resources for (a) getting informed about the most recent findings of biology education research to possibly integrate them in new personal research, and (b) studying about the teaching and learning of biology as a pre-service or in-service biology teacher. So, the main audiences for the series range from senior to early career biology education researchers and pre- or in-service biology teachers working at all educational levels. Book proposals for this series may be submitted to the Publishing Editor: Claudia Acuna E-mail: Claudia.Acuna@springer.com

Konstantinos Korfiatis • Marcus Grace
Marcus Hammann

Editors

Shaping the Future of Biological Education Research

Selected Papers from the ERIDOB 2022 Conference

 Springer

Editors
Konstantinos Korfiatis (iD)
Department of Education
University of Cyprus
Nicosia, Cyprus

Marcus Grace (iD)
Education School, Building 32
University of Southampton
Southampton, UK

Marcus Hammann
Centre for Biology Education
University of Münster
Munster, Nordrhein-Westfalen, Germany

ISSN 2662-2319   ISSN 2662-2327   (electronic)
Contributions from Biology Education Research
ISBN 978-3-031-44794-5   ISBN 978-3-031-44792-1   (eBook)
https://doi.org/10.1007/978-3-031-44792-1

This Springer imprint is published by the registered company Springer Nature Switzerland AG
The registered company address is: Gewerbestrasse 11, 6330 Cham, Switzerland

Paper in this product is recyclable.

European Researchers in Didactics of Biology (ERIDOB)

Preface

ERIDOB is the organization of European Researchers in the Didactics of Biology. It aims to bring together researchers in the didactics of biology from Europe and around the world to share and discuss their research findings and practical implications. This book of selected papers is a collection of 24 original research contributions presented at the 13th ERIDOB conference, which took place at the University of Nicosia, Cyprus, from August 29 to September 2, 2022. The contributions address a wide range of different research topics related to biology education. This also includes preparing teachers to teach biology effectively. The common theme of this collection, however, is the shared belief that research in biology education is a powerful tool to reflect upon, understand, and innovate the current practices of biology education and pre-service and in-service biology teacher training. Furthermore, the contributions in this collection are united by the belief that subject matter is important in teaching and learning.

The contributions in this collection are either theoretical or empirical in nature. Theoretical contributions describe the theoretical and conceptual underpinnings of teaching biology and research in biology education. Often, studies of this type derive insights into theoretical and conceptual frameworks by summarizing a large number of studies from the different fields of biology education. For example, one contribution in this collection addresses the concept of problematization in biology education and another contribution describes features of virtual and physical laboratory environments and their impact on learning. Empirical contributions investigate research questions related to biology education with qualitative, quantitative, or mixed methods to construct evidence-based arguments about biology education and propose solutions to problems, needs, or challenges. Empirical studies use theoretical frameworks, for example for deriving research questions and interpreting findings. For example, empirical contributions in this collection address the challenge of implementing climate change education and delivering an understanding of the anthropogenic factors affecting the loss of biodiversity. Furthermore, the boundaries between theoretical contributions and empirical contributions are both fluid and firm because studies can easily be classified as either theoretical or empirical, but, at the same time, theoretical and empirical studies are mutually dependent on one another.

ERIDOB contributes to shaping the future of research in biology education. It provides a unique platform for researchers in biology education to discuss theoretical frameworks, empirical aspects of research in biology education, for example methods of data collection and analysis, and evidence-based recommendations for innovation. The contributions in this collection refer to theoretical frameworks which guide research, draw conclusions in relation to evidence collected as part of the research, reflect on the originality of the research contributions, and discuss why it is important to gain such insights. Collectively, researchers who contribute to the ERIDOB conference – and particularly the researchers who share their research findings in this collection – advance the current understanding of effective biology teaching and learning in fields as diverse as students' conceptual understanding and reconstruction of biology, students' biology-related interests, attitudes, and motivation, environmental and health education, modelling and experimentation, and biology teacher preparation.

The chapters of this book have been organized in four parts. The first group of chapters (*Part I: Teaching Strategies and Learning Environments*) consists of seven papers focusing on a range of different aspects related to teaching strategies and learning environments. The first two chapters in this group are by authors who kindly agreed to hold keynote lectures at the conference. The chapter by Yvoni Pavlou and Zacharias C. Zacharia (Chap. 1) discusses the advantages and disadvantages of physical and virtual laboratories as means of experimentation for students in STEM+ education. The authors' main conclusion is that each mode of experimentation has specific affordances which need to be considered when making the decision to experiment in physical or virtual laboratories.

The chapter by Catherine Bruguière and Denise Orange Ravachol (Chap. 2) addresses the didactic functions of problematization, narrative, and fiction in the science classroom. The authors highlight a theoretical framework of problematization developed over more than 20 years in France and explore the more recent notion of realistic fiction by situating it in relation to work in biology didactics that borrows from the frameworks of narrative and/or fiction.

The chapter by Benjamin Stöger and Claudia Nerdel (Chap. 3) deals with mathematical modelling competence in biochemistry. The authors found that external representations and mathematical expertise have a positive effect on mathematical modelling competence in biochemistry.

The contribution of María Napal Fraile, Lara Vázquez Bienzobas, Isabel Zudaire Ripa, and Irantzu Uriz Doray (Chap. 4) focuses on the development of science process skills in pre-school children. The authors compared three types of interventions and found that the children in the adult-led intervention attained the most detailed learning of scientific concepts, while the scaffolded exploration group of children improved their scientific inquiry skills.

The next two chapters focus on informal learning environments. The contribution of Georgios Villias and Mark Winterbottom (Chap. 5) deals with educational escape rooms as a type of informal learning environment which has recently gained popularity. The authors investigated the potential of educational escape rooms to develop students' twenty-first century skills and describe design criteria for educational escape rooms.

The chapter by Anna Pshenichny-Mamo and Dina Tsybulsky (Chap. 6) focuses on the question what aspects of the nature of science (NOS) visitor guides of natural history museum address when they discuss topics related to evolution and ecology. The authors found that natural history museum guides often integrate cognitive-epistemic aspects of NOS, for example the tentativeness of scientific knowledge, which is prone to change with new findings.

The final chapter of this part is by Georgios Ampatzidis and Anastasia Armeni (Chap. 7). It deals with representations of microorganisms in biology textbooks. The authors found that microorganisms are often represented negatively so that a conceptual shift in the representations of microorganisms is needed in order to help students appreciate the importance of microorganisms in human lives and economy, as well as their role in the ecosystem.

The second group of chapters (*Part II: Students' Knowledge, Conceptions, Values, Attitudes and Motivation*) consists of seven papers focusing on a range of different cognitive and affective aspects of students' learning.

The contribution of Wilton G. Lodge, Michael J. Reiss, and Richard Sheldrake (Chap. 8) deals with students' views of the benefits of investigative school biology research projects. Specifically, participants indicated that they gained a deeper understanding of the scientific processes and the production of scientific knowledge. Furthermore, the research projects increased their science-related career aspirations.

The acceptance of evolutionary theory is an important aspect of evolution education. Andreani Baytelman, Theonitsa Loizou, and Salomi Chadjiconstantinou (Chap. 9) investigated 12th grade students' epistemological beliefs toward science and their personal beliefs in plant evolution, animal evolution, and human evolution. The authors found that students' epistemological beliefs in science predicted their personal beliefs in plant evolution and animal evolution, but not in human evolution.

The next two chapters address pupils' and students' understanding of plants and their physiology. The chapter by Alexandros Amprazis and Penelope Papadopoulou (Chap. 10) deals with plant blindness, a multidimensional construct which involves interest in plants and knowledge of plants among other variables. The authors conducted a cross-sectional study and found that students' preference for plants was lowest in high school (compared to higher preference for plants in elementary school and university students) and that correlation coefficients between interest in plants and knowledge of plants differed between the various age groups studied.

The chapter by Eliza Rybska, Joanna Wojtkowiak, Zofia Chyleńska, Pantelitsa Karnaou, and Costas P. Constantinou (Chap. 11) explores the extent to which student-constructed drawings can be used to identify how modelling skills evolve with age and how they relate to students' understanding of the mechanism of photosynthesis and plant growth. The authors found that student-constructed drawings provided insights into students' modelling skills and that the latter depended on the students' knowledge of photosynthesis.

The contribution of Malte Ternieten and Doris Elster (Chap. 12) describes findings from a study aimed at developing, testing, and evaluating a grid for the assessment of students' decision-making competencies in the context of Education for Sustainable Development. The chapter offers a description of a six-hour unit in the context of Education for Sustainable Development. Furthermore, the authors present the diagnostic grid in conjunction with findings from the evaluation study.

The chapter by Julia Holzer and Doris Elster (Chap. 13) represents an investigation of the factors influencing the students' intention to donate stem cells to leukemia patients. Based on the Theory of Planned Behavior, a specially designed educational intervention significantly impacted factors which were expected to be influential, for example beliefs, as well as the intention to donate stem cells.

The contribution of Martha Georgiou and Matina Moshogianni (Chap. 14) describes a study investigating Greek teenagers' (15–16 years old) knowledge of nutrition and healthy eating. The authors of the study found that the participants lack nutrition knowledge. Furthermore, the authors argue that the internet is the participants' main source of nutrition knowledge and that there is a gap between the students' nutrition knowledge and school-based nutrition education programs.

The third group of chapters (*Part III: Outdoor and Environmental Education*) consists of six contributions addressing aspects related to climate change, conservation, and sustainable use of biodiversity as well as school-based outdoor education.

The contribution by Nofar Naugauker, Orit Ben-Zvi-Assaraf, Daphne Goldman, and Efrat Eilam (Chap. 15) represents a case study addressing the implementation of climate change into the national curriculum in Israel. The authors provide insights into perceptions of education policymakers and education professionals. Furthermore, they describe opportunities and challenges to effective implementation, and argue that inter-sectorial collaborations need to be involved in implementing climate change education.

The contribution by Lina Varg (Chap. 16) focuses on the congruency between primary teachers' intentions to teach sustainability and their implemented teaching practices. The author observed discrepancies between primary teachers' intentions and teaching practices. Therefore, the argument is made that a greater flexibility in teaching approaches is necessary for delivering the intended educational goals.

The chapter by Chadia Rammou, Arnau Amat, Isabel Jiménez-Bargalló, and Jordi Martí (Chap. 17) examines the factors elementary school children consider important for enhancing and limiting the biodiversity of particular environments. The children considered the presence of vegetation as an important factor enhancing biodiversity, whereas human beings were considered a limiting factor of biodiversity. The authors argue that participants' views on biodiversity were rather simplistic and anthropocentric and that science education should enhance the development of a more comprehensive understanding of biodiversity.

The next two papers focus on the benefits of outdoor learning and nature experience. The chapter by Marjanca Kos, Sue Dale Tunnicliffe, Luka Praprotnik, and Gregor Torkar (Chap. 18) explores the possibilities of pre-school age children to experience biological phenomena and living organisms through unstructured play in nature. The children paid more attention to animals than plants and fungi, whereas the experiences with animals were more emotionally engaging and led to deeper learning about them. Involvement of teachers in play with young children led to longer duration of play and deeper learning.

The contribution of Marcus Hammann (Chap. 19) describes a study exploring variables related to secondary school students' perceived personal relevance of free-choice nature experiences. Furthermore, the study focused on associations between perceived personal relevance and the frequency of such nature experiences. The study found that age, gender, explicit positive evaluations, frequency of experience, and type of settings were significant determinants of perception of personal relevance. Furthermore, most participants attributed considerable importance to personal experiences with nature and advocated opportunities for nature experiences in biology instruction.

The last chapter of Part III, written by Anthi Christodoulou and Konstantinos Korfiatis (Chap. 20), examines the challenges that emerged during an action research approach applied in a school garden project. The authors found that dialogue and shared knowledge between the action research members were important to create a participatory and effective project. Furthermore, the establishment of democratic and inclusive modes of communication between teachers and students and the identification of students' personal worries and social difficulties were important factors for the success of the project.

The last part of the book (*Part IV: Biology Teachers' Professional Development*) consists of four chapters addressing the importance of biology teachers' professional development.

The chapter by Emanuel Nestler, Carolin Retzlaff-Fürst, and Jorge Groß (Chap. 21) presents findings from a study evaluating an innovative program for training biology teachers' mentors. The program aimed at enabling mentors to adapt their mentoring to the concrete situation and to the challenges of pre-service teachers in their individual development. The program promoted the mentors' dialogic skills along with their pedagogical content knowledge and content knowledge.

The contribution by Alexander Aumann and Holger Weitzel (Chap. 22) analyzes how pre-service biology teachers use their Technological Pedagogical and Content Knowledge (TPACK) to plan, implement, and reflect on a biology lesson in which students created videos explaining a specific biological topic. The authors describe two types of pre-service biology teachers who differ in their ability to combine digital technology and media with biological content knowledge. The authors suggest that it is important to enable pre-service biology teachers to use video production as a tool for content learning.

The contribution of Patrizia Weidenhiller, Susanne Miesera, and Claudia Nerdel (Chap. 23) addresses biology teachers' attitudes and self-efficacy assumptions concerning inclusion and digitalization. The chapter presents findings from an in-service teacher training study aimed at preparing teachers for the use of digital media in heterogeneous classrooms. The teachers planned and performed a digitally supported experimentation and were trained to meet the needs of diverse learners.

In the final chapter of the book (Chap. 24), Sara Großbruchhaus, Patricia Schöppner, and Claudia Nerdel describe the findings of a study investigating the potential of professional development to promote in-service biology teachers' ability to use technological equipment for the teaching of biotechnology in class. The authors describe three strategies used by the teachers to implement biotechnology experiments in the classroom. Interviews showed that a one-time professional development program can lead to sustained implementation over several years and successful school curriculum development.

Acknowledgments

We would like to thank Sophia Sergiou and Stella Petrou for helping with the logistics of the submission, review and proofreading process.

We would also like to thank the proof-reader of the chapters of the book, Dr. Daniel Wilson, for his excellent, professional work that has largely improved the quality of the written text.

We are extremely grateful to all the colleagues in the field of biology education who kindly took part in the peer reviewing of the chapters in this book:

Jeremy Airey	Daniel Olsson
Georgios Ampatzidis	Denise Orange
Petra Bezeljak	Christina Ottander
Jelle Boeve-de Pauw	Penelope Papadopoulou
Catherine Bruguiere	Michael Reiss
Graça S. Carvalho	Carolin Retzlaff-Fürst
Frederic Charles	Mikael Rydin
Ilana Dubovi	Martin Scheuch
Marida Ergazaki	Laurence Simonneaux
Niklas Gericke	Jelka Strgar
Jorge Groß	Annika Thyberg
Marcus Hammann	Iztok Tomažič
Michal Haskel-Ittah	Gregor Torkar
Lissy Jäkel	Dina Tsybulsky
Corinne Jegou	Michiel Van Harskamp
Konstantinos Korfiatis	Susanne Walan
Marjanca Kos	Holger Weitzel
Olivier Morin	Orit Ben Zvi Assaraf
Claudia Nerdel	

Finally, we would like to thank all the members of ERIDOB Academic Committee for their valuable contribution, cooperation, and support during all the stages of the preparation of the manuscripts for the current volume.

Nicosia, Cyprus   Konstantinos Korfiatis

Southampton, UK   Marcus Grace

Munster, Germany   Marcus Hammann
July 2023

Contents

Part I
Teaching Strategies and Learning Environments

Chapter 1
Using Physical and Virtual Labs for Experimentation in STEM+ Education: From Theory and Research to Practice

Yvoni Pavlou and Zacharias C. Zacharia

1.1 Introduction

Educational reform efforts in numerous countries (mostly since 2015) focus on the integration of science, technology, engineering and mathematics (STEM) policies in national curricula to promote interdisciplinary learning (Zhan et al., 2022). For example, the Next Generation Science Standards (NGSS; Next Generation Science Standards Lead States, 2013) endorse coherent and interconnected content and a coherent and interconnected approach to the STEM disciplines to support the development of students' scientific literacy. The NGSS include practices and core disciplinary ideas from engineering and the sciences that bring out the connections between the different domains, and facilitate their introduction in teaching and learning in an integrated way.

Even though inconsistencies in the definition of STEM education exist, it is nevertheless typically conceptualized as a teaching approach that incorporates skills from and knowledge of different disciplines, situated in real-world issues (Martín-Páez et al., 2019). The integration of other disciplines within STEM curricula is also a common practice in order to promote 21st century skills, such as creativity and entrepreneurship. For example, STEAM refers to the integration of arts and/or humanities in STEM curricula; STEAME additionally incorporates entrepreneurship. For the purposes of this article, the term STEM+ will be used from this point forward to represent all possible variations of STEM plus the additional disciplines that can be incorporated in STEM curricula. Educational approaches, such as inquiry-based (e.g., Pedaste et al., 2015), project-based (e.g., Capraro et al., 2013) and design-based learning (e.g., Crismond & Adams, 2012) are ideal for

Y. Pavlou (✉) · Z. C. Zacharia (✉)
Research in Science and Technology Education Group, Department of Educational Sciences, University of Cyprus, Nicosia, Cyprus
e-mail: pavlou.ivoni@ucy.ac.cy; zach@ucy.ac.cy

© The Author(s) 2024
K. Korfiatis et al. (eds.), *Shaping the Future of Biological Education Research*, Contributions from Biology Education Research, https://doi.org/10.1007/978-3-031-44792-1_1

implementing STEM+ initiatives because they are student-centered approaches that aim at addressing real-world issues through methods and practices like those of professionals (Martín-Páez et al., 2019; Thibaut et al., 2018). These approaches can facilitate the application of STEM+ concepts and practices in contexts that are relevant and interesting to students (National Academy of Sciences, 2014).

Laboratory experimentation is a fundamental aspect of science and engineering, and hence, the aforementioned educational approaches strive to provide relevant learning experiences to students. Laboratory work provides opportunities for students to implement scientific practices and skills (e.g., observation, hypothesis generation) to test theories and understand natural phenomena (de Jong et al., 2013; de Jong et al., 2014), and also to develop practical (e.g., handling equipment) and transferable (e.g., problem-solving, time management) skills (Reid & Shah, 2007). Traditionally, experimentation took place solely in physical laboratories (PL) that allowed learners to interact with real world physical/concrete materials and apparatus in order to observe and understand natural phenomena. This direct physical experience has been reported to be of pivotal importance during experimentation (e.g., Gire et al., 2010; Kontra et al., 2015; Zacharia, et al., 2012).

However, the exponential growth of technology has led to the need to rethink the practice of laboratory experimentation (Bybee, 2009; de Jong et al., 2013; National Research Council, 2006), with many studies in the past decades focusing on the use of virtual laboratories (VL) and the consequences for teaching, learning and research. Reeves and Crippen (2021) defined VL as "technology-mediated experiences in either two- or three-dimensions that situate the student as being in an emulation of the physical laboratory with the capacity to manipulate virtual equipment and materials via the keyboard and/or handheld controllers" (p. 16). The need for VL to be optimally integrated in STEM+ education became even stronger due to the COVID-19 pandemic and the subsequent rapid shift towards online/blended learning (European Commission, 2022), which inevitably promoted VL as an alternative to PL (Bazelais et al., 2022; Radhamani et al., 2021; Raman et al., 2021). Hence, it is not surprising that there is an increasing focus on the integration of technology to support learning in STEM education research (Zhan et al., 2022).

However, given that both PL and VL are available options for experimentation purposes, the dilemma persists as to which means of experimentation should be preferred for optimizing student learning. For instance, we are still looking for definite answers to questions such as, "Which means of experimentation under what circumstances optimizes student learning across grades K-16, PL or VL?" and "Should PL and VL be used alone or in combination across grades K-16?" The objective of this paper is to synthesize the theoretical and empirical perspectives emerging from the research concerning the exploration of effects of using VL and PL on students' learning during experimentation in order to contribute to the efforts of the community to answer these questions and to inform the research about and practice of laboratory experimentation in STEM+ education.

1.2 The Theoretical Perspective

Traditional views of cognition claim that the brain is made up of abstract functions and that it is a separate entity from the body, a notion that embodied theories currently challenge (Marmeleira & Duarte Santos, 2019). Embodied cognition theories differ in the degree of the effect of sensorimotor experience on cognition that is postulated (Wellsby & Pexman, 2014), but nevertheless, their overarching notion is the same: the interaction between the environment and the body influences cognition (Clark, 2008; Pouw et al., 2014; Wilson, 2002). As Pouw et al. (2014) mentioned, learning seems to depend on "gradual internalization of sensorimotor routines" (p. 65). Hence, researchers attempt to explain human motor, perception and cognition systems as dependent on the body (Farina, 2021). The mental representation of an object involves the sensory and motor regions of the brain, which are activated when the object is perceived or interacted with (Yee & Thompson-Schill, 2016). This mental representation encompasses not only the visual properties of an object, but also relevant actions (Barsalou, 2008; Gibbs, 2005). Neurological studies have shown that memory recall activates areas of the brain associated with the sensorimotor information experienced during an episode (Kiefer & Pulvermüller, 2012). For example, during the retrieval of haptically encoded stimuli, somatosensory and motor areas of the brain are activated, whereas for visually encoded stimuli, the activation of vision-related areas is observed (Stock et al., 2009). Embodied cognition theories tend to be appealing in terms of representing the organization of conceptual knowledge because they also predict how information is obtained (through sensorimotor experiences) and how and where it is processed (in the relevant sensorimotor systems; Yee et al., 2018).

For educational research, embodied cognition theories provide the opportunity to explore the impact of action on cognition throughout development and to utilize this knowledge to scaffold the teaching and learning process (Kontra et al., 2012). Hayes and Kraemer (2017) noted that embodied cognition theories can support our understanding of how students' STEM learning is enhanced, given the student-centered and hands-on nature of STEM education. In the study by Kontra et al. (2015), even brief physical experience with the forces related to angular momentum led to activation of the sensorimotor systems of the brain used to execute similar actions in the past, which resulted in the development of understanding about that concept. Rich sensorimotor experiences can support the presence of multimodal representations that facilitate learning (Kiefer & Trumpp, 2012) with a variety of ways to engage the body, from limited (e.g., gestures) to full-body movement (Skulmowski & Rey, 2018).

The provision of high-embodiment experiences does not necessarily guarantee a positive impact on learning, but bodily experiences that relate to the task at hand seem to do so (Johnson-Glenberg, 2019; Skulmowski & Rey, 2018). For example, in a study by Mavilidi et al. (2017), preschoolers were engaged in activities related to the solar system in three experimental conditions: (1) integration of related physical activities, (2) integration of irrelevant physical activities and (3) no physical activities. It became apparent after an immediate and a 6-week assessment, that the preschoolers participating in tasks that incorporated meaningful physical activities outperformed the rest of the groups, and even the preschoolers participating in the irrelevant physical activities outperformed the students who were not involved in any type of physical activities. Zohar and Levy (2021) investigated whether an increase in bodily engagement (movie, simulation, joystick and haptic device with force feedback) subsequently leads to an increase in understanding of the concept of chemical bonding. The movie, simulation and joystick conditions resulted in similar conceptual development, whereas the participants in the haptic device condition, which offered the highest bodily engagement, had a significantly higher increase in knowledge. In a study by Qi et al. (2021), providing force feedback in a simulation to students with limited prior understanding of forces also facilitated learning, but providing additional visual cues (i.e., abstract arrows) did not improve performance. Hence, not all bodily experiences can enhance learning in the same manner; alignment between the manipulation and the learning objective is needed.

The active and meaningful interaction with materials and apparatus during experimentation is what enhances learning, and not physicality in itself (Han, 2013; Klahr, et al., 2007; Pouw et al., 2014; Triona & Klahr, 2003; Zacharia & Olympiou, 2011). Hence, embodied cognition theories do not necessarily favor a specific mode of experimentation (Rau, 2020). However, haptic perspectives on learning do favor the haptic manipulation of materials because, when combined with visual stimuli, it can support memory retrieval, minimize the likelihood of cognitive overload and support the conceptual grounding of abstract concepts (Rau, 2020). As stated by Van Doorn et al. (2010, p. 813), "[t]he term haptic refers to a perceptual system that combines both input from receptors in the skin and kinesthetic information." Through touch, we can gather information (e.g., about the properties of an object) and act (e.g., lift the object), but also, based on the sensory feedback received (e.g., the force used was not enough to lift the object), we attune our actions to fit our initial intentions (e.g., use more force; Reiner, 2008).

The haptic and the visual system are complementary in nature. In a study by Reiner et al. (2006), participants used a haptic interface to lift virtual cylinders marked with the labels "heavy", "light" or "###" (neutral condition), which were compatible, incompatible or neutral with regard to the actual weight of the cylinders. It became apparent that both the reaction time and error rate were lower for the cylinders with a label that was compatible with their weight, and higher when the label was incompatible. The haptic system has a higher processing cost than the visual system and hence it will be invoked when visual information is inadequate for addressing a targeted task (Hatwell, 2003; Klatzky, et al. 1993). Given that touch has an inherent bias towards how an object "feels" (e.g., texture, material) rather than its structural properties (e.g., size, shape; Klatzky et al., 1991; Klatzky et al., 1987; Klatzky et al., 1993), the haptic system will be activated only to assist vision when exploring structural properties (Hatwell, 2003), but it can be used for discriminating objects based on their surface textures (Heller, 1989). Thus, based on haptic perspectives on learning, it is expected that haptic manipulation, as an additional available modality, will augment the development of concept-specific understanding, particularly of concepts for which haptic cues can facilitate the development of multimodal representations, such as force (Han, 2013; Zohar & Levy, 2021) and mass (Lazonder & Ehrenhard, 2014; Pavlou et al., under review; Zacharia et al., 2012). For example, in the study by Lazonder and Ehrenhard (2014) regarding free fall, students who engaged with physical materials were able to develop scientifically correct understandings because the haptic sensory feedback available facilitated the correction of students' misconceptions and revision of mass-related beliefs. This process was not evident in the demonstration or VL groups.

Haptic sensory feedback simply cannot be offered in a virtual environment. Haptic technologies for providing sensory feedback (e.g., force feedback) in VL do exist, and studies investigating their impact on learning (e.g., Bivall et al., 2011; Han & Black, 2011; Jones et al., 2006; Magana et al., 2019; Zhuoluo et al., 2019) have reported positive findings, but the sensory feedback they provide is still quite limited in comparison to the haptic and dynamic feedback available when engaging with PL. Haptically enhanced simulations are also primarily focused on developing skills (e.g., related to surgeries; Qi et al., 2021), and the integration of such technologies, especially in formal education, is still very limited (Georgiou & Ioannou, 2019; Johnson-Glenberg, 2019; Malinverni & Pares, 2014).

As Pavlou et al. (under review) pointed out, when considering comparative studies in the field of science education from the haptic encoding and embodied cognition perspectives, findings of an equal or even negative impact of PL on learning can be attributed to a lack of significant perceptual differences between the VL and PL being compared (i.e., the perceptual stimuli and the available feedback offered in the PL did not differ significantly from the virtual environment; e.g., Han, 2013) and/or the participants' prior experiences/knowledge of the concepts under investigation. As far as the latter is concerned, it is evident that most studies in the field of science education focus on the primary school years and onwards (Wörner et al., 2022; Zacharia, 2015). For older students, knowledge of the concepts under investigation could have been grounded in the early years and hence, embodied experiences might not be a prerequisite for those students. As Yee et al. (2018) mentioned, as development progresses, the reliance on abstract knowledge increases, and direct sensorimotor experience is not as necessary as for young children. However, PL can be more beneficial than VL in the early years of education when considering the reliance on grounded experience, especially through the haptic manipulation of objects (Pavlou et al., under review).

For example, in the comparative study by Zacharia et al. (2012), preschoolers with scientifically correct prior understanding of the concepts under investigation (the function of a balance beam) who interacted with either physical or virtual manipulatives outperformed students with incorrect prior knowledge who engaged with the virtual environment. Their findings indicated that haptic sensory feedback, which is a unique affordance of PL, is a prerequisite for learning if participants do not have any previous understanding of the concepts. The study by Pavlou et al. (under review) validated and also expanded the findings by Zacharia et al. (2012). This study compared the conceptual understanding of preschoolers engaged in VL or PL in three subject domains (balance beam, springs and sinking/floating). In the balance beam and springs domains, the mass of the objects (a property that should be multimodally grounded to support learning) is a causal factor affecting the experimental output, but for the sinking/floating domain, the idea that mass affects the object's behavior in water is a common misconception children have (see, e.g., Havu-Nuutinen, 2005; Hsin & Wu, 2011; Pavlou et al., 2018). Preschoolers working in the balance beam and sinking/floating domains had prior understanding of the domains, but participants working in the springs domain did not. The mode of experimentation did not affect the learning outcome for preschoolers engaged in the balance beam domain, but preschoolers who engaged with PL during experimentation in the domain of springs outperformed the participants in the VL group because the haptic sensory feedback offered in the PL group seemed to facilitate the development of understanding of the causal effect of mass. However, in the sinking/floating domain, haptic sensory feedback available during experimentation with physical materials impeded the development of a scientifically correct understanding, and the mass-related idea that "heavy objects sink/light object float" was the most dominant idea used by preschoolers to explain the phenomenon both before and after experimentation. Similar findings were also found in the preliminary study (Pavlou et al., 2018). The authors concluded that although information about other object properties was available, especially through vision, and was at times more salient than mass, haptic cues related to the mass of the objects were the most dominant perceptual cues that led to the fixation/empowerment of relevant ideas. Hence, providing haptic sensory feedback during experimentation can be detrimental (for students holding relevant misconceptions, as in the sinking/floating domain), beneficial (for students with no prior understanding, as in the springs domain) or have no significant impact (as for the preschoolers with prior knowledge of the balance beam domain). To conclude, it seems that haptic sensory feedback, which is available through engagement in PL, is not always a prerequisite for learning (see also Zacharia, 2015). In other words, VL can be used under certain conditions as a means for experimentation because they can support or even augment the development of the understanding of scientific concepts.

1.3   The Empirical Perspective

VL carry a lot of affordances that can support learning, as highlighted by many researchers (e.g., Faulconer & Gruss, 2018; de Jong et al., 2013; Olympiou & Zacharia, 2012; Potkonjak et al., 2016; Zacharia, 2015). For example, VL provide the opportunity to experiment with unobservable phenomena (e.g., radiation), to manipulate variables (e.g., light rays) and other parameters (e.g., time, spatial dimensions, data displays) and to engage with abstract concepts (e.g., symbolic representations of light). In addition, VL provide access to multiple users, are cost- and safety-efficient and minimize trial errors. Therefore, the empirical literature initially focused on exploring whether VL can support teaching and learning and specifically, whether their impact can be similar to or even greater than that of

PL. There are studies showing an advantage of VL over PL (e.g., Akpan & Andre, 2000; Bell & Trundle, 2008; Finkelstein et al., 2005) and studies showcasing opposite findings (e.g., Gire et al., 2010; Marshall & Young, 2006). However, the majority of comparative studies do not indicate that one mode of experimentation dominates over the other (e.g., Chini et al., 2012; Evangelou & Kotsis, 2019; Klahr et al., 2007; Leung & Cheng, 2021; Reece & Butler, 2017; Triona & Klahr, 2003; Zacharia & Constantinou, 2008; Zacharia & Olympiou, 2011). Inconsistencies between these studies can in part be attributed to the varying affordances carried by PL and VL that were utilized, the varying methodological approaches employed (D'Angelo et al., 2014; Faulconer & Gruss, 2018; Ma & Nickerson, 2006) and the different theoretical perspectives (or in some cases the lack thereof) adopted to predict and explain the learning outcomes (Reeves & Crippen, 2021).

Nevertheless, literature reviews conducted over the years have pointed out that VL has an effect equal to or even greater than that of PL (Brinson, 2015; D'Angelo et al., 2014; Faulconer & Gruss, 2018) or other teaching approaches (Rutten et al., 2012; Smetana & Bell, 2012). Overall, the extant empirical research studies have revealed the potential of VL experimentation for enhancing students' learning across grades K-16 (e.g., Potkonjak et al., 2016; Triona & Klahr, 2003; van der Meij & de Jong, 2006; Zacharia & Anderson, 2003; Zacharia et al., 2008; Zacharia et al., 2012). Consequently, it can be argued that VL can provide learning experiences that are just as meaningful to students as PL and, considering the many more unique affordances carried by VL as opposed to PL, some could argue that under certain circumstances, VL could be less "messy", easier to manage, and more flexible and expandable than PL (Klahr et al., 2008), especially if we consider some of the "disadvantages" of PL (e.g., space and time restrictions, absence of abstract representations). Hence, should one mode of experimentation be preferred over the other?

VL and PL have complementary affordances and thus their combination can support presentation of multiple representations of science concepts (de Jong et al., 2013; Olympiou & Zacharia, 2012; Puntambekar et al., 2021; Zacharia, 2007; Zacharia & de Jong, 2014; Zacharia & Michael, 2016). The meaningful integration of multiple representations of a concept (e.g., a physical and a virtual/abstract representation) can enhance learning to a greater extent than stand-alone representations (Ainsworth, 2008). For example, in a study by Wang and Tseng (2018), third-graders who engaged in the combination condition (VL and then PL) outperformed the students who engaged in stand-alone modes of experimentation. Kapici et al. (2019) also reported an advantage of the combination of VL and PL for seventh-grade students for the concept of electricity. Even though some studies have showcased that the stand-alone use of VL and PL has a greater impact on learning than their combination during experimentation (e.g., Gnesdilow & Puntambekar, 2022), overall, their combination seems to be more beneficial than stand-alone use (de Jong et al., 2013; Wörner et al., 2022). Teachers from different educational levels also seemed to support the combination of the two modes of experimentation for teaching and learning practice (Tsihouridis, et al., 2019).

Studies investigating the combination of VL and PL (e.g., Achuthan et al., 2017; Fuhrmann, et al., 2014; Olympiou & Zacharia, 2012, 2014; Trundle & Bell, 2010;

Yuksel et al., 2019; Zacharia, 2007; Zacharia et al., 2008) and the different ways in which they can be combined (e.g., at the same time, blended, in sequence), have showcased their complementary nature, but nevertheless provide limited information on the preeminent affordances of each mode (Lazonder & Ehrenhard, 2014). Similarly, because most of the comparative studies focused on improving students' learning (Reeves & Crippen, 2021), VL and PL experimentation were generally compared (e.g., as instructional approaches) without necessarily accounting for the potential effect of specific affordances. This is a vital underpinning in order to achieve optimal combinations for learning (Rau, 2020; Wörner et al., 2022; Zacharia et al., 2008). Thus, identifying these unique affordances of each mode of experimentation and their effect on student learning across grades K-16, as well as understanding when and under what conditions a particular mode of experimentation, along with its unique affordances, becomes more effective is still a critical issue. For instance, the community focusing on PL and VL has not yet distinguished the circumstances under which a unique affordance optimizes learning and whether such an effect holds true across grades K-16. For example, does providing abstract representations – a unique affordance of VL – support student learning the same way across grades K-16? The community also lacks information on how different unique affordances interact with each other and how this interaction impacts student learning. Moreover, the theoretical perspective through which the issue of unique affordances is approached affects the predictions and explanations articulated with regard to the impact of VL on learning (Rau & Herder, 2021). In the next section, we attempt to synthesize the theoretical and empirical perspective presented in this article.

1.4   Bridging Theory and Research with Practice in STEM+ Education

Educational approaches that facilitate STEM+ learning, such as the inquiry-based learning approach or the engineering design learning approach, call for students' active engagement and involvement in the learning process (de Jong, 2019). Laboratory work is a focal aspect of these approaches, which can be productively enacted with virtual and/or physical means (de Jong et al., 2013; Zacharia & de Jong, 2014). The role of laboratory experimentation becomes even more crucial within STEM+ education, when science is used as the dominant discipline among all of the disciplines involved.

Given that both the theoretical and the empirical evidence support the idea that both PL and VL are viable means of experimentation for students (e.g., Finkelstein et al., 2005; Triona & Klahr, 2003; van der Meij & de Jong, 2006; Zacharia, 2015; Zacharia et al., 2008), including preschoolers (Pavlou et al., 2018, under review; Zacharia et al., 2012), the palette of experimentation possibilities available to teachers (i.e., use of PL, use of VL, use of combinations of PL and VL) needs to be expanded. In the past decades, research has shown that both VL and PL can support learning and that their meaningful combination can be conducive to developing multimodal representations of concepts (de Jong et al., 2013; Olympiou & Zacharia, 2012; Puntambekar et al., 2021; Zacharia, 2007; Zacharia & de Jong, 2014; Zacharia & Michael, 2016). However, identifying the situations in which PL and/or VL can be utilized to optimize student learning across grades K-16 is still a critical issue.

Olympiou and Zacharia (2012) developed a framework summarizing a series of considerations on how to combine/blend PL and VL. Based on their framework, contemplation of the affordances of each mode of experimentation (unique or not) in conjunction with the learning objectives of each experiment and students' background (e.g., prior conceptions, skills) is vital. The framework was validated in studies concerning undergraduate students (Olympiou & Zacharia, 2012, 2014) and primary school students (Zacharia & Michael, 2016) that exhibited the advantages of blending the two modes of experimentation instead of using VL or PL alone. The framework can support the decision-making process when designing laboratory experimentation activities for STEM+ initiatives with the integration of VL and PL. Below, we discuss – and extend – some of the key considerations integrated in this framework, in conjunction with the relevant literature.

Research has shown that despite the overlapping affordances of VL and PL offered during experimentation (e.g., manipulation of material, perceptual grounding for abstract concepts), their unique affordances can affect learning in a different manner (for more details, see Olympiou & Zacharia, 2014; Zacharia, 2015). On the one hand, the ability to visualize abstract concepts (e.g., light rays, particles, current flow) and modify parameters, such as time and dimensions, is a unique affordance of VL. For example, in a study by Finkelstein et al. (2005), university students working with a virtual laboratory for electrical circuits in which they could manipulate parameters (e.g., voltage) and visualize current flow outperformed the students working with physical materials, to whom such affordances were not offered. A study by Zacharia and de Jong (2014) highlighted the learning benefits of using VL prior to PL for the understanding of complex circuits because the virtual environment offers the additional visualization of the current-flow. The appropriate consideration of these affordances for this subject domain (see, e.g., Zacharia & Michael, 2016) can facilitate the appropriate blending of both modes of experimentation to support development of multimodal understanding.

On the other hand, the availability of haptic sensory feedback during manipulation is a unique affordance of PL. According to haptic perspectives on learning and empirical studies (e.g., Han, 2013; Lazonder & Ehrenhard, 2014; Pavlou et al., under review; Zacharia et al., 2012; Zacharia, 2015), the presence/absence of haptic sensory feedback can affect learning; hence, the identification of the conditions under which it can be beneficial or detrimental during experimentation can support the optimal integration of VL and PL in STEM+ education. The development of understanding of concepts such as forces (e.g., Han, 2013; Zohar & Levy, 2021), mass/weight (e.g., Lazonder & Ehrenhard, 2014; Pavlou et al., under review; Zacharia et al., 2012) and magnetic fields (e.g., Reiner, 1999) seems to benefit from interaction with haptic stimuli. Haptic sensory feedback is possibly even more essential for students with no prior embodied experience/knowledge of a domain, for whom haptic cues can enhance concept-specific understanding when visual cues alone are not adequate for solving a task (e.g., Pavlou et al., under review; Qi et al., 2021; Zacharia et al., 2012; Zohar & Levy, 2021). However, it should be taken into account that haptic cues related to a students' misconception might also hinder learning (i.e., lead to a fixation/empowerment of the misconception) as in the sinking/floating domain of the Pavlou et al. (under review) study.

In addition, based on the theoretical perspective in this article, the lack of embodied experiences with a domain seems to be a vital consideration when selecting/combining VL and PL. The initial sensorimotor grounding of scientific knowledge in STEM classrooms through hands-on approaches can support the development of abstract knowledge (Hayes & Kraemer, 2017). When prior knowledge/embodied experience with the concepts under investigation is lacking, then the use of PL during experimentation will most likely significantly improve the learning outcome (e.g., Pavlou et al., under review; Zacharia et al., 2012). For example, if students do not have any prior experience with a domain, PL can proceed VL to provide such experiences, especially when haptic manipulation (in combination with other modalities) can enhance conceptual understanding. To amplify the effect of embodiment, there should be a strong relation between the task at hand and the bodily movements enacted by the students working in a virtual or physical environment (e.g., Johnson-Glenberg, 2019; Mavilidi et al., 2017; Qi et al., 2021; Skulmowski & Rey, 2018; Zohar & Levy, 2021).

Of course, the need to provide embodied experiences (including haptic sensory feedback) might not be as vital for older students because such experiences were probably acquired during their early years through formal and/or informal learning, an argument that can even support the stand-alone use of VL (especially when considering the importance of distance/online learning). This argument agrees with empirical research that has shown an advantage or equal effect of VL on learning for a variety of domains and disciplines for students in primary school and older (Brinson, 2015; Zacharia, 2015). Nevertheless, the reliance of younger children on sensorimotor experience indicates the importance of using PL during experimentation. However, studies with preschoolers (Pavlou et al., 2018, under review; Zacharia et al., 2012) have shown that under certain conditions, the use of VL can also facilitate learning for younger children. Given that empirical research has focused on older students (for details, see Zacharia, 2015), our understanding of the situations in which VL can support learning in early childhood education is still limited.

Another aspect that should be taken into consideration when selecting a means of experimentation for STEM+ enactments is factors related to the affective domain. For example, Justo et al. (2022) found that even though VL and PL had similar effects on the learning of basic engineering concepts, students' motivation increased with the use of physical materials. As Tsihouridis et al. (2019) also highlighted, primary students usually favor PL, whereas older students prefer VL. Hence, aspects other than learning (which were not the focus of this article, for example, the enhancement of students' skills or attitudes) can also guide the decision about which mode of experimentation is more appropriate.

The additional value of VL in education is no longer disputed, and research seems to be turning towards addressing the question of "which one is better" in what instances, how and for whom (de Jong, 2019). As noted in this article, the theoretical underpinning guiding how these questions are being examined does matter for the instructional design and the way the findings are explained (see Rau, 2020; Rau & Herder, 2021). Combining empirical and theoretical perspectives is essential. In addition, the current literature portrays the complex interplay between aspects, such as the affordances of each mode of experimentation, the concepts under investigation (e.g., haptic cues can augment concept-specific understanding) and students' prior embodied experiences/knowledge (including their misconceptions). However, the influence of this interplay on students' learning and the possible dominance of one aspect over others (e.g., whether students' misconceptions can impede concept-specific understanding in VL/PL) is yet unclear; research is still needed to develop a comprehensive framework for the optimal integration of PL and VL in STEM+ education.

References

Achuthan, K., Francis, S. P., & Diwakar, S. (2017). Augmented reflective learning and knowledge retention perceived among students in classrooms involving virtual laboratories. *Education and Information Technologies, 22*(6), 2825–2855. https://doi.org/10.1007/s10639-017-9626-x

Ainsworth, S. (2008). The educational value of multiple-representations when learning complex scientific concepts. In J. K. Gilbert, M. Reiner, & M. Nakhleh (Eds.), *Visualization: Theory and practice in science education* (pp. 191–208). Springer. https://doi.org/10.1007/978-1-4020-5267-5_9

Akpan, J. P., & Andre, T. (2000). Using a computer simulation before dissection to help students learn anatomy. *Journal of Computers in Mathematics and Science Teaching, 19*(3), 297–313.

Barsalou, L. W. (2008). Grounded cognition. *Annual Review of Psychology, 59*(1), 617–645. https://doi.org/10.1146/annurev.psych.59.103006.093639

Bazelais, P., Binner, G., & Doleck, T. (2022). Examining the key drivers of student acceptance of online labs. *Interactive Learning Environments*, 1–16. https://doi.org/10.1080/10494820.2022.2121729

Bell, R. L., & Trundle, K. C. (2008). The use of a computer simulation to promote scientific conceptions of moon phases. *Journal of Research in Science Teaching, 45*(3), 346–372.

Bivall, P., Ainsworth, S., & Tibell, L. A. (2011). Do haptic representations help complex molecular learning? *Science Education, 95*(4), 700–719.

Brinson, J. R. (2015). Learning outcome achievement in non-traditional (virtual and remote) versus traditional (hands-on) laboratories: A review of the empirical research. *Computers & Education, 87*, 218–237. https://doi.org/10.1016/j.compedu.2015.07.003

Bybee, R. W. (2009). *The BSCS 5E instructional model and 21st century skills*. BSCS.

Capraro, R. M., Capraro, M. M., & Morgan, J. R. (Eds.). (2013). *STEM project-based learning: An integrated science, technology, engineering, and mathematics (STEM) approach*. Springer Science & Business Media.

Chini, J. J., Madsen, A., Gire, E., Rebello, N. S., & Puntambekar, S. (2012). Exploration of factors that affect the comparative effectiveness of physical and virtual manipulatives in an undergraduate laboratory. *Physical Review Special Topics – Physics Education Research, 8*(1), 010113. https://doi.org/10.1103/PhysRevSTPER.8.010113

Clark, A. (2008). *Supersizing the mind: Embodiment, action, and cognitive extension*. Oxford University Press.

Crismond, D. P., & Adams, R. S. (2012). The informed design teaching & learning matrix. *Journal of Engineering Education, 101*(4), 738–797.

D'Angelo, C., Rutstein, D., Harris, C., Bernard, R., Borokhovski, E., & Haertel, G. (2014). *Simulations for STEM learning: Systematic review and meta-analysis*. SRI International.

de Jong, T. (2019). Moving towards engaged learning in STEM domains; There is no simple answer, but clearly a road ahead. *Journal of Computer Assisted Learning, 35*(2), 153–167.

de Jong, T., Linn, M. C., & Zacharia, Z. C. (2013). Physical and virtual laboratories in science and engineering education. *Science, 340*(6130), 305–308. https://doi.org/10.1126/science.1230579

de Jong, T., Sotiriou, S., & Gillet, D. (2014). Innovations in STEM education: The Go-Lab federation of online labs. *Smart Learning Environments, 1*(1), 1–16. https://doi.org/10.1126/science.1230579

European Commission. (2022). *Impacts of COVID-19 on school education*. Publications Office of the European Union. Retrieved from https://data.europa.eu/doi/10.2766/201112

Evangelou, F., & Kotsis, K. (2019). Real vs virtual physics experiments: Comparison of learning outcomes among fifth grade primary school students. A case on the concept of frictional force. *International Journal of Science Education, 41*(3), 330–348. https://doi.org/10.1080/0950069 3.2018.1549760

Farina, M. (2021). Embodied cognition: Dimensions, domains and applications. *Adaptive Behavior, 29*(1), 73–88. https://doi.org/10.1177/1059712320912963

Faulconer, E., & Gruss, A. (2018). A review to weigh the pros and cons of online, remote, and distance science laboratory experiences. *The International Review of Research in Open and Distributed Learning, 19*(2), 155–168. https://doi.org/10.19173/irrodl.v19i2.3386

Finkelstein, N. D., Adams, W. K., Keller, C. J., Kohl, P. B., Perkins, K. K., Podolefsky, N. S., et al. (2005). When learning about the real world is better done virtually: A study of substituting computer simulations for laboratory equipment. *Physical Review Special Review Special Topics-Physics Education Research, 1*(1), 010103. https://doi.org/10.1103/PhysRevSTPER.1.010103

Fuhrmann, T., Salehi, S., & Blikstein, P. (2014). *A tale of two worlds: Using bifocal modeling to find and resolve "discrepant events" between physical experiments and virtual models in biology*. International Society of the Learning Sciences.

Georgiou, Y., & Ioannou, A. (2019). Embodied learning in a digital world: A systematic review of empirical research in K-12 education. In P. Díaz, A. Ioannou, K. K. Bhagat, & J. M. Spector (Eds.), *Learning in a digital world: Perspective on interactive technologies for formal and informal education* (pp. 155–177). Springer. https://doi.org/10.1007/978-981-13-8265-9_8

Gibbs, R. W. (2005). *Embodiment and cognitive science*. Cambridge University Press.

Gire, E., Carmichael, A., Chini, J. J., Rouinfar, A., Rebello, S., Smith, G., & Puntambekar, S. (2010). The effects of physical and virtual manipulatives on students' conceptual learning about pulleys. In *Proceedings of the 9th international conference of the learning sciences* (Vol. 1, pp. 937–943). International Society of the Learning Sciences.

Gnesdilow, D., & Puntambekar, S. (2022). Comparing middle school students' science explanations during physical and virtual laboratories. *Journal of Science Education and Technology, 31*, 191–202. https://doi.org/10.1007/s10956-021-09941-0

Han, I. (2013). Embodiment: A new perspective for evaluating physicality in learning. *Journal of Educational Computing Research, 49*(1), 41–59. https://doi.org/10.2190/EC.49.1.b

Han, I., & Black, J. B. (2011). Incorporating haptic feedback in simulation for learning physics. *Computers & Education, 57*(4), 2281–2290. https://doi.org/10.1016/j.compedu.2011.06.012

Hatwell, Y. (2003). Manual exploratory procedures in children and adults. In Y. Hatwell, A. Streri, & E. Gentaz (Eds.), *Touching for knowing: Cognitive psychology of haptic manual perception* (pp. 67–82). John Benjamins Publishing Company.

Havu-Nuutinen, S. (2005). Examining young children's conceptual change process in floating and sinking from a social constructivist perspective. *International Journal of Science Education, 27*(3), 259–279. https://doi.org/10.1080/0950069042000243736

Hayes, J. C., & Kraemer, D. J. (2017). Grounded understanding of abstract concepts: The case of STEM learning. *Cognitive Research: Principles and Implications, 2*(1), 1–15. https://doi.org/10.1186/s41235-016-0046-z

Heller, M. A. (1989). Texture perception in sighted and blind observers. *Perception & Psychophysics, 45*(1), 49–54. https://doi.org/10.3758/BF03208032

Hsin, C., & Wu, H. (2011). Using scaffolding strategies to promote young children's scientific understandings of floating and sinking. *Journal of Science Education and Technology, 20*(5), 656–666. https://doi.org/10.1007/s10956-011-9310-7

Johnson-Glenberg, M. C. (2019). The necessary nine: Design principles for embodied VR and active STEM education. In P. Díaz, A. Ioannou, K. K. Bhagat, & J. Spector (Eds.), *Learning in a digital world* (pp. 83–112). Springer.

Jones, M. G., Minogue, J., Tretter, T. R., Negishi, A., & Taylor, R. (2006). Haptic augmentation of science instruction: Does touch matter? *Science Education, 90*(1), 111–123.

Justo, E., Delgado, A., Llorente-Cejudo, C., Aguilar, R., & Caber-Almenara, J. (2022). The effectiveness of physical and virtual manipulatives on learning and motivation in structural engineering. *Journal of Engineering Education, 111*(4), 813–851. https://doi.org/10.1002/jee.20482thi

Kapici, H. O., Akcay, H., & de Jong, T. (2019). Using hands-on and virtual laboratories alone or together – Which works better for acquiring knowledge and skills? *Journal of Science Education and Technology, 28*(3), 231–250.

Kiefer, M., & Pulvermüller, F. (2012). Conceptual representations in mind and brain: Theoretical developments, current evidence and future directions. *Cortex, 48*(7), 805–825. https://doi.org/10.1016/j.cortex.2011.04.006

Kiefer, M., & Trumpp, N. M. (2012). Embodiment theory and education: The foundations of cognition in perception and action. *Trends in Neuroscience and Education, 1*(1), 15–20. https://doi.org/10.1016/j.tine.2012.07.002

Klahr, D., Triona, L. M., & Williams, C. (2007). Hands on what? The relative effectiveness of physical versus virtual materials in an engineering design project by middle school children. *Journal of Research in Science Teaching, 44*(1), 183–203. https://doi.org/10.1002/tea.20152

Klahr, D., Triona, L., Strand-Cary, M., & Siler, S. (2008). Virtual vs. physical materials in early science instruction: Transitioning to an autonomous tutor for experimental design. In J. Zumbach, N. Schwartz, T. Seufert, & L. Kester (Eds.), *Beyond knowledge: The legacy of competence* (pp. 163–172). Springer. https://doi.org/10.1007/978-1-4020-8827-8_23

Klatzky, R. L., Lederman, S. J., & Reed, C. (1987). There's more to touch than meets the eye: The salience of object attributes for haptics with and without vision. *Journal of Experimental Psychology: General, 116*(4), 356–369. https://doi.org/10.1037/0096-3445.116.4.356

Klatzky, R. L., Lederman, S. J., & Matula, D. E. (1991). Imagined haptic exploration in judgments of object properties. *Journal of Experimental Psychology: Learning, Memory, and Cognition, 17*(2), 314–322.

Klatzky, R. L., Lederman, S. J., & Matula, D. E. (1993). Haptic exploration in the presence of vision. *Journal of Experimental Psychology: Human Perception and Performance, 19*(4), 726–743. https://doi.org/10.1037/0096-1523.19.4.726

Kontra, C., Goldin-Meadow, S., & Beilock, S. L. (2012). Embodied learning across the life span. *Topics in Cognitive Science, 4*(4), 731–739. https://doi.org/10.1111/j.1756-8765.2012.01221.x

Kontra, C., Lyons, D. J., Fischer, S. M., & Beilock, S. L. (2015). Physical experience enhances science learning. *Psychological Science, 26*(6), 737–749. https://doi.org/10.1177/0956797615569355

Lazonder, A. W., & Ehrenhard, S. (2014). Relative effectiveness of physical and virtual manipulatives for conceptual change in science: How falling objects fall. *Journal of Computer Assisted Learning, 30*(2), 110–120. https://doi.org/10.1111/jcal.12024

Leung, P. K. Y., & Cheng, M. M. W. (2021). Practical work or simulations? Voices of millennial digital natives. *Journal of Educational Technology Systems, 50*(1). https://doi.org/10.1177/00472395211018967

Ma, J., & Nickerson, J. V. (2006). Hands-on, simulated, and remote laboratories: A comparative literature review. *ACM Computing Surveys, 38*(3), 1–24. https://doi.org/10.1145/1132960.1132961

Magana, A. J., Serrano, M. I., & Rebello, N. S. (2019). A sequenced multimodal learning approach to support students' development of conceptual learning. *Journal of Computer Assisted Learning, 35*(4), 516–528. https://doi.org/10.1111/jcal.12356

Malinverni, L., & Pares, N. (2014). Learning of abstract concepts through full-body interaction: A systematic review. *Journal of Educational Technology & Society, 17*(4), 100–116.

Marmeleira, J., & Duarte Santos, G. (2019). Do not neglect the body and action: The emergence of embodiment approaches to understanding human development. *Perceptual and Motor Skills, 126*(3), 410–445.

Marshall, J. A., & Young, E. S. (2006). Preservice teachers' theory development in physical and simulated environments. *Journal of Research in Science Teaching, 43*(9), 907–937. https://doi.org/10.1002/tea.20124

Martín-Páez, T., Aguilera, D., Perales-Palacios, F., & Vílchez-González, J. M. (2019). What are we talking about when we talk about STEM education? A review of literature. *Science Education, 103*(4), 799–822. https://doi.org/10.1002/sce.21522

Mavilidi, M., Okely, A. D., Chandler, P., & Paas, F. (2017). Effects of integrating physical activities into a science lesson on preschool children's learning and enjoyment. *Applied Cognitive Psychology, 31*(3), 281–290.

National Academy of Sciences. (2014). *STEM integration in K-12 education: Status, prospects, and an agenda for research.* National Academies Press. https://doi.org/10.17226/18612

National Research Council. (2006). *America's lab report: Investigations in high school science.* National Academies Press.

Next Generation Science Standards Lead States. (2013). *Next generation science standards: For states, by states.* National Academies Press. Retrieved from http://www.nextgenscience.org/

Olympiou, G., & Zacharia, Z. C. (2012). Blending physical and virtual manipulatives: An effort to improve students' conceptual understanding through science laboratory experimentation. *Science Education, 96*(1), 21–47. https://doi.org/10.1002/sce.20463

Olympiou, G., & Zacharia, Z. C. (2014). Blending physical and virtual manipulatives in physics laboratory experimentation. In C. Bruguière, A. Tiberghien, & P. Clément (Eds.), *Topics and trends in current science education* (pp. 419–433). Springer.

Pavlou, Y., Papaevripidou, M., & Zacharia, Z. (2018). Can preschoolers develop an understanding of the sinking/floating phenomenon through physical and virtual experimental environments? In M. Kalogiannakis (Ed.), *Teaching natural sciences in preschool education: Challenges and perspectives* (pp. 76–95). Gutenberg.

Pavlou, Y., Zacharia, Z., & Papaevripidou, M. (under review). Comparing the impact of physical and virtual manipulatives in different science domains among preschoolers. *Science Education.*

Pedaste, M., Mäeots, M., Siiman, L. A., de Jong, T., van Riesen, S. A. N., Kamp, E. T., et al. (2015). Phases of inquiry-based learning: Definitions and the inquiry cycle. *Educational Research Review, 14*, 47–61. https://doi.org/10.1016/j.edurev.2015.02.003

Potkonjak, V., Gardner, M., Callaghan, V., Mattila, P., Guetl, C., Petrović, V. M., & Jovanović, K. (2016). Virtual laboratories for education in science, technology, and engineering: A review. *Computers & Education, 95*, 309–327.

Pouw, W. T. J. L., van Gog, T., & Paas, F. (2014). An embedded and embodied cognition review of instructional manipulatives. *Educational Psychology Review, 26*, 51–72. https://doi.org/10.1007/s10648-014-9255-5

Puntambekar, S., Gnesdilow, D., Dornfeld Tissenbaum, C., Narayanan, N. H., & Rebello, N. S. (2021). Supporting middle school students' science talk: A comparison of physical and virtual labs. *Journal of Research in Science Teaching, 58*(3), 392–419.

Qi, K., Borland, D., Brunsen, E., Minogue, J., & Peck, T. C. (2021). The impact of prior knowledge on the effectiveness of haptic and visual modalities for teaching forces. In *Proceedings of the 2021 international conference on multimodal interaction, Montréal, Canada* (pp. 203–211). https://doi.org/10.1145/3462244.3479915

Radhamani, R., Kumar, D., Nizar, N., Achuthan, K., Nair, B., & Diwakar, S. (2021). What virtual laboratory usage tells us about laboratory skill education pre-and post-COVID-19: Focus on usage, behavior, intention and adoption. *Education and Information Technologies, 26*(6), 7477–7495. https://doi.org/10.1007/s10639-021-10583-3

Raman, R., Vinuesa, R., & Nedungadi, P. (2021). Acquisition and user behavior in online science laboratories before and during the COVID-19 pandemic. *Multimodal Technologies and Interaction, 5*(8), 46. https://doi.org/10.3390/mti5080046

Rau, M. A. (2020). Comparing multiple theories about learning with physical and virtual representations: Conflicting or complementary effects? *Educational Psychology Review, 32*(2), 297–325. https://doi.org/10.1007/s10648-020-09517-1

Rau, M. A., & Herder, T. (2021). Under which conditions are physical versus virtual representations effective? Contrasting conceptual and embodied mechanisms of learning. *Journal of Educational Psychology, 113*(8), 1565–1586. https://doi.org/10.1037/edu0000689

Reece, A. J., & Butler, M. B. (2017). Virtually the same: A comparison of STEM students' content knowledge, course performance, and motivation to learn in virtual and face-to-face introductory biology laboratories. *Journal of College Science Teaching, 46*(3), 83–89.

Reeves, S. M., & Crippen, K. J. (2021). Virtual laboratories in undergraduate science and engineering courses: A systematic review, 2009–2019. *Journal of Science Education and Technology, 30*(1), 16–30. https://doi.org/10.1007/s10956-020-09866-0

Reid, N., & Shah, I. (2007). The role of laboratory work in university chemistry. *Chemistry Education Research and Practice, 8*(2), 172–185.

Reiner, M. (1999). Conceptual construction of fields through tactile interface. *Interactive Learning Environments, 7*(1), 31–55. https://doi.org/10.1076/ilee.7.1.31.3598

Reiner, M. (2008). Seeing through touch: The role of haptic information in visualization. In J. K. Gilbert, M. Reiner, & M. Nakhleh (Eds.), *Visualization: Theory and practice in science education* (pp. 73–84). Springer. https://doi.org/10.1007/978-1-4020-5267-5_4

Reiner, M., Hecht, D., Halevy, G., & Furman, M. (2006). Semantic interference and facilitation in haptic perception. In *Proceedings of the Eurohaptics conference, Paris* (pp. 41–35).

Rutten, N., van Joolingen, W. R., & van der Veen, J. T. (2012). The learning effects of computer simulations in science education. *Computers & Education, 58*(1), 136–153. https://doi.org/10.1016/j.compedu.2011.07.017

Skulmowski, A., & Rey, G. D. (2018). Embodied learning: Introducing a taxonomy based on bodily engagement and task integration. *Cognitive Research: Principles and Implications, 3*(1), 6. https://doi.org/10.1186/s41235-018-0092-9

Smetana, L. K., & Bell, R. L. (2012). Computer simulations to support science instruction and learning: A critical review of the literature. *International Journal of Science Education, 34*(9), 1337–1370. https://doi.org/10.1080/09500693.2011.605182

Stock, O., Röder, B., Burke, M., Bien, S., & Rösler, F. (2009). Cortical activation patterns during long-term memory retrieval of visually or haptically encoded objects and locations. *Journal of Cognitive Neuroscience, 21*(1), 58–82. https://doi.org/10.1162/jocn.2009.21006

Thibaut, L., Ceuppens, S., De Loof, H., De Meester, J., Goovaerts, L., Struyf, A., et al. (2018). Integrated STEM education: A systematic review of instructional practices in secondary education. *European Journal of STEM Education, 3*(1), 2.

Triona, L. M., & Klahr, D. (2003). Point and click or grab and heft: Comparing the influence of physical and virtual instructional materials on elementary school students' ability to design experiments. *Cognition and Instruction, 21*(2), 149–173. https://doi.org/10.1207/S1532690XCI2102_02

Trundle, K. C., & Bell, R. L. (2010). The use of a computer simulation to promote conceptual change: A quasi-experimental study. *Computers & Education, 54*(4), 1078–1088. https://doi.org/10.1016/j.compedu.2009.10.012

Tsihouridis, C., Vavougios, D., Batsila, M., & Ioannidis, G. (2019). The optimum equilibrium when using experiments in teaching – Where virtual and real labs stand in science and engineering teaching practice. *International Journal of Emerging Technologies in Learning, 14*(23), 67–84.

van der Meij, J., & de Jong, T. (2006). Supporting students' learning with multiple representations in a dynamic simulation-based learning environment. *Learning and Instruction, 16*(3), 199–212. https://doi.org/10.1016/j.learninstruc.2006.03.007

Van Doorn, G. H., Richardson, B. L., Wuillemin, D. B., & Symmons, M. A. (2010). Visual and haptic influence on perception of stimulus size. *Attention, Perception, & Psychophysics, 72*(3), 813–822.

Wang, T., & Tseng, Y. (2018). The comparative effectiveness of physical, virtual, and virtual-physical manipulatives on third-grade students' science achievement and conceptual understanding of evaporation and condensation. *International Journal of Science and Mathematics Education, 16*(2), 203–219.

Wellsby, M., & Pexman, P. M. (2014). Developing embodied cognition: Insights from children's concepts and language processing. *Frontiers in Psychology, 5*, 506. https://doi.org/10.3389/fpsyg.2014.00506

Wilson, M. (2002). Six views of embodied cognition. *Psychonomic Bulletin & Review, 9*(4), 625–636. https://doi.org/10.3758/BF03196322

Wörner, S., Kuhn, J., & Scheiter, K. (2022). The best of two worlds: A systematic review on combining real and virtual experiments in science education. *Review of Educational Research, 92*(6), 911–952.

Yee, E., & Thompson-Schill, S. (2016). Putting concepts into context. *Psychonomic Bulletin & Review, 23*(4), 1015–1027. https://doi.org/10.3758/s13423-015-0948-7

Yee, E., Jones, M. N., & McRae, K. (2018). Semantic memory. In J. T. Wixted & S. Thompson-Schill (Eds.), *The Stevens' handbook of experimental psychology and cognitive neuroscience* (4th ed., pp. 1–23). Wiley Online Library.

Yuksel, T., Walsh, Y., Magana, A. J., Nova, N., Krs, V., Ngambeki, I., et al. (2019). Visuohaptic experiments: Exploring the effects of visual and haptic feedback on students' learning of friction concepts. *Computer Applications in Engineering Education, 27*(6), 1376–1401. https://doi.org/10.1002/cae.22157

Zacharia, Z. C. (2007). Comparing and combining real and virtual experimentation: an effort to enhance students' conceptual understanding of electric circuits. *Journal of Computer Assisted Learning, 23*(2), 120–132.

Zacharia, Z. C. (2015). Examining whether touch sensory feedback is necessary for science learning through experimentation: A literature review of two different lines of research across K-16. *Educational Research Review, 16*, 116–137.

Zacharia, Z., & Anderson, O. R. (2003). The effects of an interactive computer-based simulation prior to performing a laboratory inquiry-based experiment on students' conceptual understanding of physics. *American Journal of Physics, 71*(6), 618–629. https://doi.org/10.1119/1.1566427

Zacharia, Z. C., & Constantinou, C. P. (2008). Comparing the influence of physical and virtual manipulatives in the context of the physics by inquiry curriculum: The case of undergraduate students' conceptual understanding of heat and temperature. *American Journal of Physics, 76*(4), 425–430. https://doi.org/10.1119/1.2885059

Zacharia, Z. C., & de Jong, T. (2014). The effects on students' conceptual understanding of electric circuits of introducing virtual manipulatives within a physical manipulatives-oriented curriculum. *Cognition and Instruction, 32*(2), 101–158. https://doi.org/10.1080/07370008.2014.887083

Zacharia, Z. C., & Michael, M. (2016). Using physical and virtual manipulatives to improve primary school students' understanding of concepts of electric circuits. In M. Riopel & Z. Smyrnaiou (Eds.), *New developments in science and technology education* (pp. 125–140). Springer. https://doi.org/10.1007/978-3-319-22933-1_12

Zacharia, Z. C., & Olympiou, G. (2011). Physical versus virtual manipulative experimentation in physics learning. *Learning and Instruction, 21*(3), 317–331. https://doi.org/10.1016/j.learninstruc.2010.03.001

Zacharia, Z. C., Olympiou, G., & Papaevripidou, M. (2008). Effects of experimenting with physical and virtual manipulatives on students' conceptual understanding in heat and temperature. *Journal of Research in Science Teaching, 45*(9), 1021–1035. https://doi.org/10.1002/tea.20260

Zacharia, Z. C., Loizou, E., & Papaevripidou, M. (2012). Is physicality an important aspect of learning through science experimentation among kindergarten students? *Early Childhood Research Quarterly, 27*(3), 447–457. https://doi.org/10.1016/j.ecresq.2012.02.004

Zhan, Z., Shen, W., Xu, Z., Niu, S., & You, G. (2022). A bibliometric analysis of the global landscape on STEM education (2004–2021): Towards global distribution, subject integration, and research trends. *Asia Pacific Journal of Innovation and Entrepreneurship, 16*(2), 171–203. https://doi.org/10.1108/APJIE-08-2022-0090

Zhuoluo, M. A., Liu, Y., & Zhao, L. (2019). Effect of haptic feedback on a virtual lab about friction. *Virtual Reality & Intelligent Hardware, 1*(4), 428–434.

Zohar, A. R., & Levy, S. T. (2021). From feeling forces to understanding forces: The impact of bodily engagement on learning in science. *Journal of Research in Science Teaching, 58*(8), 1203–1237. https://doi.org/10.1002/tea.21698

Chapter 2
Problematisation, Narrative and Fiction in the Science Classroom

Catherine Bruguière and Denise Orange Ravachol

2.1 Introduction

Several studies have already investigated and discussed the complex relationships between science and narratives, and put forward proposals for the classroom (Avraamidou & Osborne, 2009; Hadzigeorgiou, 2016, Orange Ravachol, 2017). Our research contributes to this body of work. Its originality lies in the fact that, whilst being situated in the same theoretical framework of learning through problematisation, it looks at using narratives to construct scientific knowledge from two slightly different approaches:

- An approach which considers that the use of realistic fiction can contribute to constructing problematised biological knowledge.
- An approach which considers that some forms of narrative, in particular storytelling, actually hinder the construction of biological knowledge.

These partially contradictory standpoints are useful when examining the extent to which fictional narratives can contribute to constructing problematised biological knowledge and under what didactic conditions the transposition of biological knowledge in fictional narratives can be considered to be an asset from a didactic perspective. They also lead us to question how shifting towards extra-ordinary

C. Bruguière (✉)
S2HEP Research Unit (UR 4148), University Claude Bernard Lyon 1, Lyon, France
e-mail: catherine.bruguiere@univ-lyon1.fr

D. Orange Ravachol (✉)
CIREL Université de Lille, Lille, France
e-mail: denise.orange@univ-lille.fr

K. Korfiatis et al. (eds.), *Shaping the Future of Biological Education Research*,
Contributions from Biology Education Research,
https://doi.org/10.1007/978-3-031-44792-1_2

forms of reasoning (as opposed to ordinary common-sense reasoning) allows us to move away from storytelling. In the first part of this paper, we will present the epistemological and didactic theories underpinning this work. We will then characterise the two approaches using a number of examples. Finally, we will highlight and discuss their compatibility, before looking at the possible implications for curricula.

2.2 Epistemological and Didactic Theory

2.2.1 Problematisation and Scientific Knowledge

Our research concerns learning in a rationalist framework (Bachelard, Popper, Canguilhem). The construction of scientific knowledge, by scientists or school pupils, is based on working through explanatory problems, with a view to identifying a common-sense explanation for the situation and then going beyond this. This construction does not focus on the starting points (the problem tackled) and end points (explanations seen as solutions) of the scientific work. Instead, it concentrates on what occurs between the posing of the problem and its resolution. In other words, it deploys a sort of investigation which not only aims to find a solution to the problem, but also seeks to identify the necessities which constrain this solution. This approach, which focusses on the work of explicitly stating problems in order to construct scientific knowledge, is problematisation (Fabre & Orange, 1997). It is a demanding approach which, within a specific framework (incorporating the use of different types of reasoning), explores the possible, impossible and necessary in the solutions (explanatory models) and how these can be controlled using critical thinking and empirical investigation. The knowledge built is not limited to solutions to the problems (assertoric knowledge), but is also characterised by necessity (apodictic knowledge). In the sciences, knowing is not only 'knowing that'; it is also knowing 'why it cannot be otherwise…' (Reboul, 1992).

2.2.2 Science and Narrative

The sciences (including biology) and scientific research are human and cultural endeavours in which narrative and fiction clearly have a role to play. However, "when we invent worlds which are possible in fiction, we never really leave the universe we know" (Bruner, 2002, p. 82). Popper (1985, p. 191) established a relationship of filiation between myth and science. He writes that what we call 'science' distinguishes itself from myth, not by the form it takes (in both cases it involves telling a story to explain a phenomenon), but because it introduces a correlate, critically analysing this story through discussion, comparing it with other possible stories and with empirical evidence. In this way he separates scientific knowledge from our highly subjective, common knowledge, and places it in a third world – 'a world without a knowing subject'. This is an objective and autonomous world, whose 'inhabitants' are the theoretical systems, the problems and the state of these problems (state of discussion, state of exchanging critical arguments) (Popper, 1991).

The sciences, through their proximity to and conflict with narratives, should therefore on the one hand make 'responsible use' of stories (Bruner, 1996), in order to preserve the relevance of the problem and critical depth of the work, and, on the other hand, remove themselves from common knowledge, which is so effective in terms of the shortcuts it takes in its reasoning (*ad hoc* causes, simple causality, for example). As Bachelard puts it, "in common knowledge, the facts are linked to the reasons far too quickly. From the fact to the idea, the circuit is far too short" (1938, p. 44).

2.2.3 Fictional Narratives and Scientific Problematisation: The Notion of Realistic Fiction

Through the notion of realistic fiction (Bruguière & Triquet, 2012) we have charac-terised certain fictional narratives for use as didactic objects suited to the problema-tised construction of biological knowledge. Our work differs from the more psychology and epistemology-focused studies (Egan, 1986; Hadzigeorgiou, 2016) which look at the impact of stories on students' recall, emotional involvement and motivation. Here, we assign the narrative and the fiction a cognitive function thanks to the narrative's capacity to question the objects of the world (Barthes, 1966) on the one hand, and to the fiction's capacity to represent possible worlds on the other (Eco, 1979). The fictional narratives we consider to have this didactic potential are fictional stories which do not provide any answers, never mind any scientific expla-nation as is the case for the explanatory stories (Reiss et al., 1999), but in which the plot is constrained by underlying scientific phenomena, and in which the fiction, because it creates possible worlds close to the real world of the pupil, unfolds in the pupils' spheres of reference. In these fictional narratives, neither the scientific phe-nomena which guide the unfolding plot nor the variations in scientific knowledge in the fiction's possible worlds are mentioned explicitly. These realistic-fiction narra-tives can therefore generate problems in terms of interpretation (Eco, 1979) and lead the reader to actively contribute to questioning this scientific knowledge. This involves working on the characteristics of these realistic-fiction narratives, such as the complexity of the plot and the contrast between characters' intentional and unin-tentional actions (caused by biological phenomena), all of which are levers for prob-lematisation (Bruguière & Triquet, 2012).

2.3 Narratives in Learning About Biological Phenomena

2.3.1 Reasoned Interpretation of Works of Realistic-Fiction

Here we propose a case study based on the interpretive study of the realistic-fiction picture book *La promesse* (Willis & Ross, 2010), with the aim of problematising the concept of animal metamorphosis with young primary school pupils. In this book, the animal characters undergo the metamorphoses inherent to their species, but against their will. When working on their comprehension of the reasons for the events that go against the characters' wishes, the pupils are confronted with a number of epistemological obstacles (Rumelhard, 1995, Triquet & Bruguière, 2014) inherent to the problematising of the concept of animal metamorphosis including:

– the obstacle of the non-permanence of the species.
– the obstacle of an instantaneous, radical, intentional metamorphosis.
– the obstacle of the primacy of perception over conceptualisation, which hinders their understanding of the invisible nature of the profound changes which take place during the metamorphosis.

This didactic situation is led by a teacher, who is part of our collaborative research team, with the pupils in her class who are all 6 or 7 years old. They are asked to interpret the last four pages of this picture book after looking at the first four pages. In this opening section, the tadpole character, living in a pond makes a promise to "never change" to the caterpillar character who lives on a branch of a willow tree. However, this promise made out of love is undone as first the tadpole grows front legs, then back legs, despite renewing its promise each time. In the last four pages, the characters are now adults. The butterfly and the frog no longer recognise each other and while they are looking for each other, the frog eats the butterfly. The analysis focusses on two excerpts during an interpretive reading of the end of the story when the frog eats the butterfly (Excerpt 2.1).

Excerpt 2.1
Teacher: How can we be sure that the butterfly is the same as the caterpillar at the beginning of the story?
Émilie: Because we know that at the beginning of the story the caterpillar calls the tadpole 'my shiny black pearl' and then says it again later on.
Teacher: So, it's because it uses the same nickname. Is there anything else which tells us that the butterfly is the caterpillar from the beginning of the story?
Enzo: Because the caterpillar is fat and multi-coloured.
Teacher: Oh, well, for the colours all right, we have the same colours as at the beginning, so that could be a clue; there's the nickname, then the colours. And how do you know the frog is the tadpole? It's not obvious, is it?
Younes: Because at the end the legs have grown.
Teacher: Yes, but maybe it's a different frog; we don't know what happened in the middle of the story. How do we know it's not a different frog?
Clara: Because at the end the frog is waiting for their beautiful rainbow.

In this excerpt, the pupils are invited to think about the identity of the frog and the devoured butterfly: "How can we be sure that the butterfly is the same as the caterpillar at the beginning of the story?" and "How do you know the frog is the tadpole? It's not obvious, is it?" the teacher asks. The aim is to grasp the continuity between the tadpole/frog character and the caterpillar/butterfly character despite the fact that the characters do not recognise each other at the end of the story, so that the pupils can confront the problem of the permanence of living beings during their metamorphosis. The questioning takes place on two levels. On the narrative level, it involves understanding why the characters no longer recognise each other at the end of the story, despite knowing each other at the beginning of the story. On the scientific level, it involves identifying the permanence of a living organism despite its morphology changing over time.

The pupils identify the clues that point to this continuity. The first clue concerns the nickname given to the characters, as expressed by Emilie: "Because we know that at the beginning of the story the caterpillar calls the tadpole "my shiny black pearl" and then says it again later in the book" or as Clara puts it: "The frog is waiting for their beautiful rainbow." The second clue identified is that the caterpillar and the butterfly share similar colours: "Because the caterpillar is fat and multi-coloured" (like the butterfly), says Enzo. The third clue concerns the continuity of form, as suggested by Younes: "Because at the end the legs have grown." It could be argued that the pupils agree on the continuity of the butterfly and frog characters by focussing on the similarities between the adult and larval forms. Although this allows pupils to grasp the idea of the permanence of species, it might also reinforce the idea that the adult form is derived from the larval form and mitigate the profound and invisible changes undergone during metamorphosis. This is precisely what the teacher tries to get the pupils to work on when interpreting Excerpt 2.2.

Excerpt 2.2

Teacher: At the start of the story, the tadpole and the caterpillar were in love. How you would describe their relationship now as a frog and a butterfly?

Nawel: Before, at the beginning, the tadpole lived in the water. Later they arrange to meet, and at the end, the frog didn't realise that it had eaten 'its rainbow'.

Teacher: Could the tadpole have eaten the caterpillar before?

Milo: Well, no, it isn't possible, because if the caterpillar falls into the water, it will die, and if the tadpole comes out of the water, it will die, too.

Teacher: And, therefore, can the frog eat the butterfly?

Milo: Yes, they are still in love, but the frog didn't recognise the caterpillar, so it ate it.

Émilie: In fact, it's not surprising that the frog eats the butterfly because the butterfly is an insect and frogs eat insects.

Teacher: So, for you, the end of the story is logical.

Neila: I kind of agree with Émilie because butterflies are insects, but if it had known that it was the caterpillar which had transformed itself, I'd say that it wouldn't have eaten it.

In this excerpt, the teacher turns the pupils' attention to understanding the ending to the story when the frog eats the butterfly. More specifically, she focuses on the conditions of possibility of this ending: How is the frog able to eat the butterfly at the end of the story when at the beginning of the story this predatory relationship is not possible? She does this by formulating questions which move away from the narrative towards what is possible and impossible: "Now how <u>would</u> you describe their relationship as frog and butterfly" […] "<u>Could</u> the tadpole have eaten the caterpillar before?" […] "And, therefore, <u>can</u> the frog eat the butterfly?"

The first biological reason suggested by the pupils concerns the environment: For the predation to occur, the animals must be in the same environment. As Nawel puts it: "Before, at the start, the tadpole was down below in the water," meaning the tadpole was not in the same environment as the caterpillar and "Later they arrange to meet," meaning they find themselves in the same environment. Milo goes a step further by explaining why it is impossible for the tadpole and the caterpillar to live in the same environment at the beginning of the story: "Well, no, it isn't possible, because if the caterpillar falls into the water, it will die, and if the tadpole comes out of the water, it will die, too." The second reason relates to the animals' diets. "It's not surprising that the frog eats the butterfly because the butterfly is an insect and frogs eat insects" explains Émilie, a statement confirmed by Neila. The pupils do not directly consider the profound transformations themselves, but rather their consequences, i.e. the change in environment and change in diet.

By making sense of the narrative, the pupils come to question the real-life phenomena, working off their own understanding of animal metamorphosis. The pupils work together to explore the possibilities and identify the conditions of possibility governing the transformations of the animal characters, within the restricted framework of the fictional narrative. This leads them to compare and contrast the change/permanence of living organisms' duality, as well as the visible/invisible transformations duality. One of the major benefits of these realistic-fiction picture books is that it allows this problematisation work to take place, something which is not possible when using a non-fiction work which sets out the facts and solutions with little debate.

2.3.2 Methods of Narrative Reasoning and Scientific Problematisation

From our perspective, science education should give pupils access to new ways of seeing the world, new ways of thinking which go beyond what is considered common knowledge. In order to achieve this, it is important to go beyond the teaching of scientific results – *"teaching scientific results is not science education"* (Bachelard (1938, p. 264). It is important that pupils be confronted with biology problems with real epistemological substance and asked to work on them, as this will allow for proper involvement in the process of constructing scientific knowledge.

Furthermore, building on the work by Astolfi (2008), and within the framework of problematisation, we place particular importance on putting pupils in the right conditions to provide them with reasoned access to ways of thinking which go beyond the ordinary – what we call here 'extra-ordinary reasoning'. This is achieved by going beyond the tendency to provide explanations through storytelling. Here, three case studies in nutritional biology are used to investigate the conditions in which pupils can be led to adopt extra-ordinary reasoning: One in a primary school setting and two in secondary school settings. Epistemological and didactic markers are then used to analyse the pupils' statements.

First example: breathing at primary school (pupils, 8–10 years old).

The teacher wanted to go beyond the notion of inhalation and exhalation (air going in and out of the lungs). She explains to the class that the air we breathe is a mix of oxygen and nitrogen, and that nitrogen is not useful for our body, but oxygen is a vital requirement. She shows the pupils a magnified representation of the composition of air (Fig. 2.1a). Then, in groups of three or four, the pupils are asked to answer the following question: "How do all the different parts of your body get oxygen from the air you breathe?" They are asked to produce a diagram with legends and explanations on a poster on which an outline of a human body has already been drawn. Then the groups present and discuss their posters (Orange et al., 2008).

This written work produced by the pupils (for example: group 2, fig. 2.1b) and the discussion show the pupils' explanations have two main characteristics:

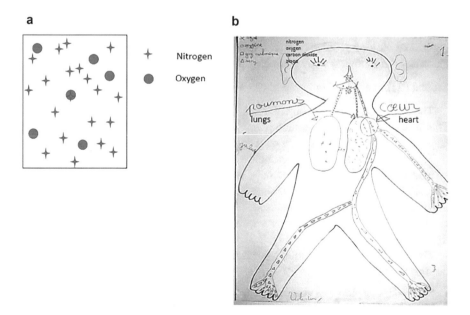

Fig. 2.1 The explanation of breathing proposed by one group of pupils (8–10 years old). (**a**) The main gas components of air (**b**) The poster produced by group 2

– A clear tendency to provide explanations using storytelling

Some of the pupils think about the distribution of blood around the body according to a schematic representation: blood production (in the heart, for example), distribution via the blood vessels and use by the different parts of the body. This is an irrigation-type explanation. It is based on a series of steps which are linked both by chronology and causality. This type of explanation by storytelling is characteristic of human thought (Bruner, 2002).

– Intra-objectal explanations

Let us take a look at what ERW, a pupil in the class, says: "And the heart, when it's too old… the old blood goes back to the heart and then the heart, well, actually…."

He disputes the basic irrigation explanation and proposes the notion that the blood returns to the heart, which leads the class to work on the notion of circulation. However, this circulation is conceived as the old blood returning to the heart to be rejuvenated. In other words, his explanation is based on the quality of the blood considered as a whole, rather than on the idea of the blood as a container or means of transportation. Moreover, we can see that this idea of the blood as a single entity is reproduced in the poster: There is no figurative representation of the oxygen (dots on the top left of the poster) inside the blood vessels, only of the blood. The pupils thus build their explanations based on the qualities assigned to the objects (the blood) rather than the relationships between objects (the blood as a carrier of oxygen): They are intra-objectal rather than inter-objectal (Piaget & Garcia, 1983).

These two types of explanation, storytelling and intra-objectal reasoning, are also proposed by older pupils when tackling problems in nutritional biology or other areas. This contrasts with the scientists' explanations for whom addressing the nutritional biology problem requires going beyond storytelling, using inter-objectal reasoning, and moving away from irrigation and towards circulation.

Second example: blood circulation (pupils, 16–17 years old).

The pupils first work on how the muscles in the body are supplied with oxygen. This is something they had already studied a few years prior. The class does not have much difficulty in agreeing on a double circulation diagram (Fig. 2.2).

The pupils are then asked to complete the diagram, modifying it as they see fit to take into account the fact that not only do the muscles need oxygen, but also nutrients, and that, as well as producing CO_2, they also produce nitrogenous waste including urea, discharged in the urine.

Here are two representative examples of the work produced by pupils when trying to answer this question (Fig. 2.3).

These two explanatory models position all the organs in series. They are based on storytelling:

> The blood is supplied with oxygen in the lungs; then it recovers nutrients in the intestines, before taking everything to the muscle, where it loads up the nitrogenous waste and CO_2 which it eliminates via the kidneys and the lungs, respectively.

Fig. 2.2 Double blood circulation

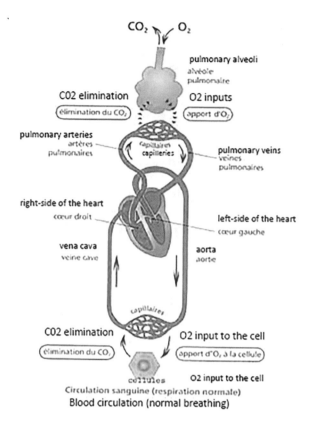

Circulation sanguine (respiration normale)
Blood circulation (normal breathing)

The pupils' attachment to explaining through storytelling is striking when they are given a circulatory model based on current scientific knowledge with which to compare their diagrams (Fig. 2.4).

Many pupils say they do not understand how this model can function. They identify several points as being particularly problematic:

- They think that it is inefficient for the nutrients absorbed into the blood in the digestive tube to go round the whole system before reaching the muscles.
- They see the situation regarding the nitrogenous waste as even more critical. They would prefer for this waste to be taken directly to the kidneys to get rid of it as quickly as possible. However, in this diagram, not only is this waste carried all the way around the system via the heart (twice) and the lungs, but also there is no guarantee it will ultimately pass through the aorta into the renal artery. They are extremely disturbed by the idea of this "dirty" blood flowing around the body.

This difficulty in thinking about the treatment of the nitrogenous waste in the model with the muscles and kidneys connected in parallel is due to the fact that the pupils cannot simply tell the story of what happens to this nitrogenous waste, as the fate of

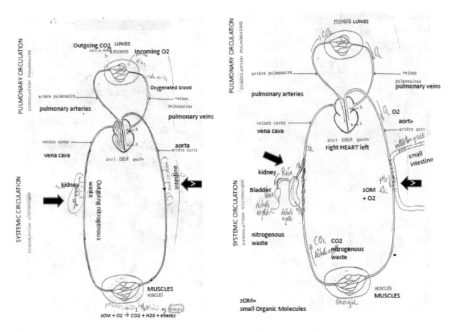

Fig. 2.3 Two examples of pupils' work (16–17 years old). (Orange & Orange, 1995)

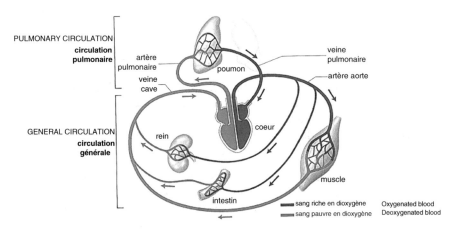

Fig. 2.4 Blood circulation (muscles, intestines and kidneys in parallel)

these molecules is not determined *a priori*. This means they have to change their point of view and use a compartmental explanation, which no longer reasons in terms of the story of what happens to the waste molecules, but in terms of inflows and outflows within the body and the concentration thresholds of the substances in the blood.

Third example: molecular renewal (pupils, 16–17 years old).

The pupils have thought individually and then in a group about how an animal cell works. Comparing the work produced allows them to create a diagram showing an overview of how a living organism functions and how any given cell within this organism works. The diagram summarises the processes (respiration and cell syntheses) referenced by the pupils, but the type of diagram imposed by the teacher is compartmental, showing inputs and outputs of matter. This aims to move away from the conventional blood circulation/cell or organ representations and thereby move away from telling the stories of molecules. The study continued by looking at molecular renewal, which allowed the pupils to complete the diagram, adding to the cell diagram an arrow moving from the big organic molecules (bOM) to small organic molecules (sOM) as the big organic molecules release small organic molecules (Fig. 2.5).

After this session, two pupils asked to see the teacher as they were having trouble making sense of the diagram. They did not understand how it could work:

> I don't know if that is what happens first.' 'So, first of all you eat the SOM, then the SOM go directly to the O_2 and make energy and then you have other SOM which arrive at the ….

These pupils want to tell a story, but as they do so, they get to a point where they cannot continue, as the story does not correspond to the diagram.

This third example again shows the pupils' tendency to explain using storytelling, in which the chronology provides a causal explanation, and intra-objectal reasoning, in which objects have specific properties. These are very real obstacles to

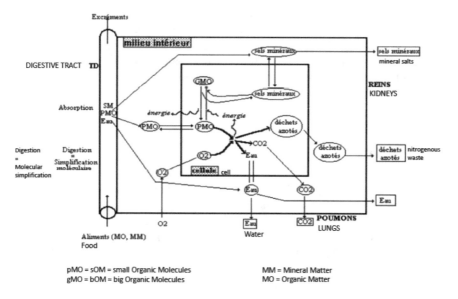

Fig. 2.5 An overview of the functioning of a living organism and its cells. (Orange, 1997)

understanding how complex systems function and to building reasoned scientific knowledge in the field of biology. It is important to take this trend seriously, as these types of ordinary explanations structure pupils' thinking and are found in numerous biology problems. When studying biological systems, two types of extra-ordinary reasoning need to be acquired, namely systemic reasoning and inter-objectal reasoning.

2.4   Discussion

In the theoretical framework of learning by problematisation, should we see narratives as a help or hinderance in the construction of reasoned scientific knowledge?

In terms of modes of reasoning, we can oppose the stories told by the pupils and the scientific explanations. Faced with the same problem, the former will mobilise objects (blood, small organic molecules, etc.), as characters with capacities and/or functions and/or their own intentions, carried along by events, the sequence of which conjugates syncretistically temporality and causality, with the limited constraints resulting in a multiplicity of possibilities. The latter will instead start with possible explanations verified against the impossibilities and necessities. It is, however, interesting to question the types of storytelling concerned. Because in addition to "journey stories" and chronicles (Orange Ravachol & Guerlais, 2005), there are also stories which lead to explanatory impossibilities and thus become levers for building apodictic knowledge.

In terms of the interpretive reading of realistic fiction, we have shown that these can encourage pupils to look for scientific explanations, or at least the biological facts required to understand the plot. In certain didactic conditions, the pupils engage in constructing problems by exploring the possibilities within the restricted framework of a fictional narrative. However, reading these works of fiction does not allow for the testing of necessities, which need to be supported by other types of scientific investigation.

2.5   Conclusion

Given the importance placed on narratives in our common knowledge, it might seem paradoxical to focus attention on them when building apodictic scientific knowledge in the science classroom. There are therefore two choices as regards learning: Research the conditions under which these narratives can be excluded, or study their complexity and look for levers to go beyond the constraints of some of their forms. We have chosen to pursue the second option. Our focus on working on problems encourages exploration and discussion of the explanatory possibilities and impossibilities, in a doubly-restrictive framework, that of the different forms of epistemology we reference, and of the school timetable. In this process, language practices (argumentation, oral productions in interaction with written productions) are fundamental, as well as teacher guidance to stop pupils falling into the trap of putting forward unquestioned proposals, standardised by true and false, and to guide them towards proposals supported with reasons and transformed through the logic of the possible/impossible/necessary. In this way, pupils are led to look objectively at both the conceptions and types of reasoning they use, and the benefits and limitations of the forms of narrative supporting their explanations.

References

Astolfi, J.-P. (2008). *La saveur des savoirs*. E.S.F.

Avraamidou, L., & Osborne, J. (2009). The role of narrative in communicating science. *International Journal of Science Education, 31*(12), 1683–1707.

Bachelard, G. (1938). *La formation de l'esprit scientifique*. PUF.

Barthes, M. (1966). Introduction à l'analyse structurale des récits. *Communications, 8*, 1–27.

Bruguière, C., & Triquet, E. (2012). Des albums de fiction réaliste pour problématiser le monde vivant. *Repères, 45*, 1–22.

Bruner, J. (1996). *Actual minds, possible worlds*. Harvard University Press.

Bruner, J. (2002). *Making stories: Law, literature, life*. Harvard University Press.

Eco, U. (1979). *Lector in fabula. Le rôle du lecteur ou la coopération interprétative dans les textes narratifs*. Grasset.

Egan, K. (1986). *Teaching as storytelling: An alternative approach to teaching and curriculum in the elementary school*. Althouse Press.

Fabre, M., & Orange, C. (1997). Construction des problèmes et franchissement d'obstacles. *ASTER, 24*, 37–57.

Hadzigeorgiou, Y. (2016). *Imaginative science education. The central role of imagination in science education*. Springer.

Orange, C. (1997). *Modèles et modélisation en biologie: quels apprentissages pour le lycée ?* PUF.

Orange, C., & Orange, D. (1995). Biologie et géologie, analyse de quelques liens épistémologiques et didactiques. *ASTER, 21*, 27–49.

Orange, G., Lhoste, Y., & Orange Ravachol, D. (2008). Argumentation, problématisation et construction de concepts en classe de sciences. In C. Buty & C. Plantin (Eds.), *L'argumentation en classe de sciences* (pp. 75–116). INRP.

Orange Ravachol, D. (2017). Récits des élèves et récits des scientifiques dans les sciences de la Nature. *Cahiers de Narratologie, 32*.

Orange Ravachol, C., & Guerlais, M. (2005). *Construction de savoirs et rôle des enseignants dans une situation de débat scientifique à l'école élémentaire: comparaison de deux cas*. Actes du 5e Colloque International Recherche(s) et Formation, Nantes.

Piaget, J., & Garcia, R. (1983). *Psychogenèse et histoire des sciences*. Flammarion.

Popper, K. (1985). *Conjectures et réfutations*. Payot.

Popper, K. (1991). *La connaissance objective*. Aubier.

Reboul, O. (1992). *Les valeurs de l'éducation*. PUF.

Reiss, M., Millar, R., & Osborne, J. (1999). Beyond 2000: Science/biology education for the future. *Journal of Biological Education, 33*(2), 68–70.

Rumelhard, G. (1995). Permanence, métamorphose, transformation. *Biologie-Géologie, 2*, 333–345.

Triquet, E., & Bruguière, C. (2014). Album de fiction, obstacles sur la métamorphose et propositions didactiques. *RDST, 9*, 51–78.

Willis, J., & Ross, T. (2010). *La promesse*. Gallimard.

Chapter 3
Using External Representations to Support Mathematical Modelling Competence in Biology Education

Benjamin Stöger and Claudia Nerdel

3.1 Models in Modern Society

Models are one of the most important tools in the natural sciences. This became very clear during the coronavirus disease 2019 (COVID-19) pandemic in 2020. Mathematical models were used to predict the course of the pandemic. Government action was considered and integrated into the forecasts. This is an example of the use of models and modelling as a central aspect of scientific inquiry. Modelling tasks are characterised by their interdisciplinarity. This becomes clear especially through problems in everyday life. Due to their interdisciplinary character, these can usually only be solved with the help of several different fields.

Therefore, a setting from the field of biochemistry is particularly suitable for teaching biological concepts and working methods, as it already represents a recognised interdisciplinary field. In addition, it provides the basis for central subject areas of biology, such as cell biology (e.g. lipid bilayers), animal and plant physiology (e.g. glycolysis and photosynthesis) or genetics (e.g. how RNA polymerase works). This shows that both chemical and biological subject concepts are necessary for this subject. Therefore, it must be considered fundamental to the teaching of modern STEM-classes to include interdisciplinary tasks. Müller et al. (2018) concluded mathematical knowledge as a predictor of success in science. Mathematics is seen as a catalyst for achieving competence in the natural sciences. This implicates the importance of mathematical knowledge for understanding models. However, at this point it must be emphasised that scientific knowledge is also necessary to derive a valid mathematical model from a scientific model. Nevertheless, mathematics is one source of students' difficulties in learning natural sciences (Müller et al., 2018).

B. Stöger (✉) · C. Nerdel
Life Science Education, Technical University of Munich, Munich, Germany
e-mail: benjamin.stoeger@tum.de; claudia.nerdel@tum.de

© The Author(s) 2024
K. Korfiatis et al. (eds.), *Shaping the Future of Biological Education Research*,
Contributions from Biology Education Research,
https://doi.org/10.1007/978-3-031-44792-1_3

3.2 Theoretical Background

3.2.1 (Mathematical) Models

Mahr (2008) distinguished between the model and the modelled object. The model focuses on the original idea; the modelled object is mostly what is understood as the model. This heterogeneity of models (in combination with models' illustration feature 'models being representations') allows the derivation of additional domain-specific functions of models. An example is the elimination of misconceptions through conceptual change (Krüger, 2007; Posner et al., 1982). Furthermore, the simplifications that are made possible by models improve the communication of technically correct information (Krell et al., 2019).

Therefore Mahr (2008) distinguished between two perspectives: model usage and model creation. The 'creation' perspective focusses on the induction of views, experiences, measurements, characteristics, findings, or rules through selection and generalisation (Grünkorn et al., 2014; Upmeier zu Belzen & Krüger, 2010). The opposite is true from the perspective of 'usage'. The deduction of a modelled object to facilitate multiple applications of the model is foregrounded (Mahr, 2008). Mathematical models are understood as models based on mathematical relations, functions or methods.

Reiss and Hammer (2012) described mathematisation as an 'essential element of working in physics,' thus spanning the important arc from mathematics to the natural sciences. The corresponding mathematical modelling has been described by Eck et al. (2017) as the translation of concrete problems in applied sciences into well-defined mathematical tasks. They described two essential aspects of mathematical models: A concrete application problem as a basis, as well as the simplification of the application problem during the modelling process.

3.2.2 Modelling Competence

The ability to construct, apply and evaluate models is understood as modelling competence (Grünkorn et al., 2014; Upmeier zu Belzen & Krüger, 2010; Blum & Leiss 2007). They distinguished specialist knowledge from modelling competence, which is assigned primarily to knowledge acquisition. They identified two dimensions of modelling competence: 'knowledge about models' and modelling' (Upmeier zu Belzen & Krüger, 2010, p. 50). Upmeier zu Belzen and Krüger (2010) defined the aspects of the 'knowledge about models' dimension as the ability to understand models as reconstructions from the creation perspective and to consider the possibility for alternative reconstructions. Whereas modelling describes the process of gaining knowledge, this process includes different steps. Therefore, they identified three aspects of the modelling dimension: The dimension 'purpose of models' refers to the understanding of models as useful reconstructions; the dimensions 'testing' and 'revision' refer to the ability to test or to apply models to phenomena and to modify them if necessary (Grünkorn et al., 2014; Upmeier zu Belzen & Krüger, 2010). They identified three levels for each sub-competence. The levels differ in the description of the relationship between the model and reality, and the role of concepts in dealing with models. At the first level, models are classified by their level of representation. At the second and third levels, the classification is based on the usage and creation perspectives (Mahr, 2008). The levels are not hierarchical (Mahr, 2008).

For the special case of mathematical modelling, Borromeo Ferri et al. (2013) developed even more precise components, which are based on the modelling competences identified by Maaß (2004) and Kaiser (2007). These result from the different steps within the mathematical modelling process that are necessary to successfully complete a process. The components are understanding, simplifying, mathematising, interpreting, validating, and communicating. The components are deduced sequentially, thereby facilitating the development of modelling cycles.

3.2.3   Modelling Cycles

Modelling cycles allow chronological splitting of the whole modelling process and provide insight into the modelling process (Grünkorn et al., 2014; Upmeier zu Belzen & Krüger, 2010).

Mathematical Modelling Cycle (Blum & Leiss, 2005)
Blum and Leiss (2005) (Fig. 3.1, blue), Borromeo Ferri (2006), and Blum and Borromeo Ferri (2009) developed the mathematical modelling cycle.

The first category is 'real situation'. It describes a macroscopic problem, task or phenomenon. Followed by creating a mental representation, the situation model, on the base of the real situation (Blum & Borromeo Ferri, 2009; Blum & Leiss, 2005). It corresponds to the constructed multiple mental representations of the presented situation, as described by Kintsch and van Dijk (1983) for texts and Schnotz (2001) for pictures and diagrams. The following real model is created through the addition of knowledge from memory to the situation model (Borromeo Ferri, 2006). The developed real model is formalised into a mathematical model through additional mathematical knowledge (Kimpel, 2018). The resulting description of the mathematical model allows the use of mathematical methods to determine a mathematical result (Borromeo Ferri, 2006). The translation or interpretation of the mathematical results into real results occurs before the final step, 'validation' (Borromeo Ferri, 2006). Validation is performed to determine the plausibility of the real results.

Chemical Modelling Cycle (Schmidt & Di Fuccia, 2012)
Schmidt and Di Fuccia (2012) applied the modelling cycle to chemistry (Fig. 3.1, green). They inserted an additional area 'chemistry' between mathematics and reality (referred to as 'the rest of the world'). Accordingly, 'the rest of the world' area represents only the macroscopic area of a subject. Furthermore, the mathematical

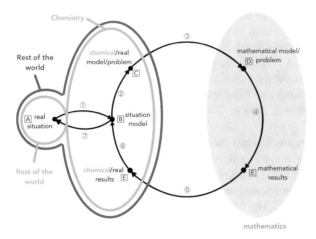

Fig. 3.1 Mathematical modelling cycle and mathematical modelling cycle for chemical contexts. (Blum & Leiss, 2005; Schmidt & Di Fuccia, 2012)

results need to be checked not only for mathematical, but also for chemical plausibility. This translation from macroscopic phenomena to mathematical symbols indicates that chemical phenomena can be described through models at different aspects according to Johnstone (1991). Thus, phenomena can be described or explained through symbolic and sub-microscopic models and representations (Kimpel, 2018).

3.2.4 Multiple External Representations

Each representation has strengths, such as organising diagrams (Larkin & Simon, 1987). Multiple representations, like simultaneous presentations of familiar and unfamiliar representations, can help to prevent misinterpretation (Ainsworth, 1999).

Schnotz and Bannert (2003) identified two channels for the cognitive analysis of descriptive representations (e.g. text) and depictional representations (e.g. diagrams). Descriptive representations are processed in two steps sub-semantically and semantically in learners' working memory. Through sub-semantic processing, words, word groups and complete sentences are identified and analysed (Schnotz, 2001, 2002). A mental representation of the text surface is then formed. The text surface, organised in cognitive schemata, can be activated during semantic processing and enables the creation of a coherent propositional representation (Schnotz, 2001, 2002). The second channel focusses on the processing of depictional representations. Information is also processed first sub-semantically, through the perception of the image as a whole, as well as its graphic design elements (Schnotz, 2001).

The mental model of the illustration is constructed from this visual image through semantic processing (Schnotz, 2001). This informed the structural model developed by Lachmayer et al. (2007) to describe the processing of diagrams. There are close interactions between mental models and propositional representations (Schnotz & Bannert, 2003). This includes mutual transference via the cognitive processes of model inspection or model construction, whereby new information can be generated in the process (Schnotz, 2001). This suggests that descriptive representations are processed internally to become propositional representations, as well as mental models, and vice versa (Schnotz & Bannert, 2003). According to Gilbert and Treagust (2009), the interaction between mental models and propositional representations connects two main types of representations: model type and symbolic type. The ability to switch between different representations is important for understanding science ('scientific literacy') (Norris & Phillips, 2003). The mediation of the relationships among representations is achieved through the translation process in which the content-related connections become apparent and understanding increases (Ainsworth, 1999).

3.3   Research Question

According to the PISA study, pupils have difficulties in representing situations mathematically (e.g. in recognising mathematical structures, regularities, relationships and patterns). In addition, there are also difficulties in evaluating mathematical solutions and placing them in the context of the modelling tasks (Edo, Putri & Hartono, 2013). Jankvist & Niss (2020) came to a similar conclusion. They were also able to identify mathematisation as one of the crucial difficulties in developing solutions. In addition, the structuring of the tasks (here called 'pre-mathematisation') was a problem for the students. These findings on student difficulties on modelling and the positive effects of multiple external representations on learning (Ainsworth, 1999) led to the following questions and hypotheses:

Question 1: Do external representations have an effect on (mathematical) modelling competence in biochemistry?

Leads to Hypothesis 1: External Representations have a positive effect on (mathematical) modelling competence in biochemistry.

Question 2: Does mathematical expertise have an effect on (mathematical) modelling competence in biochemistry?

Leads to Hypothesis 2: Mathematical expertise has a positive effect on (mathematical) modelling.

Question 3: Are external representations helpful for learners with less prior knowledge about the modelling process?

Leads to Hypothesis 3: External representations have a positive effect on modelling processes of learners with less prior knowledge.

3.4 Methods

3.4.1 Sample

Twelve individuals (M = 24.3 years old, SD = 2.06) participated in the study. Eleven were Technical University of Munich students of science and one was a high school graduate. In order to obtain the broadest possible distribution of prior knowledge, participants without a university place were also included. For the study, students who were studying both mathematics and chemistry ($n = 6$) were defined as mathematics experts. The other participants ($n = 6$) are referred to as mathematics novices (Table 3.1). Their study subjects are natural science, such as biology, chemistry or health science, but not mathematics. The students were randomly assigned to the experimental conditions.

3.4.2 Experimental and Learning Environment Design

A two-factorial experimental design was used. The first independent variable, learning environment design, was systematically modified regarding the used representations (IV1, triple-staged; see IV1.a [text and symbol], IV1.b [text, symbol and diagram], and IV1.c [text, symbol and image]). The participants' mathematical expertise was checked as a second factor (double-staged; see IV2.a [mathematic experts] and IV2.b [mathematic novices]; Table 3.1). The participants worked on a task related to enzyme kinetics using pen and paper. The learning environment included examples for the participants to elaborate on. Using methanol poisoning as the task topic, the students had to determine the amount of ethanol needed to treat poisoning. The task was divided into six subtasks based on the steps in Schmidt and Di Fuccia's (2012) chemical modelling cycle (see also Sect. 3.2.4).

Table 3.1 Research design

	IV2.a Mathematics experts	IV2.b Mathematics novices
IV1.a Text + symbol (TS)	2 participants	2 participants
IV1.b Text + symbol + diagram (TDS)	2 participants	2 participants
IV1.c Text + symbol + image (TBS)	2 participants	2 participants

3.4.3 Survey Method

The relevant participant characteristics were obtained from a questionnaire. The participants had to use the 'thinking aloud' method (Sandmann, 2014) while working on the learning environment. Their verbal utterances were recorded and transcribed in accordance with the guidelines by Kuckartz (2008). Qualitative content analysis was performed in accordance with Mayring (2010) on the basis of the elaboration profiles generated by the participants.

Frick's (2019) category system was used (Table 3.2). This was originally used to gather students' mental models while solving mathematical modelling tasks. The category system was deductively extended in accordance with the modelling competences (category 3: modelling competence) identified by Upmeier zu Belzen and Krüger (2010) and Grünkorn et al. (2014). First- and second-level subcategories were also created. These were developed on the basis of the model by Upmeier zu Belzen and Krüger (2010). Accordingly, there were two subcategories based on the dimensions 'knowledge about models' (3.1) and 'modelling' (3.2). For those two, concrete subcategories were developed based on the levels described in the model (e.g., 'statements about the relationship between the model object and the initial object' (3.1.2)). A total of three such categories were developed for 3.1 and eight for 3.2. The 'monitoring' category concluded with statements of understanding, misunderstanding, astonishment, doubt, and joy.

3.4.4 Evaluation

For the qualitative analysis, the independent variables, experts and novices, as well as TS, TDS and TBS, grouped the participants. The dependent variable, i.e. the participants' 'elaboration profiles', was evaluated by qualitative content analysis based on Mayring (2010). The profiles were analysed with MAXQDA 12 (version 12; VERBI). For this purpose, the statements from the elaboration protocols were assigned to the corresponding categories of the category system. The frequencies for the individual and cumulative subcategories were considered. For every category, contingency tables were generated. They were the base for chi-square tests of independence for selected categories.

Table 3.2 Main categories

1. Adding knowledge by retrieving related knowledge
2. Adding knowledge by generating knowledge through inference
3. Modelling competence
4. Monitoring

3.5 Results

3.5.1 Categories

In order to examine model competence, category 3 was divided into two subcategories: 'knowledge about models' and 'modelling' (see Sects. 3.2.3 and 3.4.3). The individual sub-competences within a dimension were used together with the levels to differentiate within the category. To examine modelling competence, a distinction was made between knowledge about models and modelling (Upmeier zu Belzen & Krüger, 2010). Therefore category 3.1 'knowledge about models' focusses on comparisons of reality and the model at an abstract level (3.1.1) and a concrete level (3.1.2). The simplification of a real situation (category 3.2.1) was identified as the 'purpose of models' sub-competence. The use of models to describe phenomena and to evaluate results was characterised by two categories (3.2.2 & 3.2.3). The comparison of the model to reality and the validation of the model were understood as model testing (category 3.2.5). The ability to make revisions on the basis of the test results was a sub-competence (3.2.6).

3.5.2 Elaboration Profiles and Modelling Strategies

To explore elaboration, the elaboration profiles (superficial + deep) developed by Lind et al. (2005) were applied to the relevant categories. Superficial elaboration includes (re-)reading a text, unsuccessful attempts at remembering, retrieving information in the text and the creation of inferences to develop a text base (Lind et al., 2005). Deep elaboration is characterised by the creation of inferences that facilitate the development of a situation model and the retrieval of knowledge from raw content.

Analogous modelling strategies were constructed on the basis of the levels identified by Upmeier zu Belzen and Krüger (2010). Level 1 was defined as *recognition and application of models*, and Levels 2 and 3 were defined as *modelling* (Table 3.3).

3.5.3 Qualitative Analysis

3.5.3.1 Comparison of the Effect Different Representations

Some of the frequencies for the individual categories related to the usable external representations were noteworthy, e.g., '2.1.4 description of solutions' (Category which includes all descriptions of solutions). The absolute values for 'additional representations' were twice as large as those for the control condition

Table 3.3 Assignment rules for modelling strategies

Modelling strategy	Recognition and application of models: M_n	Modelling M_e
Assigned categories	*Assessment of the level of abstraction and similarity to the modelled phenomenon (3.1.1)* *Explanation of the relationship between the model and the initial object (3.1.2)* *Use of model to describe a phenomenon (3.2.2)*	*Estimates and simplification of the modelling process (3.2.1)* *Check the model for fit with the original object (3.2.5)* *Revise the model to reflect new knowledge about the phenomenon (3.2.7)* *Revise model because of falsified hypotheses (3.2.8)*

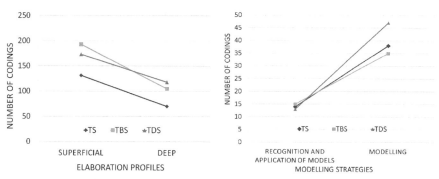

Fig. 3.2 Frequencies (Total number of expressions made by students in these categories) for superficial and deep elaboration profiles (left) and modelling strategies (right) by representation

$[h_{TBS}(2.1.4) = 48, h_{TDS}(2.1.4) = 44, h_{TS}(2.1.4) = 27]$. The explication of the mathematical reasoning (Categories 2.5) improved with additional external representations (cumulative frequencies in Category 2.5: $h_{TS}[2.5] = 67$, $h_{TBS}[2.5] = 101$, $h_{TDS}[2.5] = 113$).

Regarding elaboration (Categories 1.1.1 ('Successful recall of knowledge from the information text'), 1.3 ('Explicit use of external sources of knowledge'), 2.1 ('Inferences that serve to build a text base'), 1.1.3 ('Successful recall of knowledge from memory'), and 2.5 ('Inferences that serve to build an integrated situation model'), all the participants engaged more often in superficial elaboration $[E_s(TS) = 131, E_s(TBS) = 193, E_s(TDS) = 173]$ rather than deep elaboration $[E_d(TS) = 70, E_d(TBS) = 105, E_d(TDS) = 118]$. There were differences between the groups (Fig. 3.2). The TBS group often engaged in superficial elaboration; the TDS group in deeper elaboration.

The frequencies for 'recognition and application of models' were nearly the same in all three test groups $[M_n(TS) = 14, M_n(TBS) = 13, M_n(TDS) = 15]$. However, there were differences in the frequencies for modelling $[M_e(TS) = 38, M_e(TBS) = 35, M_e(TDS) = 47]$. The participants with access to additional diagrams created models more often.

3.5.3.2 Expertise

Experts paraphrased mathematically (Category 2.1.4) twice as high as the novices $[h_E(2.1.4) = 41, h_N(2.1.4) = 18]$. There was also a difference in the quality of the paraphrasing. Novices paraphrased mainly concrete numerical values ('If I see this correctly, we have an alpha of 1 plus 559.9, i.e. 560 approximately or 561 betters' (TS & N [novice])). Experts tended to paraphrase qualitatively ('The Michaelis–Menten equation is … for very high concentrations, the Km in the denominator is [negligible], and it can be considered void in the limit value [of the Michaelis–Menten equation]' (TDS & E)). Novices orientated themselves through quantitative data: 'To reach the value 1, c(inhibitor) would have to be 1×10 to the power of -5'. In contrast, the expert group used mathematical paraphrases much more frequently ('So [in] Michaelis–Menten kinetics, the current velocity v is equal to Vmax times c of the substrate divided by the specific Km value plus c of the substrate' (TBS & E)).

Expertise seemed to have a positive effect on the ability to integrate representations ($h_E(2.3) = 18, h_N(2.3) = 4$). The examination of the elaboration profiles and modelling strategies (Fig. 3.3) revealed different tendencies, regarding the participants' expertise.

Both groups more likely engaged in superficial elaboration $[E_s(\text{Novices}) = 166, E_d(\text{Novices}) = 77; E_s(\text{Experts}) = 331, E_d(\text{Experts}) = 216]$. It was evident that the expert group more likely engaged in elaboration. The same picture emerged for the modelling strategies. Again, the values for the novices were much lower. $[M_N(\text{Novices}) = 1, M_N(\text{Experts}) = 41, M_E(\text{Novices}) = 24, M_E(\text{Experts}) = 79]$. Therefore, experts used modelling strategies much more frequently than the novices.

3.5.3.3 Chi-Square Test

A chi-square test of independence was performed on the numbers of verbal utterances of all categories to identify correlation between variables. A selection of categories that showed statistical correlation within the expertise groups are listed in Table 3.4.

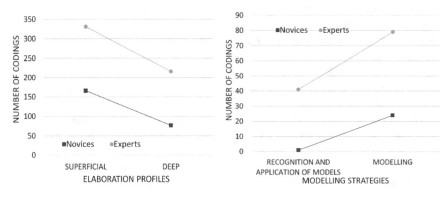

Fig. 3.3 Frequencies (Total number of expressions made by students in these categories) for superficial and deep elaboration profiles (left) and modelling strategies (right) by expertise

Table 3.4 Test of independence

Category	Degree of freedom	n	χ^2	p
2.1.4 (Description of solution paths)	2	119	12.277	.0021[*]
2.5.4 (Mathematical reasoning)	1	70	9.131	.0025[**]
2.5.5 (Calculating sought-after quantities)	1	70	4.536	.033[*]
E_s (Superficial elaboration)	2	497	4.935	.084
E_d	2	293	12.861	.0016[**]
M_N	1	103	15.482	.00008[***]

Significance levels: [*] = p < .05 ; [**] = p < .01 ; [***] = p < .001

The results revealed a statistically highly significant correlation between representation and the participants' expertise in the description of solutions (2.1.4) and the calculation of the sought-after variables (2.5.5). For 'mathematical reasoning' (2.5.4), there was also a highly significant correlation between the participants' expertise and the use of additional external representations.

The elaboration profiles yielded different results. There was no statistically significant correlation for superficial elaboration (E_s). Analysis of deep elaboration (E_d) indicated a highly significant correlation between expertise and external representations. The correlation between mathematical expertise and additional external representation was highly significant regarding modelling strategy M_N (Recognition and Application of models).

3.6   Discussion

3.6.1   Method

It should be noted that there are limitations regarding the generalisability of the results. These results were produced in a biochemical context and are meaningful only in that context. Another limitation is the small sample, which emphasises its qualitative character. To make quantitative conclusions, the sample would need to be larger and more representative (Kroß & Lind, 2001). This would make the results of frequency analyses and inferential statistics more meaningful.

3.6.2   Results

Hypothesis 1 The results indicate that additional external representations can have a positive influence on modelling competence, but learning with diagrams requires greater cognitive effort because of the abstract structural similarity. Learners might lack the additional cognitive capability for text–diagram integration (Schnotz, 2001, 2002). Looking at the sample as a whole, a positive effect is nevertheless evident. The situation is different when the test groups (Experts vs. Novices) are considered individually (See hypothesis 3).

The hypothesis regarding mathematical conclusions was confirmed. The learning environment with images contained the highest number of codes. This implies that the participants with additional images modelled more than the group without additional representations. Since the learning environment differs only through the representations it contains, the additional modelling work can be explained by the representations. Category 3.1.1 (Assessment of the degree of similarity and abstraction of the model) revealed an advantage for the variant without additional representation. A possible explanation is that the images contained implicit estimates or simplifications that did not require prior elaboration. The number of codes with deep elaboration and different representations was at least half that of the number of codes with superficial elaboration. This indicates differences in the effectiveness of images and diagrams. Images seem to invite superficial elaboration because of their concrete similarity to the represented object, and diagrams require more cognitive power and stimulate deep elaboration (Lowe, 1993; Schnotz, 2002).

In summary, pictorial representations were found to have a positive effect on modelling competence in biochemistry. This is primarily based on the increased frequency of modelling (category M_e), as well as the superficial and deep elaborations in all categories relevant to modelling.

Hypothesis 2 The positive influence of mathematical expertise on mathematical modelling competence was also confirmed. On the one hand, this can be explained by the frequency of the statements in the modelling categories (Fig. 3.3 right). On the other hand, it can also be explained from the significant chi-square test carried out for the category M_N.

The experts argued mainly on a mathematical level. The novices avoided mathematical abstraction. Experts included mathematical concepts in task definitions and contributed additional mathematical knowledge.

Experts were more actively involved in the creation of situation models than novices. It was evident that mathematical conclusions required appropriate mathematical expertise to achieve text–image integration that fosters learning. In contrast, novices experienced difficulty in applying external representations to mathematical conclusions. This corresponds to the intrinsic cognitive load component of the cognitive load theory (Sweller et al., 1998). The examination of the knowledge about models indicated that the experts made statements more frequently (Experts 39 times; Novices 2 times). The recognition and description of the model in relation to the phenomenon under study are an important step in model construction (Grünkorn et al., 2014; Upmeier zu Belzen & Krüger, 2010). The experts in all the test groups described the relationship between the real object and the model. This indicated a deeper understanding of the mathematical model. There were also differences in validation strategies. The experts tended to estimate their results on the basis of their prior knowledge (Category 1 'Adding knowledge from memory', experts 147 times; novices 70 times). The novices tended to make their assessments intuitively. This is consistent with validation strategies formulated by Borromeo Ferri (2006).

The results of the chi-square test indicated that in modelling, representation and expertise seemed to be mutually influential. Experts used external representations twice as often as novices. This suggests that external representations help experts to generate estimates and simplifications.

Hypothesis 3 The general positive effect of additional external representation from hypothesis 1 could only be observed for the expert. The hypothesised positive effect of representations on novices was not confirmed. The findings indicate that external representations can be fully used only with appropriate expertise. Experts used additional representations more frequently and effectively. Kozma and Russell (1997) found that experts in chemistry were more likely to use multimedia representations than novices. This contrasts with Schnotz (2001), who found that learners with little prior content-specific knowledge achieved greater success in multimedia learning environments (Schnotz, 2001). The reason was that the variety in the presentation of information stimulated the construction of multiple mental representations. Therefore, Category 3.1.1 indicated that external representations were helpful, especially for experts, in the generation of estimates and simplifications.

3.7 Conclusions

The results suggest that external representations and mathematical expertise have a positive influence on mathematical modelling competence with the limitation that it is mainly experts who benefit from additional representations. Because this mixed-methods approach was used in combination with a small sample, these results serve only as a source of inspiration for future studies. Studies with larger samples that allow for quantification would be the logical next step. Participants could be classified as mathematics, biology or chemistry experts and novices. Other subject areas, such as epidemiology or physical chemistry, could be studied to determine the role of prior discipline-specific knowledge.

The application of the results to competence-oriented academic tasks would reveal the positive effects of multimedia representations. It should also be noted that the use of additional representations for performing mathematical modelling tasks had a positive effect, especially for high-performing students; however, this was not the case for underachieving students.

It is therefore important to determine appropriate designs to promote the success of low-achieving learners in understanding intensive modelling strategies. The use of dynamic representations might better emphasise the model character. This could provide additional support to students as they learn about modelling. A further possibility for instructional design would be to explain and demonstrate the mathematical modelling cycle over time to clearly define the modelling process.

The equally low rate on modelling M_e for the experts and novices indicated that the defined modelling strategy, especially model criticism and modification, was seldom used in the school environment. There, learners seem to get a wrong picture of models as something absolute. However, modelling, the relevant working method, is hardly ever taught. Perhaps different teaching methods can eliminate the misconception of the 'absolute model' through conceptual change (Posner et al., 1982). This could be achieved by active engagement in model criticism.

References

Ainsworth, S. (1999). The functions of multiple representations. *Computers & Education, 33*(2–3), 131–152.

Blum, W., & Borromeo Ferri, R. K. (2009). Mathematical modelling: Can it be taught and learnt? *Journal of Mathematical Modelling and Application, 1*(1), 45–58.

Blum, W., & Leiss, D. (2005). Modellieren im Unterricht mit der 'Tanken'-Aufgabe. *Mathematik Lehren, 128*, 18–21.

Blum, W., & Leiss, D. (2007). Investigating quality mathematics teaching: The DISUM project. Developing and researching quality in mathematics teaching and learning. *Proceedings of MADIF, 5*, 3–16.

Borromeo Ferri, R. (2006). Theoretical and empirical differentiations of phases in the modelling process. *ZDM – Mathematics Education, 38*(2), 86–95.

Borromeo Ferri, R., Greefrath, G., & Kaiser, G. (Eds.). (2013). *Realitätsbezüge im Mathematikunterricht. Mathematisches Modellieren für Schule und Hochschule: Theoretische und didaktische Hintergründe.* Springer Fachmedien. https://doi.org/10.1007/978-3-658-01580-0

Eck, C., Garcke, H., & Knabner, P. (2017). *Mathematical modeling* (Springer undergraduate mathematics series). Springer.

Edo, S. I., Putri, R. I. I., & Hartono, Y. (2013). Investigating secondary school students' difficulties in modeling problems PISA-model level 5 and 6. *Journal on Mathematics Education, 4*(1), 41–58.

Frick, D. E. (2019). *Statische und dynamische Repräsentationen als Unterstützung bei mathematischen Modellierungsaufgaben in der Biologie.* University Library of TU Munich.

Gilbert, J. K., & Treagust, D. F. (2009). *Multiple representations in chemical education* (Vol. 4, pp. 1–8). Springer.

Grünkorn, J., Upmeier zu Belzen, A., & Krüger, D. (2014). Assessing students' understandings of biological models and their use in science to evaluate a theoretical framework. *International Journal of Science Education, 36*(10), 1651–1684. https://doi.org/10.1080/09500693.2013.873155

Jankvist, U. T., & Niss, M. (2020). Upper secondary school students' difficulties with mathematical modelling. *International Journal of Mathematical Education in Science and Technology, 51*(4), 467–496.

Johnstone, A. H. (1991). Why is science difficult to learn? Things are seldom what they seem. *Journal of Computer Assisted Learning, 7*(2), 75–83.

Kaiser, G. (2007). Modelling and modelling competencies in school. In *Mathematical modelling (ICTMA 12): Education, engineering and economics*, pp. 110–119.

Kimpel, L. (2018). *Aufgaben in der Allgemeinen Chemie: Zum Zusammenspiel Von Chemischem Verständnis und Rechenfähigkeit* (Studien Zum Physik- und Chemielernen Ser: v.249). Logos Verlag.

Kintsch, W., & van Dijk, T. A. (1983). *Strategies of discourse comprehension.* Academic.

Kozma, R. B., & Russell, J. (1997). Multimedia and understanding: Expert and novice responses to different representations of chemical phenomena. *Journal of Research in Science Teaching, 34*(9), 949–968.

Krell, M., Walzer, C., Hergert, S., & Krüger, D. (2019). Development and application of a category system to describe pre-service science teachers' activities in the process of scientific modelling. *Research in Science Education, 49*(5), 1319–1345. https://doi.org/10.1007/s11165-017-9657-8

Kroß, A., & Lind, G. (2001). Einfluss des Vorwissens auf Intensität und Qualität des Selbsterklärens beim Lernen mit biologischen Beispielaufgaben. *Unterrichtswissenschaft, 29*(1), 5–25.

Krüger, D. (2007). Die conceptual change-Theorie. In D. Krüger & H. Vogt (Eds.), *Theorien in der biologiedidaktischen Forschung: Ein Handbuch für Lehramtsstudenten und Doktoranden* (pp. 81–92). Springer.

Kuckartz, U. (2008). *Qualitative evaluation: Der Einstieg in die Praxis* (2., aktualisierte Aufl.). VS, Verlag für Sozialwissenschaften.

Lachmayer, S., Nerdel, C., & Prechtl, H. (2007). Modellierung kognitiver Fähigkeiten beim Umgang mit Diagrammen im naturwissenschaftlichen Unterricht (Modelling of cognitive abilities regarding the handling of graphs in science education). *Zeitschrift Für Didaktik der Naturwissenschaften, 13*, 161–180.

Larkin, J. H., & Simon, H. A. (1987). Why a diagram is (sometimes) worth ten thousand words. *Cognitive Science, 11*(1), 65–100.

Lind, G., Friege, G., & Sandmann, A. (2005). Selbsterklären und Vorwissen. *Empirische Pädagogik, 19*(1), 1–27.

Lowe, R. K. (1993). Constructing a mental representation from an abstract technical diagram. *Learning and Instruction, 3*(3), 157–179.

Maaß, K. (2004). Mathematisches modellieren im Unterricht—Ergebnisse einer empirischen studie. *Journal für Mathematik-Didaktik, 25*(2), 175–176.

Mahr, B. (2008). Ein Modell des Modellseins. Ein Beitrag zur Aufklärung des Modellbegriffs. In U. Dirks & E. Knobloch (Eds.), *Modelle* (pp. 187–218). Peter Lang.

Mayring, P. (2010). Qualitative Inhaltsanalyse. In G. Mey & K. Mruck (Eds.), *Handbuch Qualitative Forschung in der Psychologie*. VS Verlag für Sozialwissenschaften. https://doi.org/10.1007/978-3-531-92052-8_42

Müller, J., Stender, A., Fleischer, J., Borowski, A., Dammann, E., Lang, M., & Fischer, H. E. (2018). Mathematisches Wissen von Studienanfängern und Studienerfolg. *Zeitschrift für Didaktik Der Naturwissenschaften, 24*(1), 183–199.

Norris, S. P., & Phillips, L. M. (2003). How literacy in its fundamental sense is central to scientific literacy. *Science Education, 87*(3), 224–240.

Posner, G. J., Strike, K. A., Hewson, P. W., & Gertzog, W. A. (1982). Accommodation of a scientific conception: Toward a theory of conceptual change. *Science Education, 66*(2), 211–227.

Reiss, K., & Hammer, C. (2012). *Grundlagen der Mathematikdidaktik: eine Einführung für den Unterricht in der Sekundarstufe*. Springer.

Sandmann, A. (2014). Lautes Denken–die Analyse von Denk-, Lern-und Problemlöseprozessen. In D. Krüger, I. Parchmann, & H. Schecker (Eds.), *Methoden in der naturwissenschaftsdidaktischen Forschung* (pp. 179–188). Springer.

Schmidt, I., & Di Fuccia, D.-S. (2012). Mathematical models in chemistry lessons. *Giornale Di Didattica E Cultura Della Società Chimica Italiana, 34*(3), 331–335.

Schnotz, W. (2001). Sign systems, technologies, and the acquisition of knowledge. In *First international seminar on using complex information systems*, pp. 9–29.

Schnotz, W. (2002). Wissenserwerb mit Texten, Bildern und Diagrammen. In L. J. Issing & P. Klimsa (Eds.), *Information und Lernen mit Multimedia und Internet: Lehrbuch für Studium und Praxis* (3rd ed., pp. 65–81). Beltz PVU.

Schnotz, W., & Bannert, M. (2003). Construction and interference in learning from multiple representation. *Learning and Instruction, 13*(2), 141–156.

Sweller, J., van Merrienboer, J. J. G., & Paas, F. G. W. C. (1998). Cognitive architecture and instructional design. *Educational Psychology Review, 10*(3), 251–296.

Upmeier zu Belzen, A., & Krüger, D. (2010). Modellkompetenz im Biologieunterricht. *ZeitschriftfFür Didaktik Der Naturwissenschaften, 16*(1), 41–57.

Chapter 4
The Effect of Adult Intervention in the Development of Science Process Skills

María Napal Fraile, Lara Vázquez Bienzobas, Isabel Zudaire Ripa, and Irantzu Uriz Doray

4.1 Introduction

Policy-documents in science education internationally emphasize that there is great value in teaching science to children already in preschool (Delserieys et al., 2018; Akerblom & Thorstag, 2021). Several studies have shown that 6-year-old children are able to think scientifically, using deductive reasoning and hypothesizing, and demonstrating at least an incipient command of the skills that scientists use in their work (Canedo-Ibarra et al., 2010). Obviously, these abilities are still in development in children, and this development requires that children be provided with various opportunities and contexts for learning (Sutton-Smith, 1970; Gelman & Brenneman, 2004). In that way, intellectual development follows manipulation and interaction with objects and phenomena, but also with other people (peers or adults).

4.1.1 Science Process Skills (SPS)

Notwithstanding the relevance of scientific concepts (phenomena, theories and models), science must be understood as a tool to understand the way the world works, with joy and pleasure. As such, the teaching of science at school should seek to equip children with skills to continue exploring and learning throughout life, which is achieved through the development of basic skills, by exposing them to situations in which science is done (Harlen, 1999). Indeed, doing science involves unique process skills, which can be defined as "a set of broadly transferable

M. Napal Fraile (✉) · L. Vázquez Bienzobas · I. Zudaire Ripa · I. Uriz Doray
Department of Sciences, Public University of Navarre (UPNA), Pamplona, Spain
e-mail: maria.napal@unavarra.es; vazquez.122538@e.unavarra.es;
mariaisabel.zudaire@unavarra.es; iranzu.uriz@unavarra.es

abilities, appropriate to many science disciplines and reflective of the behaviour of scientists" (Padilla, 1990), and which include actions such as critical thinking, hypothesizing, manipulating and reasoning skills. The basic (simpler) process skills – observing, inferring, measuring, communicating, classifying, predicting – provide a foundation for learning the integrated (more complex) skills – controlling variables, defining operationally, formulating hypotheses, experimenting, formulating models – and SPS are inseparable in practice from the conceptual understanding that is involved in learning and applying science (Harlen & Gardner, 2010; Ilma et al., 2020).

The development of SPS depends on the maturation of the child, which may constrain or limit the development of more complex process skills. Some authors recommend that only the basic SPS be taught to young children (Rezba et al., 2007, cited in Vartianen & Kumpulainen, 2019). Others, however, such as Ergül et al. (2011), suggest that children develop thinking by using the different SPS. Furthermore, Ergül et al. (2011) suggests that the SPS are hierarchical so that children need to acquire the basic SPS first before they can acquire the more advanced SPS. In conclusion, the learning of SPS should never be neglected with excuses, such as lack of time or overloaded programmes.

4.1.2   Facilitating the Learning of Science in the Classroom

Undoubtedly, free play provides a variety of opportunities for learning and developing SPS. However, the intervention of the adult can be pivotal in moving children from their actual development zone to the proximal development zone (*sensu* Vygotski), overcoming the limitations of discovery learning. This adult intervention may include a range of actions, from providing adequate environments that facilitate the development of their experiences (Santer et al., 2007; van Limped et al., 2020), to guiding exploration with timely clues.

The first factors supporting the development of science concepts and skills are adequate spaces and materials, i.e. materials that are safe, natural, open-ended, scientifically rigorous and quotidian (Pedreira & Márquez, 2017) in stimulating spaces that promote exploration (Santer et al., 2007).

The second factor is language, which is foundational to science learning. Not only technical vocabulary allows the learner to be more precise; also, dialogic talk, in which there is an interchange with the aim of exploring an event in depth. This provokes thinking and can help students to construct new representations that they did not have yet (Canedo-Ibarra et al., 2010). Furthermore, depending on how the questions are formulated and when the teacher intervenes, their effect can vary greatly, from promotion to inhibition of exploratory learning (Harlen, 2018).

In this vein, asking a good question is the first step towards a good answer because that question invites the students to continue exploring where the answer can be found. As Fine and Desmond (2015) state, open-ended questions that follow the learning objectives are capable of directing the child's thinking. In this sense, productive questions (Martens, 1999) engage the children in deep exploration, increase the time they spend focusing their attention on the task that is being done, or prompt them to establish relationships among concepts and ideas.

All things considered, the attitude of the teacher, and the degree and the intent of adult intervention may determine the outcome of the instruction, and more specifically the development of skills. Indeed, the scaffolding provided – including the degree of intervention, timing and format as depending on when the instruction is given – determines the orientation and intensity of exploratory behaviour (Bonawitz et al., 2010). Productive teaching should support children in generating powerful ideas and coherent arguments (Granja, 2015). This involves finding a careful trade-off between giving direct instructions and letting children explore.

According to Pedreira (2018), adult intervention can be based on the level of directivity (adult's choice as to whether to give new ideas or not) and also in tune with the child considering what he or she is doing. If these two dimensions are interrelated, the different types of standard intervention appear (Fig. 4.1).

4.1.3 Objectives

The objectives of the research were two-fold: first, to describe the effect of a proposal intended to develop SPS in the learning of science among young children; and second, to describe the impact of adult intervention on the engagement with science tasks and their impact on learning.

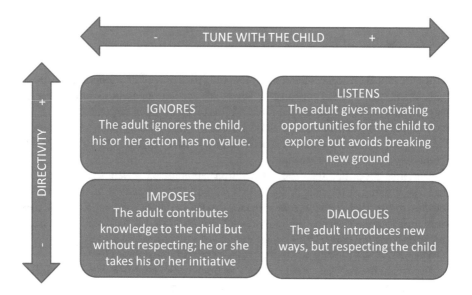

Fig. 4.1 Styles of adult intervention

More precisely, we aimed to answer the following two research questions:

- Does a proposal built around basic SPS also improve knowledge on the selected topics?
- How does the style of adult intervention impact the development of SPS and content acquisition?

4.2 Research design and method

4.2.1 Participants

The participants included a convenience sample of 42 children, aged four to six, belonging to three classes (A: $n = 14$; B: $n = 13$; C: $n = 15$), all attending the same public school. The institution approved the execution of the study.

The leader teacher was a pre-service teacher, who had received specific training on the development of Inquiry-Based Science Education and productive questioning, accompanied by three teachers (female; aged 38–47; 14–21 years of experience). Both the in-service and the pre-service teachers were instructed on the style of adult guidance to apply during each intervention.

Each group was divided into three subgroups, which followed three different types of adult intervention: adult-led (directive; 'imposed' in Fig. 4.1), children-led (discovery; 'listening' in Fig. 4.1) and guided (exploration guided by productive questions; 'dialogue' in Fig. 4.1).

4.2.2 Design of the Intervention

Groups A and B (second course, 4–5 years old) participated in a proposal related to magnetism, and C (third course; 5–6 years old) carried out an observation of ants. Despite the different themes, the three proposals specifically targeted the development of process skills and included phenomena that are within reach and belong to the everyday context of children. Therefore, they served as a context to attempt different degrees of adult intervention.

The first proposal, the 'ant observation', comprised seven sessions, which involved looking at the ants in the school yard and taking care of an anthill inside the classroom, as well as some sessions for discussion about facts and findings. SPS were mainly observing and communicating, but also inferring, predicting, measuring or interpreting data.

The activities were adapted to the styles of adult intervention (Table 4.1).

The second proposal, 'magnetism', included six activities, which included exploring the properties of magnetic and non-magnetic materials, and magnetic

Table 4.1 Adaptation of the activities in the 'ant observation' according to the different styles of adult intervention

	Adult-led (imposed)	Guided exploration (dialogue)	Children-led (listening)
Act 3. Observe ants	Prompted to observe specific structures, using technical vocabulary *Look, they take their food with the front part; they have clamps like a jaw*	Guided with questions such as: *Have you seen? Have you figured out? Have you noticed? How many? Why do you think?*	Given freedom to observe for a while; then asked to explain what they have observed
Act 4. How can we make ants deviate from their path?	Carry out the action of passing the hand without the previous action questions	Ask action questions to help them make predictions: *What will happen if you rub your hand on the path? And if you move the leaf?*	Probe question, only leaving them to be the ones that direct their exploration
Act 5. What do ants eat?	Directing the exploration	Helping with questions to focus their attention/ reason	Let explore

interactions. SPS included observing, communicating, comparing, inferring and predicting, but also interpreting data and measuring.

4.2.3 Data Collection and Analysis

Annotations and audio recordings were transcribed and analysed by applying techniques of content analysis to ascertain learning gains and the presence and degree of development of SPS. Each of the literal transcripts (see, for example, Table 4.4) were skimmed for evidence of SPS (instances of observation, description, making inferences, etc.) and then graduated according to the level of complexity. For example, for communication, judgements were made on the precision in the use of language, meaningful introduction of technical vocabulary; for predicting, making predictions based on the observed phenomena or models, and not just on everyday experience or intuition. In addition, drawings and written productions were taken at different moments of the process. These served to measure the progression in knowledge.

Given the small sample size and the difficulty in gathering systematic data from children at this age, the variety of evidence collected was used just to illustrate the range of answers produced in each situation. No quantitative comparisons were made.

When analysing the drawings, attention was given to the anatomy (number and proportion of body parts, number of legs, and presence of antennae), behaviour (number of ants, gregarious behaviour) and environment (anthills).

4.3 Findings

Following the ant observation sequence, the three groups improved the knowledge of the anatomy and behaviour of these insects (Table 4.2).

Table 4.2 Sample drawings, before and after the intervention, for each group

ADULT-LED (imposed)

Body with 4 parts. 8 legs (including 4 in the other side)	Body with 3 parts; 6 legs. Ants form a row. The bee has antennae.

CHILD-LED (listening)

 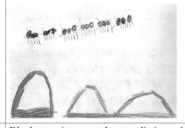

Body with 3 parts, 6 legs (instructed), colourful ant hill	Black ants, in a row, three realistic ant hills.

GUIDED EXPLORATION (dialogue)

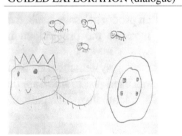 

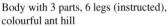

The queen wears a crown, huge head, no antennae. Ant hill with windows.	Ants with antennae, long body not segmented, ant hill with a hole. The queen is bigger.

When we analysed the knowledge gains, in terms of both contents and SPS, it became clear that pupils in the child-led group got lost more frequently, and even had difficulties persisting with the exploration. In turn, students in the adult-led group demonstrated the highest retention of concepts, with no improvement in SPS (inferring, predicting, interpreting data, etc.). Last, children in the guided exploration group advanced most in SPS, although not all the concepts were mastered. In this last group, teachers' productive questions were decisive to trigger processes (Table 4.3).

In the same vein, in the proposal about magnetism, all the groups acquired the basic concepts: there are magnetic and non-magnetic materials, magnets have two poles, oppositely charged magnetic poles repel each other, magnetic attraction (or repulsion) is a weak force that acts at a certain distance, and it is possible to temporally turn some metallic objects into magnets. Children showed some degree of development in certain basic and integrated SPS, including predicting, making inferences or posing hypotheses (Table 4.4).

Table 4.3 Productive questions and answers. Guided exploration group

Reasoning questions	*(T) Which ant type are these? Is it the queen? (S) NO (T) Why? (S) Because they're small (T) And what are those? (S) Looking for food*
Questions to focalise attention	*(T) Where do the ants go? (S) They go on both sides (T) I'm going to rub my hand, what's happening? (S) This one has changed direction!*
Questions to promote observation	*(T) Have you seen how the ant has got the food? (S) With the mouth and the forelegs (T) How many legs do they have? (S) 6 legs and 2 horns (T) Horns? (S) I cannot see its face. Ah, look! This one only has one antenna*

T teacher, *S* student

Table 4.4 Examples of the development of SPS

OBSERVE	Describe properties of materials and objects: (T) *What happens when you hold the magnet near to the iron filings inside the bottle? (S) They move, it's magic!*
INFER	*(T) Will they stick? (S) Yes, because it's like silver*
PREDICT	A magnet breaks. *(T) Will we be able to put together this blue with this blue? (S) No, both are the same colour* *(T) At which distance will they stick? (S1) At 3 centimetres. (S2) No, at 5 – it's bigger*
COMMUNICATE	New concepts are learned: *(T) What are you doing? (S1) Looking for things that stick (S2) Yes, magnetic things!*
CLASSIFY	Two objects stick to the magnet *(T) Do they have something in common? (S) Yes, both are made of iron* (= they're metallic).
MAKE HYPOTHESES	Purposeful actions are initiated: *(S) Can I try in the golden part?* [The pupil wants to put the magnet close to a golden object.]
INTERPRET DATA	They organise the data to extract conclusions. *(S) Here, there are little things, and here many. There are more metallic things. (T) Very good, girls (S) Only the girls guess the answer*

T teacher, *S* student

Table 4.5 Comparative outcomes following open and productive questions

	OPEN QUESTION (adult-led)	PRODUCTIVE QUESTION (guided exploration)
OBSERVE	*It sticks*	*… only to iron.*
INFER	*Because it is magnetic*	*… and magnetic items are attracted by the magnet.*
PREDICT	*It won't stick*	*… because equal poles repel each other.*
MEASURE	*It sticks at 3 cm*	*… because of their different power.*
COMUNICATE	*The magnet sticks*	*… to the iron filings because they are attracted.*
CLASSIFY	*Sticks to magnet or not*	*Objects that stick to the magnet are made of iron; the rest are made of other materials.*

T teacher, *S* student

Regarding the effect of adult intervention, pupils in the adult-led intervention proved less autonomous and more dependent on adult supervision. In the children-led group, conceptual development was scarce, while those in the guided investigation made better observations and better inferences. Indeed, making productive questions increased the detail of the answers provided (Table 4.5).

4.4   Discussion

Contrary to the commonly held belief, this proposal has shown that infant children are able to think scientifically and progress in their scientific thinking (Canedo-Ibarra et al., 2010; Robbins, 2005). Specifically, they have demonstrated understanding of some *a priori complex* concepts, such as magnetic attraction or the anatomy and social organisation of insect communities. Throughout the proposal, the children progressed in the use of SPS, confirming that children may develop their SPS when given appropriate contexts (Sutton-Smith, 1970). This requires that children be exposed to suitable environments and materials (Santer et al., 2007) that are safe and supportive, but also challenging.

The type of questions that are asked is crucial for the development of good thinking (Elstgeest, 1985), with productive questions that prompt the children to focus their attention or find a solution leading to much more accurate and complete answers. The choice of the topic and contexts for the research, and the instruments and materials provided, were key to the development of SPS: The development of scientific skills has increased, especially in the guided exploration group (dialogue). The key factors that supported this improvement were: to start from an interesting topic for the students, to ask them productive questions, to ensure an adequate classroom climate in which to carry out their inquiry, and, last, to provide them with instruments and materials, such as magnifying glasses and marbles of magnets, which increase their interest.

Maximum development of SPS was achieved by the guided exploration group, following a higher interest in exploration, due to the questions and challenges that were being set (Bonawitz et al., 2010).

Indeed, experience does not directly generate knowledge (Hodson, 1994; Kite et al., 2021), unless there are some intellectual interplays happening (Couso, 2014). For this, the intervention of the adult guiding the class talk is crucial (Harlen, 2018). Better results in the guided exploration group (similar content acquisition with better SPS development) can be explained by a better and more sustained exploration, following the questions and challenges set (Bonawitz et al., 2010). However, SPS had a limited development in the adult-led group: While they participated in observation, they only rarely formed hypotheses or made adequate predictions. In turn, students in the child-led group had difficulties following the processes and thus did not go deeper into the properties and behaviour of the materials.

Also, the type of questions proved decisive to promote thinking (Elstgeest, 1985). Simple, closed questions that aim at knowledge reproduction do not have an effect on the development of SPS. On the contrary, questions that help focus attention result in answers that are much more accurate and complete. Preschool teachers have a big responsibility to help students open their minds, hook them with open but productive questions following the scientific method and education by inquiry, as it makes children critical and empowered in their learning (Fine & Desmond, 2015).

All this confirms that at preschool, the adult plays a central role, not only in the acquisition of concepts, but essentially in the development of SPS that equip the students for learning throughout life. However, in-service preschool teachers are often reluctant to engage in science projects that they perceive as too difficult for their students. Hence the importance of teacher trainings, which make teachers feel confident, prepared, open-minded and capable of reaching new horizons (Elstgeest, 1985).

4.5   Conclusion

When provided with adequate support and stimulating situations, children at the preschool age show at least an incipient command of some SPS. The style of adult intervention strongly determined the outcome of the activity, with guided exploration (autonomous exploration guided by productive questions) producing more insightful observations and interpretations, as compared to adult-led and discovery approaches.

References

Åkerblom, A., & Thorshag, K. (2021). Preschoolers' use and exploration of concepts related to scientific phenomena in preschool. *Journal of Childhood, Education & Society, 2*(3), 287–302. https://doi.org/10.37291/2717638X.202123115

Bonawitz, E., Shafto, P., Gweon, H., Goodman, N. D., Spelke, E., & Schulz, L. (2010). The double-edged sword of pedagogy: Instruction limits spontaneous exploration and discovery. *Cognition, 120*(3), 322–330. https://doi.org/10.1016/j.cognition.2010.10.001

Canedo-Ibarra, S. P., Castello-Escandell, J., Garcia-Wehrle, P., & Morales-Blake, A. R. (2010). Precursor models construction at preschool education: An approach to improve scientific education in the classroom. *Review of Science, Mathematics and ICT Education, 4*(1), 41–76.

Couso, D. (2014). De la moda de "aprender indagando" a la indagación para modelizar: una reflexión crítica [From the fashion of "learning by inquiry" to Model-Based Inquiry: a critical revisión]. In R. Jiménez, A. M. Wamba, M. A. de las Heras, A. A. Lorca, & B. Vázquez (Eds.), *26EDCE. Investigación y Transferencia Para Una Educación En Ciencias: Un Reto Emocionante* [Research and transference for science education].

Delserieys, A., Jégou, C., Boilevin, J. M., & Ravanis, K. (2018). Precursor model and preschool science learning about shadows formation. *Research in Science & Technological Education, 36*(2), 147–164. https://doi.org/10.1080/02635143.2017.1353960

Elstgeest, J. (1985). The right question at the right time. In W. Harlen (Ed.), *Primary science: Taking the plunge* (pp. 33–40). Heinemann Educational Books.

Ergül, R., Şımşeklı, Y., Çaliş, S., Özdılek, Z., Göçmençelebı, S., & Şanli, M. (2011). The effects of inquiry-based science teaching on elementary school students' science process skills and science attitudes. *Bulgarian Journal of Science and Education Policy (BJSEP), 5*(1), 48–69.

Fine, M., & Desmond, L. (2015). Inquiry-based learning: Preparing young learners for the demands of the 21st century. *Educator's Voice, 7*, 2–11.

Gelman, R., & Brenneman, K. (2004). Science learning pathways for young children. *Early Childhood Research Quarterly, 19*(1), 150–158.

Granja, D. O. (2015). Constructivism as theory and teaching method. *Sophia, 19*(2), 93–110. https://doi.org/10.17163/soph.n19.2015.04

Harlen, W. (1999). Purposes and procedures for assessing science process skills. *Assessment in Education: Principles, Policy & Practice, 6*(1), 129–144.

Harlen, W. (2018). *The teaching of science in primary schools*. David Fulton Publishers. https://doi.org/10.4324/9781315398907

Harlen, W., & Gardner, J. (2010). *Developing teacher assessment*. McGraw- Hill.

Hodson, D. (1994). Hacia un enfoque más crítico del trabajo de laboratorio [Towards a more critical approach to laboratory work]. *Enseñanza de Las Ciencias, 12*(3), 299–313.

Ilma, S., Al-Muhdhar, M. H. I., Rohman, F., & Saptasari, M. (2020). The correlation between science process skills and biology cognitive learning outcome of senior high school students. *Journal Pendidikan Biologi Indonesia, 6*(1), 55–64. https://doi.org/10.22219/jpbi.v6i1.10794

Kite, V., Park, S., McCance, K., & Seung, E. (2021). Secondary science teachers' understandings of the epistemic nature of science practices. *Journal of Science Teacher Education, 32*(3), 243–264. https://doi.org/10.1080/1046560X.2020.1808757

Martens, M. L. (1999). Productive questions: Tools for supporting constructivist learning. *Science and Children, 36*(8), 24.

Padilla, M. J. (1990). *The science process skills* (Research matters – To the science teacher, No. 9004). National Association for Research in Science Teaching (NARST). http://www.narst.org/publications/research/skill.cfm

Pedreira, M. (2018). Intervenir, no interferir: el adulto y los procesos de aprendizaje [To intervene, not to interfere: the adult and the learning processes]. *Aula de Infantil, 96*, 9–13.

Pedreira, M., & Márquez, C. (2017). Espacios de ciencia de libre elección: posibilidades y límites [Elective science spaces: possibilities and limits]. In M. Pedreira & C. Márquez (Eds.), *Enseñanza de las Ciencia e Infancia. Problemáticas y avances de teoría y campo desde Iberoamérica* [Science teaching and childhood: Problems and advances of theory and field from Iberoamerica] (pp. 151–169). Bellaterra. Sociedad Chilena de Didáctica, Historia y Filosofía de las Ciencias.

Robbins, S. P. (2005). *Essentials of organizational behavior* (8th ed.). Prentice Hall.

Santer, J., Griffi, C., & Goodall, D. (2007). *Free play in early childhood. A literature Review*. National Children's Bureau. https://www.bl.uk/collection-items/free-play-in-early-childhood-a-literature-review

Sutton-Smith, B. (1970). *A descriptive account of four modes of children's play between one and five years*. Columbia Univeristy College Teachers College. https://eric.ed.gov/?id=ED049833

van Liempd, I. H., Oudgenoeg-Paz, O., & Leseman, P. P. (2020). Do spatial characteristics influence behavior and development in early childhood education and care? *Journal of Environmental Psychology, 67*, 101385.

Vartiainen, J., & Kumpulainen, K. (2019). Promoting young children's dynamic practice. In K. Kumpulainen & J. Sefton-Green (Eds.), *Multiliteracies and early years innovation: Perspectives from Finland and beyond* (pp. 77–94). Routledge. https://doi.org/10.432 4/9780429432668-5

Chapter 5
Facilitating the Practice of 4C Skills in Biology Education Through Educational Escape Rooms

Georgios Villias and Mark Winterbottom

5.1 Introduction & Theoretical Background

Educational escape rooms (EERs) are game-based activities that adopted the initial concept from the escape room industry and adapted it appropriately for use in an educational context. By offering immersive experiences that promote students' active participation, EERs have emerged as promising alternative approaches to fostering students' conceptual and skill-based learning (Nicholson, 2018). A deeper look inside the design principles and the conceptual framework of an EER reveals their connection to several well-established educational methodologies (e.g., problem-based, inquiry-based, experiential, game-based, narrative-based, etc.), as well as motivational theories (e.g., self-determination theory, self-efficacy theory, etc.), justifying their learning potential. EERs have been widely used with the intention to create an active learning environment, motivate students, simulate conditions from real-life scenarios for health-care professionals, and facilitate the learning of content knowledge and the development of 21st century skills (Lathwesen & Belova, 2021; Taraldsen et al., 2022; Veldkamp et al., 2020a, b).

The popular and overarching term '21st century skills', refers to a set of skills and competencies (learning, social and cultural, life and career, literacy, etc.) that are considered to be of vital importance for both present and future, in order to confront the challenges, introduced to the new generation with the arrival of the third millennium (Bapna et al., 2017). Many attempts have been made by various initiatives and educational organisations to outline and classify the most important 21st century skills, resulting in several different skill frameworks. Among them are included frameworks from the P21: Partnership for 21st Century Learning (P21, 2009),

G. Villias (✉) · M. Winterbottom (✉)
Faculty of Education, University of Cambridge, Cambridge, UK
e-mail: gv283@cam.ac.uk; mw244@cam.ac.uk

© The Author(s) 2024
K. Korfiatis et al. (eds.), *Shaping the Future of Biological Education Research,*
Contributions from Biology Education Research,
https://doi.org/10.1007/978-3-031-44792-1_5

63

the Assessment and Teaching of 21st Century Skills – Cisco/Intel/Microsoft (http://www.atc21s.org/), the World Economic Forum, and OECD initiatives (DeSeCo and PISA). Although there are many overlapping competencies in these frameworks, there is a lack of consensus between them about which 21st century skills should be regarded as the most essential. Depending on their focus (e.g. studies, work, literacy, social life etc.), they categorise skills differently (Ananiadou & Claro, 2009; Bapna et al., 2017; Bialik et al., 2015; Lai & Viering, 2012). Bialik et al. (2015, p. 3) compared these frameworks to identify their commonalities and concluded that the learning skills which are present in most of them are the following four: Critical thinking, Creativity, Collaboration and Communication (4Cs).

The 4Cs seem to have a central role in modern proposals for 21st century curriculum re-design (Fadel et al., 2015) and school networks for 21st century learning (e.g. EdLeader21). Each of these skills encompasses different performance areas (abbreviations in brackets) that should be considered when trying to measure or develop them. Critical thinking, according to Sternberg (1986, p. 3), "comprises the mental processes, strategies, and representations people use to solve problems, make decisions, and learn new concepts." These include the ability to discover information (IFD), interpret and analyse collected data (IPA), make and support claims with valid reasoning (RES), and propose adequate, applicable solutions when needed (PRB). Regarding Creativity, Sternberg and Lubart (1998, p. 3) define it as "the ability to produce work that is both novel (i.e. original, unexpected) and appropriate (i.e. useful, adaptive concerning task constrains)." That means being able to brainstorm and generate ideas (IDG), articulate and refine these ideas (IDR), but also to effectively select and integrate materials to develop a unique product or finish a specified task (CPI). Collaboration is a "coordinated, synchronous activity that is the result of a continued attempt to construct and maintain a shared conception of a problem" (Roschelle & Teasley, 1995, p. 70). While collaborating, individuals are expected to take initiatives or even lead the group (LDI), follow appropriate norms, avoid conflicts, and share their insights (CPF), be responsible and productive (RPR), and show responsiveness to others through the provision and acceptance of feedback (RSP). Finally, Communication is described by Qian and Clark (2016, p. 51) as "the ability to articulate thoughts and ideas in a variety of forms, communicate for a range of purposes and in diverse environments, and use multiple media and technologies." Engaging in fruitful conversations and discussions entails the use of verbal and non-verbal language, empathy, conveying of emotion, and consensus building (ENG).

Over the past five years, educational research on EERs has been gaining momentum. Researchers' increasing interest in this trending phenomenon is reflected in the growing number of papers being published each year (Fotaris & Mastoras, 2019). Most of these studies were conducted on university level students, coming from several different fields (STEM, medicine, nursing, computer science). Nevertheless, a great number of STEM-oriented EERs have also been developed, implemented, and studied in secondary schools (Lathwesen & Belova, 2021). Knowledge about the structure and design of EERs is accumulating fast since their first appearance, leading to the development of several design frameworks to guide the early adopters

(Clarke et al., 2017; Fotaris & Mastoras, 2022; Nicholson, 2018; Nicholson & Cable, 2021; Veldkamp et al., 2020a, b). EERs' educational and design aspects have been systematically reviewed (Veldkamp et al., 2020a, b), offering insights into their learning mechanism. Nevertheless, evidence-based research on EERs' learning effectiveness has still not provided conclusive results. Several researchers have assessed students' cognitive (e.g., content knowledge, skills) or affective (e.g., motivation, interest, engagement) learning outcomes after their participation in EERs. The research designs they used range from quantitative (pre-/post-participation surveys and tests), to qualitative (interviews, informal feedback and observations), and sometimes mixed methods (Taraldsen et al., 2022). However, there are some issues in reference to the applied research methodology. Only a few of these researchers have actually adopted a control and treatment group design. Fotaris and Mastoras (2019) also highlight the importance of using larger sample sizes to avoid questioning of these studies' results. According to Lathwesen and Belova (2021, p. 10), there is not enough empirical evidence on the short and long-term effectiveness of these activities in comparison to the traditional, lecture-based approach. As they stress, "to research whether escape games have a long-lasting learning effect, multiple post-tests need to be undertaken at different time intervals." Taraldsen et al. (2022) also acknowledge the need for systematic, longitudinal studies in primary and secondary schools. They suggest that researchers should adopt more complex research designs to evaluate EERs' learning gains, emphasising 21st century skills and school subjects' content learning.

Attempting to bridge the gap in knowledge presented above, this study examined EERs' learning impact on secondary school students, focusing on the practice and development of their 21st century skills. The study's main research questions are the following:

(a) What 4C skills are practised by students when engaging in an EER? Do they develop?
(b) What structure and features of an EER enable students to practise and possibly develop their 4C skills?

Some of the present study's characteristics and the added value it brings in this particular field of educational research are outlined below:

- adopts a mixed-method research design; thus, it capitalises on the benefits from both quantitative and qualitative approaches.
- uses several different data collection and analysis methods; thus, it facilitates the validation of its data through triangulation.
- is applied on a large number of students from different schools; thus, it increases the findings' reliability.
- is based on a control and treatment group design; thus, it evaluates EERs' learning impact on students' 21st century skills in comparison to other didactic approaches.
- is longitudinal with multiple data collection points; thus, it offers insights into the long-term effectiveness of EERs.

- approaches knowledge holistically; thus, it combines information provided by different sources (e.g. interviews, tests, observation, questionnaires) and different perspectives (e.g. student, facilitator, researcher).
- uses a design-based research methodology; thus, it reviews and optimises the developed EERs' design during its three successive iterations.
- provides design guidelines for developing puzzles which promote the practice and development of 4C skills; thus, it offers workable solutions and ideas to education practitioners for biology teaching and learning.

5.2   Materials and Methods

5.2.1   Design of the Study and Data Collection

The work presented here is the result of a nine-month, longitudinal, mixed-method research study that adopted a design-based research methodology and followed an iterative process of three meso-cycles. Three EER interventions (EER1, EER2, EER3) were designed and developed according to a series of criteria that were consistent with the students' age group, expected cognitive abilities, pre-existing knowledge, but also the lessons' content, learning objectives and connection to common misconceptions. These three EER activities were embedded in the teaching of the biology science course, as part of the normally expected-to-be-taught Greek national curriculum and lasted between 45 and 60 min. Even though the content knowledge of these EER activities was different, all of them retained the same focus on developing students' 4C skills and thus they are connected. Each EER activity was built upon the previous one. Therefore, all the theoretical and practical findings that were produced during the empirical micro-cycles that preceded (Analysis/Exploration) and followed (Evaluation/Reflection) the implementation of the first EER activity acted as layers that informed and added value to the respective stages of the next meso-cycle (second EER). The same approach was applied to the third meso-cycle (third EER), as well. A total of 209 Year 10 students, from three different Greek (main, secondary, and back-up unit) schools and one Cypriot school (pilot unit) divided into ten different classes, participated in the study. However, we only conducted an in-depth analysis of 125 full datasets that were collected from students enrolled in the main and secondary school unit. Acknowledging the importance of using a control group to contradict plausible counterfactual inferences as argued by Marsden and Torgerson (2012) led us to the adoption of a control group research design. An effort was also made so that both groups were comparable in terms of student socioeconomic background, gender ratio and prior performance in science. Nevertheless, in all schools, students' allocation to classrooms was non-random, but based on criteria predefined by the schools' administrations.

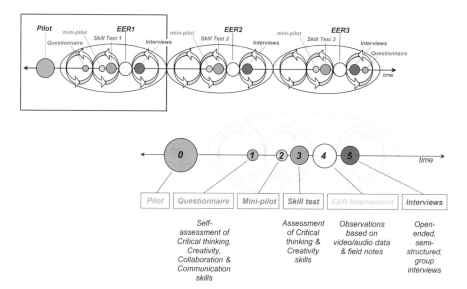

Fig. 5.1 The overall research plan of this study and data collection methods as applied on the first meso-cycle

Several different data collection methods were employed, including question-naires, skill tests, observations, as well as interviews (Fig. 5.1). Considering that each of these methods has inherent weaknesses that may affect the data's validity and provide inaccurate results (e.g. observer bias, lack of recording equipment, omission of participants' less noticeable behaviour, difficulty in analysing collected data, interviewees' introversion or reluctance to answer, skills being overrated, poor self-awareness, miscomprehension of questions), we did not rely exclusively on one of the aforementioned data collection methods, but we triangulated the findings of many, so as to have an objective and unbiased assessment of participants' skills. The main data collection method that we applied during the implementation of the EERs was the use of video/audio recordings. For each implementation, there were four groups of students, consisting of between three and six, dependent on the group, working on large benches simultaneously.

Students were usually standing or sitting around the bench, with all the resources needed in the middle. Using 360-degree cameras, placed on the students' benches, we collected 88 video/audio recordings (each video's average duration: 40 min). Field observations were also collected, assessing participants' performance from the facilitator's perspective. Regarding interviews, we conducted six informal, exploratory interviews to assess and optimise the EERs' design (after the mini-pilots), as well as 15 semi-structured, in-depth, group or small-group interviews to gain insights into the structure, design, and learning impact of EERs, from the per-spective of 33 students (after the interventions). In reference to the study's written assessment tools (tests, questionnaires), we constructed and validated our own. Three skill assessment tests that contained different but equivalent items were designed to measure and compare objectively students' critical thinking and creativity skills. Critical thinking test items were based on the format of standardised situational judgement tests (SJTs), while for creativity we used items adapted from a standardised scientific creativity test, developed for secondary school students (Hu & Adey, 2002). Finally, at the beginning and at the end of the research study, we administered a 7-point Likert scale and self-assessment questionnaires, in order to measure students' self-perception of all their 4C skills.

5.2.2 Qualitative Data Analysis

A major part of the analysis was devoted to observational data from video/audio recordings. Despite initially collecting a vast amount of data, less than half of them were analysed in-depth. Nevertheless, the student groups whose recordings were not analysed exhibited the same patterns of behaviour in the 4C skills practice. In order to analyse these rich data in a methodical and efficient way, we used a set of rubrics designed for a performance-based assessment of the 4Cs. These rubrics were revised and customised according to the EERs' special design features. Their two-dimensional design consisted of a vertical axis which evaluated the sub-categories of different performance areas of each skill, and a horizontal, 4-level, performance rating scale that evaluated more accurately the mastery level of each sub-category (Fig. 5.2).

Based on the revised 4C skills rubrics' design presented above, we developed an analytical coding framework of 12 double indicators. Each of them corresponded to one of the 4C skills performance areas mentioned in the introduction section (i.e. IFD, IPA, RES, PRB, IDG, IDR, CPI, LDI, CPF, RPR, RSP, ENG). The aforementioned indicators were used to code students' practice of the 4C skills during the observational analysis of the video/audio recordings. Adopting this strategy of double indicators allowed us to have a coding scheme that was flexible enough to describe observational variability. At the same time, we restrained the number of its basic codes to an operational degree (Table 5.1). This double system also allowed us to conduct a 'two-axes' observational analysis. The first 'axis' analysed the skills' frequency of appearance, while the second one described the observed progress of the skills' level of practice. Considering qualitative data analysis tools, we used version 22 of ATLAS.ti software. Its powerful analysis tools helped us to calculate each code's absolute frequency, and focus on the co-occurrence of codes for specific skills and puzzle designs. Coloured heat maps were used to visualise these quantified observational data for each EER's puzzle and showcase their ability to facilitate the practice of certain aspects of the 4C skills (Table 5.2).

As regards observational data from field notes and interview data, they were also systematically analysed after firstly being classified into categories, based on their content (structure of the game, teams' operation, acquisition of knowledge).

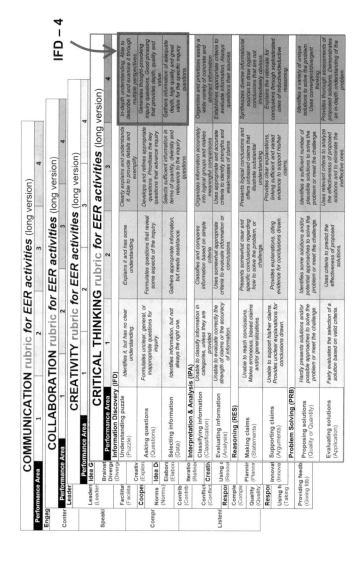

Fig. 5.2 Connection of a 4C skills rubrics' design to the analytical coding framework for EER activities (IFD-4 corresponds to Information Discovery – Level 4)

Table 5.1 Double indicators for observational data analysis on 4C skills

			Performance area	Indicators' second part (xxx-X) Level of Skill mastery			
				1 Poor level	2 Average level	3 High level	4 Expert level
Indicators' first part (XXX-x) Performance area	CT	IFD	Information discovery	IFD-1	IFD-2	IFD-3	IFD-4
		IPA	Interpretation & Analysis	IPA-1	IPA-2	IPA-3	IPA-4
		RES	Reasoning	RES-1	RES-2	RES-3	RES-4
		PRB	Problem-solving	PRB-1	PRB-2	PRB-3	PRB-4
	CR	IDG	Idea generation	IDG-1	IDG-2	IDG-3	IDG-4
		IDR	Idea refinement	IDR-1	IDR-2	IDR-3	IDR-4
		CPI	Creative production & Innovation	CPI-1	CPI-2	CPI-3	CPI-4
	CB	LDI	Leadership & Initiative	LDI-1	LDI-2	LDI-3	LDI-4
		CPF	Cooperation & Flexibility	CPF-1	CPF-2	CPF-3	CPF-4
		RPR	Responsibility & Productivity	RPR-1	RPR-2	RPR-3	RPR-4
		RSP	Responsiveness	RSP-1	RSP-2	RSP-3	RSP-4
	CM	ENG	Engaging in discussions	ENG-1	ENG-2	ENG-3	ENG-4

CTCR critical thinking & creativity, *CB* collaboration, *CM* communication, *Dark colour* initial questionnaire, *Light colour* final questionnaire, *Purple* experimental group, *Green* control group

Table 5.2 Cross tabulation of 4Cs' performance area codes per puzzle in EER1

	P1	P2	P3	P4	P5	P6
IFD	(15)	(15)	(36)	(12)	(31)	(33)
IPA	(6)	(2)	(5)	(5)	(20)	(12)
RES	(2)	(21)	(8)	(8)	(23)	(36)
PRB	(8)	(8)	(16)	(6)	(21)	(16)
IDG	(8)	(4)	(19)	(4)	(10)	(19)
IDR	(5)	(7)	(8)	(1)	(5)	(8)
CPI	(8)		(11)	(9)	(5)	(22)
LDI	(8)	(9)	(9)	(6)	(9)	(13)
CPF	(16)	(11)	(22)	(7)	(18)	(28)
RPR	(11)	(2)	(4)	(8)	(6)	(7)
RSP	(3)	(13)	(7)	(1)	(12)	(9)
ENG	(15)	(15)	(20)	(11)	(26)	(27)

IFD information discovery, *IPA* interpretation & analysis, *RES* reasoning, *PRB* problem solving, *IDG* idea generation, *IDR* idea refinement, *CPI* creative production & innovation, *LDI* leadership & initiative, *CPF* cooperation & flexibility, *RPR* responsibility & productivity, *RSP* responsiveness, *ENG* engagement in discussions, *P1-P6* puzzles

5.2.3 Quantitative Data Analysis

After calculating the numerical scores of the study's written assessments (skill tests, questionnaires), we analysed them using descriptive statistics in SPSS statistical software (version 28). Regarding the skill assessment tests, for each single test item we calculated the responses' mean score and standard deviation. These measures were also calculated for the tests' sub-scores of critical thinking, creativity and originality. Apart from the two basic groups, i.e. the experimental and the control group, during our quantitative data analysis, we divided students into more groups, selecting them based on several other criteria (by school, by gender, by game performance) and compared their responses. Depending on the responses' distribution, parametric or non-parametric tests were applied so as to check for statistically significant changes between the items' ratings. In reference to the Likert-scale questionnaires, we followed the same analytical process, using different measures of central tendency and dispersion (median, mode, interquartile range).

5.3 Results

5.3.1 Observational Data

In order to explore which of the 4C skills were practised more by participating students during each puzzle of the three developed EER activities, we applied a code co-occurrence analysis tool to their coded data. Using data from several implementations each time (9 for EER1, 10 for EER2, and 8 for EER3), we calculated the absolute frequencies of all the skills' codes per puzzle and EER, i.e. the aggregating scores of the codes' combined presence. Visualising these quantified data with three coloured heatmaps facilitated us to detect similarities, differences or repeating patterns among these data, allowing us to associate the practice of specific 4Cs skills with the puzzles' properties (i.e. duration, type, difficulty) and their unique design features (i.e. provided resources, required tasks, complexity).

Analysing these heatmaps (Table 5.2), it became clear that some puzzles favoured more the practice of specific 4C skills compared to others (the higher the frequency of practice, the darker the heatmap's boxes). *Critical thinking* skills were mostly promoted by puzzles that contained several items or offered multiple options that students had to select from. The resources' complexity, i.e. having different information in one place that needed to be combined, also activated students' analytical thinking. By linking puzzles to challenging parts of the syllabus, where students already had some prior knowledge, it motivated them to make claims and use argumentation to support them. Last but not least, puzzles that did not offer straightforward solutions troubled students and forced them to apply different problem-solving strategies in order to come up with a solution. *Creativity* skills were practised less by students compared to the other 4C skills, either because of students' difficulty to express them in such time-pressuring activities, or because of the inappropriate design of the developed EERs' puzzles. What our analysis showed is that puzzles that included hidden elements (e.g. symbols, letters), or coded messages that were difficult to decipher, ignited students' divergent thinking and encouraged their brainstorming. *Collaboration* and *Communication* skills, as expected, were dominant and inextricably linked with each other during these team-based activities. Easy puzzles or puzzles with a very limited number of items were often solved by a single individual; thus they did not favour these skills. On the contrary, puzzles of medium to high difficulty, that consisted of a considerable number of resources, involved the majority of students in solving them. Students collaborated actively by sharing their ideas, undertaking certain tasks, and contributing to the team through independent and team work. In general, puzzles of higher difficulty that troubled students for a greater amount of time resulted in a broader practice of all 4C skills.

Another interesting finding emerged after comparing students' game performance between the first two EER activities. The relative frequency of codes' labels that corresponded to higher levels of demonstrated skill mastery, slightly increased in activity EER2 compared to EER1; the communication, collaboration, and critical thinking skills in particular. That was also inferred from descriptive evaluations that showed improved performance both at a team level (8 out of 12 teams), as well as at an individual level (15 out of 62).

5.3.2   Self-Assessment Questionnaires

Two identical self-assessment questionnaires were administered to the study's participants, one at the beginning and the other at the end of a six-month period. 100 students answered both questionnaires. A comparison of their scores revealed that the experimental group students, compared to the control group students, improved significantly in terms of their communication skills (CM: 4.715 → 5.221, $t(47) = -3.157, p = 0.003$) (Fig. 5.3).

5.3.3   Skill Assessment Tests

Considering that the experimental group students participated in all three EERs, while the control group students did not participate in EER1, we focused our test scores' comparison mainly on tests ST1 (pre-test) and ST2 (post-test), and less on test ST3 (delayed post-test). For this comparison, we selected only those students that had participated in at least two EER activities and had fully completed all three skill assessment tests (80 students in total). In both groups, students' test scores fluctuated in a similar manner (Table 5.3). Critical thinking scores initially decreased (ST1 → ST2) and increased afterwards (ST2 → ST3). The exact opposite happened with their creativity scores. While analysing different student groups' test scores, we observed the same pattern occurring repeatedly, irrespective of the applied

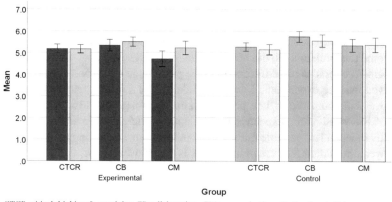

CTCR critical thinking & creativity, *CB* collaboration, *CM* communication, *Dark colour* initial questionnaire, *Light colour* final questionnaire, *Purple* experimental group, *Green* control group

Fig. 5.3 Bar charts of the responses' means (7-point Likert scale) for all scale items that derived from students' initial and final questionnaires (by group)

Table 5.3 Descriptive statistics and statistically significant differences (SSD) between the main overall scores of students' skill assessment tests (by group)

All 3 tests	Experimental group School A + B (n = 40) ST1 ST2 ST3			* (p ≤ 0.05) ** (p ≤ 0.01)		
	Mean & Observed changes			ST1-ST2	ST2-ST3	ST1-ST3
Critical Thinking	49.90^n	42.20^n	42.60^n	** $t(39) = 7.240$ $p < 0.001$	No SSD $p = 0.784$	** $t(39) = 5.730$ $p < 0.001$
	↓ 7.7 / 15.4%		↑ 0.4 / 0.9%			
Creativity	42.40	58.00	50.53^n	** $Z = -4.286$ $p < 0.001$	* $Z = -2.307$ $p = 0.021$	** $Z = -2.877$ $p = 0.004$
	↑ 15.6 / 36.8%		↓ 7.5 / 12.8%			

All 3 tests	Control group School A + B (n = 40) ST1 ST2 ST3			* (p ≤ 0.05) ** (p ≤ 0.01)		
	Mean & Observed changes			ST1-ST2	ST2-ST3	ST1-ST3
Critical Thinking	51.80^n	40.20	44.00^n	** $Z = -5.316$ $p < 0.001$	* $Z = -2.536$ $p = 0.011$	** $t(39) = 6.488$ $p < 0.001$
	↓ 11.6 / 22.4%		↑ 3.8 / 9.5%			
Creativity	53.43^n	62.75	53.03^n	** $Z = -3.009$ $p = 0.003$	** $Z = -3.159$ $p = 0.002$	No SSD $p = 0.887$
	↑ 9.3 / 17.4%		↓ 9.7 / 15.5%			

n normally distributed data, *ST1* 1st skill assessment test, *ST2* 2nd skill assessment test, *ST3* 3rd skill assessment test, *No SSD* no statistically significant differences criteria (e.g. by school, by group, by gender, by performance). Therefore, we concluded that the selected skill tests' items were not as equivalent in terms of difficulty as we had initially thought. In order to practically cancel the items' inequality and overcome this problem, we did not focus on the exact scores, but we calculated and compared the tests' score difference and the percentage of the observed change. We observed that the experimental group's students had a much greater average score improvement compared to their peers from the control group (+15.6 vs +9.3 points, or + 36.8% vs +17.4%, respectively). Students' participation in activity EER1 was the only independent variable that changed in that case, suggesting that the observed improvement in their ST2 creativity scores is somehow related.

5.3.4 Interviews

Based on their personal experience, interviewed students stated that participating in EER activities required them to have a cooperative disposition, to show empathy, to respect others' opinions and avoid conflicts (collaboration skills), but also to establish good communication for the sharing of information and the exchange of views and ideas (communication skills). They also acknowledged the importance of being observant, having ingenuity (critical thinking skills) and using divergent thinking (creativity). When asked if their participation in the EERs facilitated them to develop any of these skills, they claimed that they did not notice any significant change. However, it is worth noting that several of them found themselves to be more alert, more observant, and able to work more efficiently with their teammates after the first EER activity.

5.4 Discussion

5.4.1 Connection Between Puzzle Types and the Practice of 4C Skills

During this study's three EER activities, we tested in total fifteen different puzzles that were developed based on ten different puzzle designs. Our study's observational data analysis revealed that each of these designs facilitated the practice of 4C skills, to a greater or lesser extent, dependent on their unique features. Trying to make inferences based on the broader categories that these puzzle designs belonged to, we associated them with specific puzzle types (word, observation, logic, deduction, cryptography and meta-puzzles), or combinations of them, as presented in the work of Nicholson and Cable (2021). Word puzzles (crossword), combined with text-based resources, facilitated students' text data-mining and collaboration skills. Observation puzzles (pattern recognition, image datasets) were usually enriched with multiple items or other resources, favouring the active involvement and collaboration of most team members. Logic puzzles (matching up items) encouraged the practice of several critical thinking skills, with an emphasis on reasoning. Deduction puzzles (narrative-based questions) and cryptography (encrypted or hidden information) ignited learners' creativity and divergent thinking. Finally, meta-puzzles usually required and fostered the practice of all the 4Cs.

5.4.2 Design Guidelines for EERs that Foster the 4Cs

The learning outcome that an EER activity is capable of delivering depends greatly on its overall design. Since education practitioners develop EERs with different intended goals (e.g. content knowledge, skills, motivation, interest, engagement), applying on each of these occasions a specialised design framework, appropriately adapted to their distinct learning objectives, could benefit more of their learners. Taraldsen et al. (2022, p. 9) stressed in their review article that "researchers and educators have started to look for frameworks for designing escape rooms for educational purposes and for evaluating both 21st century skills and subject matter competence on an individual level." According to one of the latest and most complete design frameworks proposed by Fotaris and Mastoras (2022), there are several elements and parameters that need to be considered when designing an EER activity (e.g. demographic information about the participants' background, skill level, needs, and motivation; the activity's goal, learning objectives, constraints, required knowledge, group size, game type, playtime length, curriculum position, theme, setting, narrative, characters, puzzle types, puzzle designs, puzzle path, game flow, game assets, room layout, hint system, scoring system, introduction, rules, and reflection). Exploring thoroughly some of the features mentioned above, reflecting on the insights which we gained from the findings of the present research study, and building upon design guidelines provided by previous studies (Clarke et al., 2017; Fotaris & Mastoras, 2019, 2022; Nicholson, 2018; Nicholson & Cable, 2021; Veldkamp et al., 2020a, b, 2022), we recommend the following set of practical guidelines for the design of EERs that focus both on the development of participants' 4C skills and content learning:

1. *forming teams of four or five members.* Teams of that size collaborate more effectively during gameplay and offer their members adequate opportunities to access the puzzle resources, engage actively and learn.
2. *using escape boxes.* This particular game type has proven to be ideal for use in schools, in terms of practicality, time efficiency, cost, and facilitation. Regardless of the educational environment in which the activities are implemented (class, science lab, auditorium), escape boxes and the existence of a fixed working space encourage the team members to gather all together, brainstorm, discuss, and solve the game puzzles.
3. *including physical objects as puzzle components.* By increasing students' engagement and visualising theoretical concepts, these items arouse students' curiosity, allow them to learn by doing, and facilitate them to practise their 4C skills while using them.

4. *adopting a combination of linear and multi-linear puzzle pathways.* While linear puzzle pathways ensure that all team members are exposed to the same amount of knowledge, multi-linear puzzle pathways force the less engaged or introvert students to take action and participate more actively.

5. *designing self-guided puzzles that align well with the EERs' learning objectives, the curriculum, and the game narrative.* Appropriately designed puzzles that immerse learners in the storytelling experience and do not require (substantial) scaffolding to be solved have a good chance of increasing their cognitive, behavioural and affective engagement, and foster learning.

6. *adding layers to puzzles.* Instead of using puzzle components that provide all needed information in a direct and clear way, encrypting part of that information or cleverly 'hiding' it by making it seem trivial, can offer more depth, increase the puzzle's difficulty, and boost players' creativity and analytical thinking.

7. *increasing the number and the complexity of puzzle components.* Apart from making the puzzle more difficult, the increased number of available items enables, and sometimes requires, the engagement of more team members, creating some sort of social interdependence. Effective collaboration and communication become a prerequisite for solving the puzzle. Selecting, sorting, or matching items also foster the learners' observational, analytical and reasoning skills.

8. *limiting the provision of scaffolding and hints to a minimum.* Most players' observation, creativity and critical thinking usually sharpen when they reach an impasse, as long as they remain at a state of flow. Reducing the amount of available information makes the puzzle's solution less straightforward and increases the time players spend on it, thereby extending the practice of 4C skills.

9. *incorporating meta-puzzles.* Meta-puzzles are usually placed on the convergence point of complex or multi-linear puzzle paths. They offer an excellent opportunity for synthesising findings from previous puzzles, but also for re-examining information more carefully and discover something new that might have been overlooked. Their advanced complexity requires higher order thinking skills and effective collaboration between the members of a team.

10. *challenging learners' pre-existing knowledge.* Puzzles that deal with part of the syllabus that students find challenging are very useful in revealing students' weaknesses and misconceptions. At the same time, they can easily ignite discussion among students, fostering the practice of their analytical thinking and reasoning skills.

5.5   Conclusions

Practitioners and entrepreneurs alike have been claiming for a long time now that students utilise the 4C skills while engaging in EER activities. Educational researchers have also investigated this matter, considering that EERs have been widely used with the learning intention of practising and developing several types of skills, including the 4Cs. Our longitudinal study provided strong evidence that verified these claims. The application of a meticulous observational analysis on rich qualitative data that derived from numerous video/audio recordings and a control-based research design, as opposed to the methods applied in previous studies, offered a more reliable, detailed and accurate documentation of these skills. Apart from the practice of the 4C skills, indications of their development were also found. Data collected from several different methods (skill tests, questionnaires, interviews) corroborated these indications. Furthermore, we identified some connections between specific puzzle types and the practice of certain 4C skills. Based on the study's findings and informed by the existing literature, we created a list of practical design guidelines that the early adopters could utilise to develop EERs that can foster these skills combined with content learning.

Teaching biology in the 21st century is much more than a sheer transfer of content knowledge. Using this knowledge effectively requires certain skills that are equally important. Among other things, biology students are expected to: (a) critically analyse biological data in order to understand them and propose creative solutions for real-life problems; (b) care about socio-scientific issues in biology, be able to express their opinion and take action; (c) use scientific reasoning to communicate their knowledge; (d) collaborate with others towards common goals. Bearing in mind the importance that educational reforms place on developing students' 4C skills, it is necessary to investigate further the design and long-term effectiveness of appealing educational activities that can deliver this outcome, like the EERs.

References

Ananiadou, K., & Claro, M. (2009). *21st century skills and competences for new millennium learners in OECD countries* (OECD education working papers, 41, p. 33). https://doi.org/10.1787/218525261154

Bapna, A., Sharma, N., Kaushic, A., & Kumar, A. (2017). *Handbook on measuring 21st century skills.* https://doi.org/10.13140/RG.2.2.10020.99203

Bialik, M., Fadel, C., Trilling, B., Nilsson, P., & Groff, J. (2015). Skills for the 21st century: What should students learn?, *117*(July 2017). Retrieved from http://curriculumredesign.org/wp-content/uploads/CCR-Skills_FINAL_June2015.pdf

Clarke, S., Peel, D. J., Arnab, S., Morini, L., Keegan, H., & Wood, O. (2017). escapED: A framework for creating educational escape rooms and interactive games for higher/further education. *International Journal of Serious Games, 4*(3), 73–86.

Fadel, C., Bialik, M., & Trilling, B. (2015). *Four-dimensional education: The competencies learners need to succeed.* Center for Curriculum Redesign.

Fotaris, P., & Mastoras, T. (2019, October). Escape rooms for learning: A systematic review. In *Proceedings of the European conference on games-based learning*, pp. 235–243. https://doi.org/10.34190/GBL.19.179

Fotaris, P., & Mastoras, T. (2022). Room2Educ8: A framework for creating educational escape rooms based on design thinking principles. *Educational Sciences, 12*(11), 768. https://doi.org/10.3390/educsci12110768

Hu, W., & Adey, P. (2002). A scientific creativity test for secondary school students. *International Journal of Science Education, 24*, 389–403. https://doi.org/10.1080/09500690110098912

Lai, E. R., & Viering, M. (2012, April). Assessing 21st century skills: Integrating research findings. In *Annual meeting of the national council on measurement in education*, p. 66.

Lathwesen, C., & Belova, N. (2021). Escape rooms in stem teaching and learning – Prospective field or declining trend? A literature review. *Educational Sciences, 11*(6). https://doi.org/10.3390/educsci11060308

Marsden, E., & Torgerson, C. J. (2012). Single group, pre- and post-test research designs: Some methodological concerns. *Oxford Review of Education, 38*(5), 583–616. https://doi.org/10.1080/03054985.2012.731208

Nicholson, S. (2018). Creating engaging escape rooms for the classroom. *Childhood Education, 94*(1), 44–49. https://doi.org/10.1080/00094056.2018.1420363

Nicholson, S., & Cable, L. (2021). *Unlocking the potential of puzzle-based learning: Designing escape rooms and games for the classroom* (1st ed.). SAGE.

P21. (2009). *P21 framework definitions*. Retrieved from http://www.p21.org/documents/P21_Framework_Definitions.pdf

Qian, M., & Clark, K. R. (2016). Game-based learning and 21st century skills: A review of recent research. *Computers in Human Behavior, 63*, 50–58. https://doi.org/10.1016/j.chb.2016.05.023

Roschelle, J., & Teasley, S. D. (1995). The construction of shared knowledge in collaborative problem solving. In C. E. O'Malley (Ed.), *Computer-supported collaborative learning* (pp. 69–197). Springer. https://doi.org/10.1007/978-3-642-85098-1_5

Sternberg, R. J. (1986). *Critical thinking: Its nature, measurement and improvement* (p. 37). National Institute of Education.

Sternberg, R., & Lubart, T. (1998). The concept of creativity: Prospects and paradigms. In R. Sternberg (Ed.), *Handbook of creativity* (pp. 3–15). Cambridge University Press. https://doi.org/10.1017/CBO9780511807916.003

Taraldsen, L. H., Haara, F. O., Lysne, M. S., Jensen, P. R., & Jenssen, E. S. (2022). A review on use of escape rooms in education–touching the void. *Education Inquiry, 13*(2), 169–184. https://doi.org/10.1080/20004508.2020.1860284

Veldkamp, A., Daemen, J., Teekens, S., Koelewijn, S., Knippels, M. C. P. J., & van Joolingen, W. R. (2020a). Escape boxes: Bringing escape room experience into the classroom. *British Journal of Educational Technology, 51*(4), 1220–1239. https://doi.org/10.1111/bjet.12935

Veldkamp, A., van de Grint, L., Knippels, M. C. P. J., & van Joolingen, W. R. (2020b). Escape education: A systematic review on escape rooms in education. *Educational Research Review, 31*(January), 100364. https://doi.org/10.1016/j.edurev.2020.100364

Veldkamp, A., Rebecca Niese, J., Heuvelmans, M., Knippels, M. C. P. J., & van Joolingen, W. R. (2022). You escaped! How did you learn during gameplay? *British Journal of Educational Technology, 53*, 1430–1458. https://doi.org/10.1111/bjet.13194

Chapter 6
Museum Guides' Views on the Integration of the Nature of Science While Addressing Visitors' Experiences: The Context of Ecological and Evolutionary Issues

Anna Pshenichny-Mamo and Dina Tsybulsky

6.1 Theoretical Background

Understanding nature's animate and inanimate parts will help us face local and global challenges, such as environmental deterioration, loss of biological variability, and climate change. Therefore, natural history is a necessary component of biological education (King & Achiam, 2017). Natural History Museums (NHMs) are considered important natural history learning environments due to the research, educational, and curating activities that take place there. NHMs' collections present visitors with natural history's essential content, practices, and discourse in a tangible form (Marandino et al., 2015). NHMs present both the history of nature and related scientific research and, in this sense, shape visitors' views of scientific activity and the scientific community. Thus, natural history museums have a distinctive advantage in promoting evolution education (Diamond & Evans, 2007; King & Achiam, 2017; Piqueras et al., 2022) and education on ecology-related topics, such as environmental issues (King & Achiam, 2017), climate change (McGhie et al., 2018), etc. NHMs have a critical role in making science, knowledge, and the narratives that facilitate the accessibility of scientific learning to the public, as they are research institutions that interact with a diverse audience (Mujtaba et al., 2018). As centers of informal education, NHMs are open to all ages, to a wide range of cultures, and to all walks of society. They also have the power to contribute to the education of citizens who have science capital and base their decisions on scientific knowledge. One of the means of communication between the scientific community and the public is through NHM exhibits (Achiam et al., 2014). To establish visitors' connection to the science presented in the exhibits, it is crucial to highlight the

A. Pshenichny-Mamo (✉) · D. Tsybulsky (✉)
Faculty of Education in Science and Technology, Technion – Israel Institute of Technology, Haifa, Israel
e-mail: panna@campus.technion.ac.il; dinatsy@technion.ac.il

© The Author(s) 2024
K. Korfiatis et al. (eds.), *Shaping the Future of Biological Education Research*,
Contributions from Biology Education Research,
https://doi.org/10.1007/978-3-031-44792-1_6

79

intricacy of scientific processes and social implications (Hine & Medvecky, 2015). This approach can contribute to shaping the visitors' understanding of the nature of science (NOS).

NOS refers to the fundamental aspects of science, such as its definition, methodology, ability to address certain types of questions, and limitations in addressing others. One of the frameworks for science education that incorporates NOS is the Family Resemblance Approach (FRA), which was proposed by Erduran and Dagher (2014) and builds upon the framework established by Irzik & Nola (2011). The development of FRA was based on overlaps, similarities, and differences among the scientific disciplines. The FRA highlights that NOS has cognitive-epistemic and social-institutional aspects which depend on each other, and cannot exist without each other. Together, these aspects portray a comprehensive picture of science and underline its vital importance. Each aspect is followed by categories that describe the various characteristics of the sciences. The epistemic-cognitive aspects include the categories of aims and values, practices, methods and methodological rules, and scientific knowledge. The categories of social-institutional aspects are: professional activities, scientific ethos, social certification and dissemination, social values of science, social organizations and interactions, political power structures, and financial systems (Erduran & Dagher, 2014). Recently, Irzik & Nola (2022) proposed a new category to be added to the social-institutional aspects of NOS: the reward system of science. Only a few studies have examined the context of NOS in NHMs (e.g., Piqueras et al., 2022). These studies indicate that visitors' direct interaction with the exhibits they are exposed to in NHMs may contribute to their understanding of several aspects of NOS. Along with their contribution to visitors' understanding of NOS, visits to NHMs may generate rewarding, memorable and satisfying experiences (Pekarik et al., 1999; Tsybulskaya & Camhi, 2009). In the professional literature, it is noted that positive experiences for museum visitors can be classified into four categories: object experiences (originate from encounters with significant exhibit objects), cognitive experiences (originate from interpretive or intellectual opportunities provided by the experience), introspective experiences (originate from private perceptions and experiences) and social experiences (originate from social engagement). These experiences are uniquely able to engage visitors with the content discussed and may influence the process of the development of understanding of NOS during the guided tour.

Many studies in the field of informal education have focused on visitors and their learning outcomes (Achiam et al., 2016; Bamberger & Tal, 2007, 2008; Nesimyan-Agadi & Ben Zvi Assaraf, 2021; Piqueras et al., 2008; Shaby et al., 2019b; Tal & Morag, 2007). Recent studies have shown increased awareness of the importance of the guides in these environments (Pshenichny-Mamo & Tsybulsky, 2023), and have focused on guides' professional development (Lederman & Holliday, 2017; Piqueras & Achiam, 2019; Tran & King, 2007, 2011) and the interactions between guides and visitors (Pattison & Dierking, 2013; Plummer et al., 2021; Shaby et al., 2019a). Despite the fact that guides – as educators – play a key role in connecting the museum's agenda with the needs and interests of its visitors, and are essential in designing museum visitors' experiences (Tran, 2007), the research largely overlooks their perspectives. The goal of this study was to examine guides' views regarding the integration of NOS, with attention to visitor satisfaction, during instruction on relevant biological topics in the areas of ecology and evolution. The decision to focus on the context of topics pertaining to ecology and evolution stemmed from the importance of NHMs to the didactics of biological topics. Similar to King and Achiam (2017), we too believe that it is necessary to develop comprehensive natural history education in order to develop a well-rounded understanding of science. Natural history education encompasses a thorough understanding of plants and animals in their current and historical natural environments, as well as the scientific methods and practices used in the field. Such an education would provide learners with the necessary knowledge, skills, and reasoning abilities to actively engage in discussions and initiatives related to biological issues, such as evolution and ecology.

6.2   Methods

This was a qualitative study involving 15 guides working in four different NHMs located in various regions in Israel. We chose the term "guides", as is customary in the Hebrew language, to refer to paid employees at the museum who accompany and guide visitors through exhibitions and galleries in the museum (Bamberger & Tal, 2008; Tsybulskaya & Camhi, 2009). The guides were approached via a request to the manager of education programs at each museum, who helped to put the primary researcher in contact with the guides. The guides who agreed to participate in the study did so on a voluntary basis. The guides were diverse in terms of their number of years of experience as museum guides, hours worked per week, academic degree, and gender (10 women and 5 men).

Data were collected using individual semi-structured interviews, which were recorded and transcribed. In the interviews, the guides were asked to describe their practice during guided tours from a reflective and retrospective point of view. Examples of questions asked during the interviews are: (1) How do you refer to the characteristics of science and scientific research during your guidance? (2) What are the topics of the guided tours at the museum? What are the main points in these guided tours, and how do you relate to them?

The data were then submitted for content analysis from the perspective of FRA (Erduran & Dagher, 2014), with the addition of the social-institutional aspect reward system (Irzik & Nola, 2022) and satisfying visitors' experiences (Pekarik et al., 1999; Tsybulskaya & Camhi, 2009; Tsybulsky, 2019). The analysis was performed in three stages. In the first stage, after the transcription and repeated reading of interviews, the analysis focused on expressions related to cognitive-epistemic and social-institutional aspects of NOS. After that, in the second stage, statements identified as connected to aspects of NOS were categorized into four categories that describe the visitors' experiences: object experiences, cognitive experiences, introspective experiences, and social experiences. The third stage included analysis of guides' phrases related to evolutionary or ecological topics.

During data analysis, our aim was to identify the primary aspect of NOS in the guides' utterances. However, in some cases it was not feasible to pinpoint a single primary aspect, as certain statements touched upon multiple aspects of NOS. Thus, a single statement might encompass mentions of various NOS aspects.

In the following, we illustrate the process of data analysis using several examples:

Utterances such as *"[...] the most that I use is some stories about Darwin [...]"* were not included in the analysis. Although the above includes a reference to evolution, it does not refer to any NOS aspects or any of the categories of visitors' experiences.

The following utterance was included in our analysis:

> *It is important to me to make the other person feel like a researcher. To make him think, for example, about how Darwin, who lived in a time when there was not so much technology, had to look at all the animals and imagine the evolutionary connection between them in his mind's eye. And really, when I try to connect these things, I'm more trying to connect through explaining that research is something that develops, it progresses, and things that we once knew are different today; today we know better. It's something that I think really, really characterizes my work as a guide when it comes to elements of research.*

This guide talks about Darwin's thinking process and the development of scientific knowledge in the context of evolution. She refers to the tentativeness of scientific knowledge, an epistemic-cognitive aspect of NOS, and science as a human endeavor, a social-institutional aspect of NOS. This is also a good example of the way in which a single utterance from a guide can relate to several aspects of NOS. With this utterance, the guide tries to get visitors to connect to Darwin's journey and imagine how research was done in the days when technological instruments were less developed than they are today. In doing so, she essentially tries to evoke emotions and to give the visitors an opportunity to imagine a different time from the one in which we live today, thus relating to the visitors' introspective experiences.

6.3   Findings

Our findings indicated that out of 173 references related to NOS aspects, 57 were in the context of evolution and ecology. We found that the museum guides addressed NOS mostly in terms of epistemic-cognitive aspects (43 utterances), and less in terms of social-institutional aspects (14 utterances). In the context of the museum visitors' experience, most of the utterances reflected cognitive experiences together with epistemic-cognitive aspects of NOS (10 utterances). The guides tended to reflect less on the social experiences of museum visitors. Only one guide reflected on this experience in the context of social-institutional NOS aspects. The rest of the findings are shown in Figs. 6.1 and 6.2.

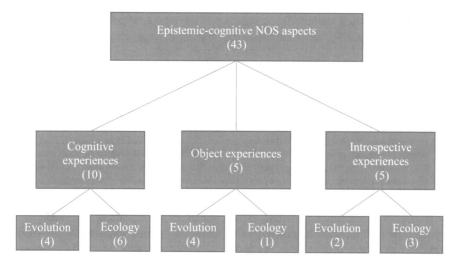

Fig. 6.1 The flow chart shows the distribution of epistemic-cognitive aspects of NOS mentioned by the guides into categories linked to visitors' experience and topic (the number of references appears in parentheses). The total number of references to epistemic-cognitive aspects is higher than the total number of references to visitor experience, as a single guide's utterance encompasses mentions of various NOS aspects in several cases

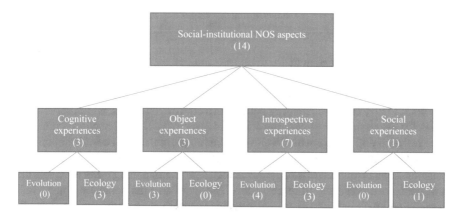

Fig. 6.2 The flow chart shows the distribution of social-institutional aspects of NOS mentioned by the guides into categories linked to visitors' experience and topic (the number of references appears in parentheses)

6.3.1   Epistemic-Cognitive Aspects of NOS

6.3.1.1   Cognitive Experiences

In the context of evolution, one guide talks about aspects of NOS by mentioning that scientific knowledge is tentative and changes with new findings. In her words, the guide tries to enrich visitors' understanding of the way in which we gain scientific knowledge. Thereby, she references the cognitive experience of the museum visitors:

> *I like to talk about the aspects of archeological findings. If the subject is "What Makes Us Human?" I talk about Homo erectus, which was believed to be the first evidence of a hominid walking upright and not a human being as we know it today. And years later, Australopithecus afarensis, known as Lucy, was discovered: a species that was much more ancient and able to walk upright. Then suddenly, the scientific world looks at it in astonishment and realizes that they made a mistake and so all the books need to be changed. That is part of the beauty of science; there is this great humility to say we were wrong, but look – now we know something new and much more significant, and who knows what else we will discover? It is always possible to un-prove what those before me found. And there is always room for doubt.*

Another reference to the cognitive experience of museum visitors was made by a guide who talked about the ecology-related issue of stream pollution. While mentioning that scientists observe things and test their observations, he made a reference to the epistemic-cognitive aspect of scientific practices related to NOS. The guide aims to enhance visitors' understanding of how research on stream pollution is conducted and facilitates their acquisition of new information and knowledge:

> *[…] I think that the museum guide has an important role in making science accessible […] It's very important that people understand more than just the bottom line. It's not enough to point and say that the streams in Israel are polluted; rather, they need to understand the research being done, how it is conducted, and that people are, in fact, working on this issue.*

6.3.1.2 Object Experiences

One guide refers to ecology in the context of a food chain exhibit. The guide draws attention to an exhibit showcasing pellets derived from owl food, which serves as a basis for discussing the underlying scientific principles. He underscores the extensive research and experimentation involved in acquiring scientific information and highlights that the museum researchers engage in such studies. The focus on the visitors' object experience is apparent as they see "the real thing"- the food that the owl consumed presented in its real form. These are also the authentic objects that researchers study in the museum. By guiding visitors to observe the exhibit in detail, the guide highlights its distinctiveness:

> *[…] There are numerous exhibits highlighting research, specifically how the research is conducted. One example of such exhibits is the owl pellet display, where visitors can observe the disassembly of pellets like a puzzle. These are things that are connected to the studies done at the museum. And these are also things that draw people's attention, people suddenly go see this table with a lot of partridges that look very bad (pellets), they're just spread out there, and they don't understand why it's displayed like that, and that's precisely the point that gives us an opening to talk about the research and why it's like that, what goes on behind the curtains […].*

Another guide talks about the technique of ringing birds as one of a variety of scientific methods that serve scientists in ecological research. The guide discusses multiple aspects of object experiences. First, she allows visitors to see "the real thing" – an exhibit that shows an authentic research method currently in use in ecological research. Second, she shows them an exhibit that illustrates the transition – from hunting animals in order to research them, to the use of research methods that do not include hunting, but rather the tracking of living animals. Thus, the guide shows visitors an important exhibit that they are likely to find meaningful:

> We have a ringed stork. That's a point where I can start talking about everything related to research, including from the past few years, about ringing and tracking migratory birds. The general direction is to take some exhibit and talk about it. Say, to talk about the development of research of bird migration, how it started with ringing, how it looks today, you know… something entirely different […].

This guide also notes that she integrates discussion of scientific research, and the nature thereof, by using exhibits, and thus emphasizes her attention to the visitors' object experiences.

6.3.1.3 Introspective Experiences

In the context of plant ecology, one of the guides mentions studies that were done in the museum's laboratory focused on incubator-like products. She notes that visitors may overlook their existence despite passing by every day. The guide mentions NOS, saying that science is a way of knowing. Attention is shown to the visitors' introspective experience in the attempt to reflect visitors' memories of this fruit body – did they ever notice it? She also emphasizes the significance of the fruit body as a subject of research:

> One researcher here deals with the product that develops when a plant mosquito bites a plant and a kind of fruit body is created – like an incubator. These incubators are hard for people to imagine, although they may pass by it every day. And I can tell them that a few floors above, a researcher is studying in her laboratory how this thing affects the ecology of the plants.

6.3.2 Social-Institutional Aspects of NOS

6.3.2.1 Cognitive Experiences

The guide discusses the cognitive experience of visitors and describes how he enriches their knowledge about the dance of honeybees. This topic is related to communication in the world of honeybees, which falls under the category of ecology. The guide highlights that the discovery of the honeybee dance earned a

Nobel Prize for the researcher, which relates to the social-institutional aspect of NOS, specifically the reward system.

> *[...] For example, I talk about the bees' dance and about communication in the world of bees. Then I mention the researcher Karl von Frisch, who discovered the bees' dance and won a Nobel prize [...].*

6.3.2.2 Object Experiences

One of the guides discusses the divergent evolution of partridges in different regions of Israel, highlighting science as a human effort noting that the findings presented in the exhibits are the work of researchers. Thus, she mentions the social-institutional aspects of NOS. Additionally, the guide draws attention to the visitors' object experiences, as they are able to see the partridges that underwent divergent evolution and observe authentic objects from nature that were used in the research:

> *In one of our galleries we have a collection of partridges from the south and partridges from the north, and you can see the differences. So I can talk about that, and I can show them, 'look, this is what researchers found, this is the process (divergent evolution) that operates on a lot of animals, it's not just partridges, it's also tigers, it's rabbits,' and I explain the topic. It's something that really connects to research done by scientists here [...].*

6.3.2.3 Introspective Experiences

A guide mentions respect for the environment – which falls under the scientific ethos aspect of NOS – while speaking about the state of the environment as an ecological issue. In her words, she tries to get visitors to connect to the problem that exists and to reflect by recalling past ideas and experiences which may be less harmful to the environment:

> *Regarding ecology, I think it is essential to talk about how we brought on this environmental situation. The most critical exhibition for understanding this is "The Human Impact". This exhibition shows the effects of the changes we have made as a result of a very difficult process of industrialization and its consequences. [...] many people are trying to find technological solutions, such as substitutes. I think it's a good idea, but we also need to find solutions for the harm that we have caused and go back to the less harmful ways used in the past.*

Another guide discusses the importance of respecting nature in the context of ecological problems, highlighting it as a core social value of science. Specifically, she focuses on the issue of species extinction and aims to raise visitors' awareness and encourage introspection. By prompting visitors to reflect on the meaning of the exhibit and encouraging them to feel a spiritual connection to the issue, she hopes to inspire them to carry this knowledge with them into the future. Through this approach, she engages with the introspective experiences of visitors, inviting them to consider their own values and actions in relation to the environment:

[…] I try to alert visitors to the fact that the animal they are observing is at risk of becoming extinct, that the other animal they were observing has already become extinct, and that the animals that they see in their natural environment are also at risk of becoming extinct. I try to show them that this is directly related to their everyday lives. I emphasize that there is something they can do about it, which is to ask questions even after they leave the museum.

6.3.2.4 Social Experiences

In the following interview, the guide emphasizes the social values of science in the context of ecology. He introduces visitors to the collection of Father Ernst Johann Schmitz and sparks a conversation on the topic of respecting the environment. This topic is discussed in connection with the research method of hunting animals that was commonly used in the past. The guide aims to generate a group discussion about the impact of these methods and the importance of ethical considerations in scientific research. In this way, he attends to the social experience of museum visitors by creating a shared activity related to the exhibit and fostering social learning:

Say, the exhibit of Father Schmitz, there we talk about him and about the fact that all of ecology and zoology was based on hunting, and people are really horrified, 'People who researched animals hunted them?' And then I try to start a discussion, a shared learning.

6.4 Conclusions and Discussion

This study explored the views of NHM guides regarding the integration of NOS aspects, while attending to museum visitors' satisfying experiences, when discussing topics related to evolution and ecology. Our study found that when guiding visitors through the museum and addressing evolution and ecology topics, the guides tend to emphasize the epistemic-cognitive aspects of NOS. Also, they pay attention primarily to the cognitive experiences of visitors in the museum. We believe these findings complement each other. When the guides integrate aspects of NOS in the epistemic-cognitive context, they try to educate their audience about the characteristics of science and the scientific process. Thus, in essence, the guides usually relate to visitors' cognitive experience during their time at the museum, allowing the visitors to gain new information and knowledge that enriches the previous understanding they may have had of various topics in evolution and ecology.

Through attention to the objects in an exhibit, the guides relate to the visitors' object experiences. Attention to this experience is shown primarily through giving visitors the opportunity to see "real things" in terms of the authenticity of the object for scientific research. Similarly, this experience allows museum visitors to see objects that are rare and highly valuable to the world of science, and perhaps to be amazed by the various exhibits in the museum. Our findings support the view presented by King and Achiam (2017) that exhibits in NHMs can serve as "windows" into the practice of natural history and can enhance science education, especially biology education, for visitors. Additionally, our study indicates that the object experiences provided by guides during museum tours can expose visitors to both the epistemic-cognitive and social-institutional aspects of NOS.

Attention to introspective experiences during museum tours occurs primarily in the context of stories about the history of science in which the guides lead visitors to identify with the researcher's story and imagine the spirit of the times in which he or she was active. Furthermore, these stories reveal the human component of science and scientific research, which has a direct connection to the social-institutional aspects of NOS. This exposure can enhance visitors' understanding of these aspects. The introspective experiences enable visitors to recall their own past experiences at different points during the tour and to feel how these experiences take on new meaning within the context of the museum. Additionally, our findings suggest that when museum guides prompt visitors' introspective experiences, it can result in visitors becoming aware of previously unfamiliar aspects of scientific research and knowledge. This heightened awareness can then deepen visitors' understanding of the epistemic-cognitive aspects of NOS, which in turn can lead to a reinterpretation of the exhibits viewed during the guided tour.

Museum visitors' social experience was mentioned in the interviews by only one guide, who spoke about the subject of ecology and mentioned a social-institutional aspect of NOS – social values of science. We believe that during the tours the guides *do* pay attention to visitors' social experience, by creating a discussion between visitors or conducting one-on-one discussions related to ecology or evolution issues at various points. It could be that during their narrations the guides ask the visitors questions or answer their questions about scientific knowledge, or conversations arise about the struggles that the scientists faced during their scientific path. These interactions provide opportunities for visitors to discuss relevant NOS aspects and gain additional relevant knowledge. However, during the interviews, the guides did not share many of these situations. We may interpret this as indicating that they did not find these situations to be suitable answers to the questions we asked.

Prior research in informal education environments has suggested that interactions between museum guides and visitors primarily consist of technical explanations (Shaby et al., 2019a). However, our study offers new insight by revealing that, from the viewpoint of the guides themselves, there is an effort to engage with visitors in a manner that extends beyond simply explaining how to operate exhibits or objects. Instead, guides aim to enhance visitors' overall experience at the NHMs and integrate aspects of NOS by incorporating relevant scientific knowledge, and knowledge about science related to ecology and evolution issues. Moreover, the interviews with the guides show that the efforts to integrate different aspects of NOS into the tours are done in an attempt to enhance visitors' overall museum experience. Bamberger and Tal (2008) suggested that the expertise and passion exhibited by guides may play a role in the positive interactions between guides and visitors, which may contribute to their heightened engagement in the learning process during the tour. Our findings show that the guides tend to incorporate anecdotes and narratives that capture visitors' interest and encourage their participation in the content being presented. Ultimately, the guides' efforts serve to enrich and enliven the museum experience for visitors.

According to Holliday and Lederman's (2014) findings, guides in informal educational settings possess a deep understanding of NOS. Our study builds on this conclusion by demonstrating that these guides also possess the pedagogical knowledge required to effectively integrate NOS into their guided tours. This knowledge includes a focus on visitors' satisfying experiences, which are valued for their ability to inform and shape visitors' expectations and preferences during museum visits or activities (Pekarik et al., 1999).

In conclusion, the focus of this study is on museum guides and their views. The objective is to gain a better understanding of the role of guides during the guided tours they give to museum visitors, building upon previous studies (Pshenichny-Mamo & Tsybulsky, 2023; Shaby et al., 2019a; Tran, 2007). Acquiring a deeper understanding of the diverse practices used in NHMs can aid museum educational staff in improving the design of educational programs and guide training programs that integrate evolutionary and ecological content and explicit instruction on NOS. This can facilitate the development of tours and exhibits that provide visitors with more satisfying and meaningful experiences, while enhancing visitors' comprehension of different aspects of NOS. A follow-up study should examine these findings in comparison with what happens in practice during guided tours and in comparison with the experiences reported by the visitors themselves during these tours. Therefore, conducting observations of guided tours and examining interactions and visitors' experiences with a focus on the integration of NOS could provide more insights and build upon our findings.

References

Achiam, M., May, M., & Marandino, M. (2014). Affordances and distributed cognition in museum exhibitions. *Museum Management and Curatorship, 29*(5), 461–481. https://doi.org/10.108 0/09647775.2014.957479

Achiam, M., Simony, L., & Lindow, B. E. K. (2016). Objects prompt authentic scientific activities among learners in a museum programme. *International Journal of Science Education, 38*(6), 1012–1035. https://doi.org/10.1080/09500693.2016.1178869

Bamberger, Y., & Tal, T. (2007). Learning in a personal context: Levels of choice in a free choice learning environment in science and natural history museums. *Science Education, 91*(1), 75–95. https://doi.org/10.1002/sce.20174

Bamberger, Y., & Tal, T. (2008). Multiple outcomes of class visits to natural history museums: The students' view. *Journal of Science Education and Technology, 17*(3), 274–284. https://doi. org/10.1007/s10956-008-9097-3

Diamond, J., & Evans, E. M. (2007). Museums teach evolution. *Evolution, 61*(6), 1500–1506. https://doi.org/10.1111/j.1558-5646.2007.00121.x

Erduran, S., & Dagher, Z. R. (2014). *Reconceptualizing the nature of science for science education.* Springer.

Hine, A., & Medvecky, F. (2015). Unfinished science in museums: A push for critical science literacy. *Journal of Science Communication, 14*(2), 1–14. https://doi.org/10.22323/2.14020204

Holliday, G. M., & Lederman, G. N. (2014). Informal science educators' views about nature of scientific knowledge. *International Journal of Science Education, Part B, 4*(2), 123–146. https://doi.org/10.1080/21548455.2013.788802

Irzik, G., & Nola, R. (2011). A family resemblance approach to the nature of science for science education. *Science & Education, 20*(7), 591–607. https://doi.org/10.1007/s11191-010-9293-4

Irzik, G., & Nola, R. (2022). Revisiting the foundations of the family resemblance approach to nature of science: Some new ideas. *Science & Education.* https://doi.org/10.1007/s11191-022-00375-7

King, H., & Achiam, M. (2017). The case for natural history. *Science & Education, 26*(1–2), 125–139. https://doi.org/10.1007/s11191-017-9880-8

Lederman, J. S., & Holliday, G. M. (2017). Addressing nature of scientific knowledge in the preparation of informal educators. In P. G. Patrick (Ed.), *Preparing informal science educators: Perspectives from science communication and education* (pp. 509–525). Springer. https://doi.org/10.1007/978-3-319-50398-1_25

Marandino, M., Achiam, M., & de Oliveira, A. D. (2015). The diorama as a means for biodiversity education. In S. D. Tunnicliffe & A. Scheersoi (Eds.), *Natural history dioramas: History, construction and educational role* (pp. 251–266). Springer. https://doi.org/10.1007/978-94-017-9496-1_19

McGhie, H., Mander, S., & Underhill, R. (2018). Engaging people with climate change through museums. In W. Leal Filho, E. Manolas, A. M. Azul, U. M. Azeiteiro, & H. McGhie (Eds.), *Handbook of climate change communication* (Vol. 3, pp. 329–348). Springer. https://doi.org/10.1007/978-3-319-70479-1_21

Mujtaba, T., Lawrence, M., Oliver, M., & Reiss, M. J. (2018). Learning and engagement through natural history museums. *Studies in Science Education, 54*(1), 41–67. https://doi.org/10.1080/03057267.2018.1442820

Nesimyan-Agadi, D., & Ben Zvi Assaraf, O. (2021). How can learners explain phenomena in ecology using evolutionary evidence from informal learning environments as resources? *Journal of Biological Education,* 1–14. https://doi.org/10.1080/00219266.2021.1877784

Pattison, S. A., & Dierking, L. D. (2013). Staff-mediated learning in museums: A social interaction perspective. *Visitor Studies, 16*(2), 117–143.

Pekarik, A. J., Doering, Z. D., & Karns, D. A. (1999). Exploring satisfying experiences in museums. *Curator, 42*(2), 152–173.

Piqueras, J., & Achiam, M. (2019). Science museum educators' professional growth: Dynamics of changes in research–practitioner collaboration. *Science Education, 103*(2), 389–417. https://doi.org/10.1002/sce.21495

Piqueras, J., Hamza, K. M., & Edvall, S. (2008). The practical epistemologies in the museum: A study of students' learning in encounters with dioramas. *Journal of Museum Education, 33*(2), 153–164. https://doi.org/10.1080/10598650.2008.11510596

Piqueras, J., Achiam, M., Edvall, S., & Ek, C. (2022). Ethnicity and gender in museum representations of human evolution: The unquestioned and the challenged in learners' meaning making. *Science & Education, 0123456789.* https://doi.org/10.1007/s11191-021-00314-y

Plummer, J. D., Tanis Ozcelik, A., & Crowl, M. M. (2021). Informal science educators engaging preschool-age audiences in science practices. *International Journal of Science Education, Part B: Communication and Public Engagement, 11*(2), 91–109. https://doi.org/10.1080/21548455.2021.1898693

Pshenichny-Mamo, A., & Tsybulsky, D. (2023). Natural history museum guides' conceptions on the integration of the nature of science. *Science & Education,* 1–19. https://doi.org/10.1007/s11191-023-00469-w

Shaby, N., Assaraf Ben-Zvi, O., & Tal, T. (2019a). An examination of the interactions between museum educators and students on a school visit to science museum. *Journal of Research in Science Teaching, 56*(2), 211–239. https://doi.org/10.1002/tea.21476

Shaby, N., Assaraf Ben-Zvi, O., & Tal, T. (2019b). Engagement in a science museum – The role of social interactions. *Visitor Studies, 22*(1), 1–20. https://doi.org/10.1080/10645578.2019.1591855

Tal, T., & Morag, O. (2007). School visits to natural history museums: Teaching or enriching? *Journal of Research in Science Teaching, 44*(5), 747–769. https://doi.org/10.1002/tea.20184

Tran, L. U. (2007). Teaching science in museums: The pedagogy and goals of museum educators. *Science Education, 91*(2), 278–297.

Tran, L. U., & King, H. (2007). The professionalization of museum educators: The case in science museums. *Museum Management and Curatorship, 22*(2), 131–149. https://doi.org/10.1080/09647770701470328

Tran, L. U., & King, H. (2011). Teaching science in informal environments: Pedagogical knowledge for informal educators. In D. Corrigan, J. Dillon, & R. Gunstone (Eds.), *The professional knowledge base of science teaching* (pp. 279–293). Springer. https://doi.org/10.1007/978-90-481-3927-9

Tsybulskaya, D., & Camhi, J. (2009). Accessing and incorporating visitors' entrance narratives in guided museum tours. *Curator, 51*(1), 81–100.

Tsybulsky, D. (2019). Students meet authentic science: the valence and foci of experiences reported by high-school biology students regarding their participation in a science outreach programme. *International Journal of Science Education, 41*(5), 567–585. https://doi.org/10.1080/09500693.2019.1570380

Chapter 7
Friends or Foes? Microorganisms in Greek School Textbooks

Georgios Ampatzidis and Anastasia Armeni

7.1 Introduction

Research on children's ideas about microorganisms reveals that children frequently hold alternative ideas about aspects such as microorganisms' size and morphology, their status as living entities, the location where they can be found, and their activity (Byrne, 2011). For example, researching the ideas of students aged 11–14, Bandiera (2007) reported that when asked about the presence of microorganisms indoors, half of her study participants placed microorganisms in specific rooms of the home, primarily the bathroom and the kitchen. Furthermore, when asked whether microorganisms could be found in a healthy human body, 38% of them answered that healthy humans are free of microorganisms. When asked where microorganisms can be found, students aged 7–14 in Byrne's (2011) study largely focused on the human body. The same finding is reported by Jones and Rua (2006) as well; among their participants, high school and middle school students believed that microorganisms could be present in the mouth or on the hands, whereas elementary school students thought that germs were located in particular areas of the body, such as saliva and skin. Other areas where microorganisms may be found according to students are filthy and unsanitary places, such as waste, dustbins, sewage treatment plants, compost piles and soil (Byrne, 2011).

According to Byrne et al. (2009), the most frequently mentioned microbial activity by students of various ages was as agents of human disease. In another study, Byrne (2011) found that students 7–14 years old held similar beliefs

G. Ampatzidis (✉)
University of Thessaly, Volos, Greece
e-mail: gampatzidis@uth.gr

A. Armeni (✉)
University of Patras, Patras, Greece
e-mail: armeni@upatras.gr

© The Author(s) 2024
K. Korfiatis et al. (eds.), *Shaping the Future of Biological Education Research*,
Contributions from Biology Education Research,
https://doi.org/10.1007/978-3-031-44792-1_7

about microorganisms, such as the belief that they are the cause of disease. Byrne and Grace (2010) also acknowledged that negative aspects of microbial activity dominated children's ideas. They noted that the participants of their study appeared to know substantially more about the connection between microorganisms and disease or food spoilage than they did about the beneficial functions that they serve in matter cycling, food production and medicine. The lack of knowledge of these aspects of microbial activity among children is alarming; learning about these applications would not only serve as a foundation for later lessons on more complex microbial technology and biotechnology, but it might also help to balance out the students' largely unfavourable views of microorganisms (Byrne & Grace, 2010).

In an effort to track the origin of students' thoughts about microorganisms, Simonneaux (2000) noted that the media appear to have a crucial part in forming students' relevant understanding, along with school lessons, personal experience, and family life. According to Mafra and Lima (2009), children often develop their understanding of microorganisms as a result of the diseases that they, their parents or other family members experience; thus the concept of microorganisms is associated with a negative meaning. Questioning 836 primary school students on their major source of information about microorganisms, Karadon and Şahin (2010) reported that the most common response was the media, such as TV shows (39.2%), with school being the main source of information about microorganisms for 21.1% of students. Simard (2021) suggested that since the 1990s, the media have emphasized the association of microorganisms with disease, while ignoring information on their beneficial roles and functions. The media greatly fuel prejudices against microorganisms by focusing on human illnesses in the news; the public gets the impression that microorganisms are only causes of disease and mortality as a result of the focus on illnesses such as SARS, AIDS and H1N1 (Simard, 2021).

Thus, it seems that (a) students considerably express negative ideas about microorganisms, mainly related to the threat they pose to human health, (b) it is of paramount importance for students to build an understanding of the positive activities of microorganisms, and (c) students' ideas about microorganisms are significantly influenced by the media, and school seems to play a rather important role as well. Focusing on school education, we note that it relies heavily on the usage of textbooks, which are essential for both teaching and learning (Liu & Khine, 2016). Over 90% of secondary science teachers in the United States use textbooks to help them plan and deliver lessons (Weiss et al., 2001). Abd-El-Khalick et al. (2008) argue that textbooks may influence what is taught and learned in the classroom to a greater extent than what educators would like.

Considering the above, we decided to explore Greek school textbooks in regard to the representation of microorganisms. Therefore, the research question addressed here is: How are microorganisms represented in Greek school textbooks?

7.2 Methods

For this study, we investigated the textbooks used for the teaching of biology in secondary education (six textbooks) and the textbooks used for the teaching of natural sciences in primary education (two textbooks) in Greece. We note here that there are no courses specialized in biology in the curriculum of Greek primary schools; thus biological concepts are discussed in natural sciences classes for Grade 5 and Grade 6 students. The textbooks investigated were the following:

- Natural Sciences for the fifth grade of Primary School (Apostolakis et al., 2015a) – Grade 5.
- Natural Sciences for the sixth grade of Primary School (Apostolakis et al., 2015b) – Grade 6.
- Biology for the first grade of Gymnasium (Mavrikaki et al., 2017a) – Grade 7.
- Biology for the second and third grade of Gymnasium (Mavrikaki et al., 2017b) – Grades 8–9.
- Biology for the first grade of Lyceum (Kastorinis et al., 2011) – Grade 10.
- Biology for the second grade of General Lyceum-General Education (Kapsalis et al., 2013) – Grade 11.
- Biology for the second and third grade of General Lyceum (Adamantiadou et al., 2013) – Grades 11–12.
- Biology for the third grade of General Lyceum-Health Studies Specialization (Aleporou-Marinou et al., 2013) – Grade 12.

The first author identified the words "microorganism/microorganisms" (in Greek: mikroorganismos/mikroorganismoi) and "microbe/microbes" (in Greek: microvio/mikrovia) in the textual corpus of the study – i.e. the whole text included in the eight textbooks of interest, apart from illustrations legends, tables of contents, chapter titles and assessment activities. We note that we transliterated the Greek terms to Latin characters above according to ELOT 743 standard which is equivalent to ISO 843 standard. For the reader's convenience, we will refer to the English words in the rest of the text, i.e. microorganism/s and microbe/s. Each paragraph including the word "microorganism/s" and/or the word "microbe/s" at least once was determined as unit of analysis.

Taking into consideration the above, our analysis resulted in 187 relevant paragraphs included in seven of the textbooks investigated:

1. Natural Sciences for the sixth grade of Primary School.
2. Biology for the first grade of Gymnasium.
3. Biology for the second and third grade of Gymnasium.
4. Biology for the first grade of Lyceum.
5. Biology for the second grade of General Lyceum-General Education.
6. Biology for the second and third grade of General Lyceum.
7. Biology for the third grade of General Lyceum-Health Studies Specialization.

Table 7.1 The coding scheme

Microorganisms as a part of the living world	Microorganisms and health	Microorganisms and food	Microorganisms in industry and technology
Biological diversity	Human pathogens	Food production	Microorganisms cultivation
Physiology	Human microbiome		Cleaning water
	Pathogens of organisms other than humans		Enzymes and antibiotics production
			Vaccines and serum production

The paragraphs were coded in mutually exclusive categories formed by drawing on a scheme proposed by Mafra and Lima (2009). As shown in Table 7.1, the coding scheme consists of four categories:

- Microorganisms as a part of the living world: the text refers to ideas, such as the variety of living organisms on earth, the different environments that microorganisms are found in, and elements of their physiology.
- Microorganisms and health: the text refers to microorganisms as human pathogens or part of the human microbiome, as well as their role as pathogens of organisms other than humans.
- Microorganisms and food: the text refers to the role of microorganisms in food production (e.g. yeasts involved in bread, yoghurt and alcohol production).
- Microorganisms in industry and technology: the text refers to the role of microorganisms in industrial and technological applications, such as the production of enzymes, antibiotics, vaccines or serum, and cleaning water.

In contrast to Mafra and Limas's (2009) scheme, we decided to include subordinate categories as shown in Table 7.1. The main reasons for this decision were (a) to distinguish between the positive and negative roles of microorganisms concerning health (i.e. microorganisms as pathogens and microorganisms as part of the microbiome) and (b) to illustrate the different technological and industrial applications linked with microbial activity in the textbooks investigated. We also have to note that, while Mafra and Lima (2009) classified the role of microorganisms in vaccine development under the "Microorganisms and health" category, we decided to classify relevant paragraphs under the "Microorganisms in industry and technology" category, arguing that in this case their role in the application of technology should be highlighted.

Both authors coded independently 93 (i.e. about 50%) randomly chosen paragraphs and the rater agreement was about 90%. The cases of disagreement were reviewed and discussed by the two coders and the rest of the analysis was carried out by the first author. We should note that our research may not be considered an exhaustive account of how microorganisms are represented in the textbooks investigated, since our results are limited by the methodology used to collect our data. For instance, searching for the words "microorganism/s" and "microbe/s", we do not trace appearances of microorganisms mentioned by the name of specific taxa (e.g. bacteria, fungi) or the name of specific species.

7.3 Results

As shown in Table 7.2, in most cases microorganisms are mentioned in relation to health (94/187). More specifically, in 85/187 paragraphs microorganisms are mentioned as human pathogens. For example, in textbook-4 we read: "Pathogenic microorganisms are transmitted from person to person, by coughing, sneezing, etc. The respiratory tract, of course, is protected by the mucus and cilia of the epithelial tissue. However, if the number of pathogenic microorganisms is high and/or the body's resistance is reduced, then these microorganisms (bacteria and viruses) can cause various diseases such as pneumonia, tuberculosis and acute bronchitis. To reduce the possibility of contracting such diseases, we must avoid indoor spaces where many people are crowded together" (Kastorinis et al., 2011, p. 83).

Moreover, 6/187 paragraphs refer to microorganisms as pathogens of organisms different than humans. For instance (textbook-2): "If, for example, we are interested in studying how water affects the growth of a plant, we should experiment with identical plants keeping all other known factors constant. That is, the plants should have the same height, we should place them in similar pots, with the same quality soil, we should provide them with the same lighting and in general we should keep all the necessary conditions constant except for the amount of water we pour on them. Of course, there are other factors that may affect the growth of the plant, but which cannot be precisely controlled, such as some microorganisms found in the soil of the pot or some insects that may harm them" (Mavrikaki et al., 2017a, p. 13).

Under the "Microorganisms and health" category we traced 3/187 paragraphs where a positive role of microorganisms is mentioned. What follows is an excerpt discussing microorganisms as part of the human microbiome (textbook-6): "Other microorganisms, such as the bacterium *Escherichia coli* that lives in the intestines, are beneficial for humans when they are in small numbers and do not migrate to other tissues and organs; they constitute the microbiome of humans, which produce useful chemical substances that humans cannot synthesize on their own (e.g. vitamin K from *E. coli*) or contribute to the body's defense" (Adamantiadou et al., 2013, p. 11).

Table 7.2 Frequencies of categories of microorganisms' representations

Categories	Frequencies	Subcategories	Frequencies
Microorganisms as a part of the living world	40	Biological diversity	16
		Physiology	24
Microorganisms and health	94	Human pathogens	85
		Human microbiome	3
		Pathogens of organisms other than humans	6
Microorganisms and food	12	Food production	12
Microorganisms in industry and technology	41	Cultivation	12
		Enzymes and antibiotics production	11
		Vaccines and serum production	7
		Cleaning water	11

The second most popular category is "Microorganisms in industry and technology" (41/187). In 12/187 paragraphs the cultivation of microorganisms is discussed without focusing on particular applications. For instance, in textbook-7 we read: "Biotechnology is a combination of science and technology aimed at using living organisms to produce useful products on a large scale. It relies mainly on recombinant DNA and microorganisms' cultivation techniques. Microorganisms grow under controlled conditions in which a range of suitable nutrients is available. In a large-scale cultivation, suitable devices, called bioreactors, are used" (Aleporou-Marinou et al., 2013, p. 116).

Moreover, 11/187 paragraphs refer to the production of antibiotics or enzymes through the treatment of microorganisms. For instance (textbook-7): "Microorganisms' products are usually produced in small quantities, sufficient to cover their metabolic needs. It is obvious that for a microorganism to be practically useful, its products, such as enzymes or antibiotics, must be produced in large quantities. This is achieved in two main ways: by regulating the cultivation conditions and by genetically modifying the organisms" (Aleporou-Marinou et al., 2013, p. 151).

Furthermore, 11/187 paragraphs refer to the development of vaccines or serum through treating microorganisms. What follows is an excerpt discussing the role of microorganisms in the development of vaccines and serum (textbook-3): "A small amount of dead or inactive microorganisms or their parts is introduced into our body with vaccination. The content of the vaccine is sufficient to activate the immune system response while it is usually not capable of causing disease. In this way, the organism develops 'memory' cells for the particular microorganism" (Mavrikaki et al., 2017b, p. 86–87).

Under the "Microorganisms in industry and technology" category we finally traced 11/187 paragraphs referring to the role of microorganisms in cleaning water e.g. from oil. The following text is an example of such a reference (textbook-7): "Marine microorganisms can contribute in breaking down oil spills: After the volatile fractions of the oil evaporate, its organic chemical compounds are broken down by bacteria and fungi belonging to more than 70 genera, resulting in their metabolism into carbon dioxide" (Aleporou-Marinou et al., 2013, p. 162).

The number of paragraphs classified under the "Microorganisms as a part of the living world" category is 40/187. Most of them (24/187) deal with elements of microorganisms' physiology. For example, in textbook-5 we read: "Many microorganisms perform anaerobic respiration, that is glucose is oxidized to produce ATP in the absence of oxygen. Of course, there are also cells of multicellular organisms, which occasionally, when they are forced to produce energy and there is not enough oxygen in their environment, also perform anaerobic respiration. Muscle cells are a typical example" (Kapsalis et al., 2013, p. 110).

The rest of the paragraphs classified under the "Microorganisms as a part of the living world" category mention microorganisms through discussing the diversity of living organisms or their habitat. For instance (textbook-1): "A trip to the countryside or a walk in the forest is enough to feel the relaxation and pleasure offered by observing the natural environment, but also to see the great variety of microorganisms, plants and animals, as well as the relationships that these living organisms develop among themselves" (Apostolakis et al., 2015b, p. 74).

Finally, 12/187 paragraphs refer to the role of microorganisms in food and drink production. The following excerpt of textbook-3 is an example of such a reference: "For thousands of years, humans have been taking advantage of the properties of certain organisms, with the aim of improving their lives. They grow plants, breed animals and, by the method of selective breeding, create organisms with desirable properties-phenotypes. They use various organisms (e.g. herbs) as raw material for the preparation of medicines and cosmetics. In addition, with the help of specific microorganisms, they develop various useful products (food and drinks), such as bread, beer, wine, cheese and yogurt" (Mavrikaki et al., 2017b, p. 120).

As shown in Fig. 7.1, microorganisms are mostly mentioned in relation to health in 4/7 textbooks. The category "Microorganisms and health" is the second most popular in textbook-1, the least popular in textbook-7, while in textbook-5 the microorganisms' link with health is not mentioned at all. The category "Microorganisms as a part of the living world" is the most popular in 2/7 textbooks, while the category "Microorganisms in industry and technology" is the most popular in 1/7 textbooks.

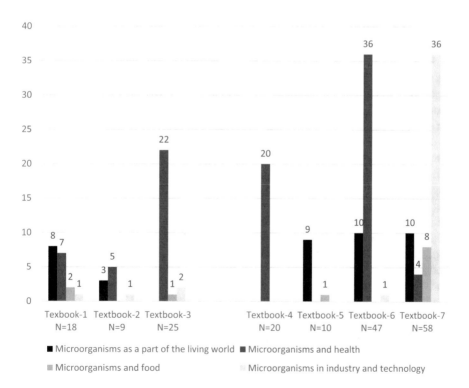

Fig. 7.1 Frequencies of categories of microorganisms' representations in textbooks

In most textbooks some categories are not represented. In 2/7 textbooks there are paragraphs classified under all four categories, 3/7 textbooks include paragraphs classified under three of the categories, in 1/7 textbooks two categories are represented, while in 1/7 textbooks microorganisms are mentioned only in regard to their link with health. In sum, the category "Microorganisms as a part of the living world" appears in 5/7 textbooks, the category "Microorganisms and health" appears in 6/7 textbooks, the category "Microorganisms and food" appears in 4/7 textbooks and the category "Microorganisms in industry and technology" appears in 5/7 textbooks.

7.4   Discussion

Reviewing eight textbooks used in biology and natural sciences classes in secondary and primary education in Greece respectively, we noticed that microorganisms are most frequently mentioned in relation to health; about 50% of the paragraphs including the word "microorganism/s" and/or "microbe/s" are classified under the "Microorganisms and health" category. It should be noted that this is not the case for each textbook separately; in 3/7 textbooks microorganisms are not mostly mentioned in relation to health. For instance, in textbook-7 the most popular category is the "Microorganisms in industry and technology", while in textbook-1 and textbook-5 the most popular category is the "Microorganisms as a part of the living world". We argue that this is because of curriculum priorities. For instance, textbook-5 focuses on cell biology, thus discussing the physiology and anatomy of unicellular organisms, while textbook-7 is a specialization textbook which focuses on applications of biotechnology. Specific curriculum priorities are also the reason why microorganisms are not mentioned in "Natural Sciences for the fifth grade of Primary School", which focuses on concepts of physics and chemistry rather than biology.

At this point we should discuss two methodological choices we made. The first one is distinguishing between pathogens of humans and pathogens of organisms other than humans. Today there is an increased awareness of the possibility and essential need to address health concerns and to attain health objectives by refocusing health management more on the interaction between ecosystem health, animal health, and human health (Evans & Leighton, 2014). The One Health concept, which is defined as the "collaborative effort of multiple health science professions, together with their related disciplines and institutions – working locally, nationally, and globally – to attain optimal health for people, domestic animals, wildlife, plants, and our environment" (One Health Commission, as cited in Gibbs, 2014), is receiving attention. Although health tends to be approached in a holistic way, we decided to distinguish between human and other organisms' pathogens in our analysis in order to get a better understanding of the role of textbooks in the conceptualization of microorganisms as a threat to human health.

The second choice we should discuss is, unlike Mafra and Lima (2009), distinguishing between the role of microorganisms as pathogens and their role as part of the human microbiome. Microorganisms have several positive roles as part of the human microbiome. For instance, the gut microbiome is known to contribute to the renewal of gut epithelial cells, the development of the immune system, harvesting nutrients that would not otherwise be available and synthesizing vitamins (Turnbaugh et al., 2007). By illustrating the presentation of microorganisms as part of the human microbiome in the textbooks we investigated, we intended to trace possible instances of mentioning microorganisms in a positive way that would otherwise go unnoticed.

Returning to the results of our analysis, we noted that microorganisms are presented in a negative way in 91/187 paragraphs (subcategories: "Human pathogens" and "Pathogens of organisms other than humans"), while they are presented in a positive way in 56/187 paragraphs (subcategories: "Human microbiome", "Cultivation", "Enzymes and antibiotics production", "Vaccines and serum production", "Cleaning water"). There are 40/187 paragraphs where microorganisms are presented in neither a positive, nor negative way (subcategories: "Biological diversity" and "Physiology"). It seems that the instances where microorganisms are presented in an unfavorable way (i.e. as pathogens for humans and other organisms) are more frequent than the instances where microorganisms are presented in a favorable way (i.e. as part of the human microbiome or as having a role in beneficial applications and procedures).

All public and private schools in Greece use the same textbooks. Textbooks are government mandated and the curriculum relies heavily on their use by the teachers of all grades. Thus, in our research we are looking at textbooks as a reflection of what is taught in the Greek classroom. To the best of our knowledge, the representation of microorganisms in Greek textbooks of biology and natural sciences currently in use has not been investigated before. The discussion of our analysis findings (a) contributes to the literature on how students think and learn about microorganisms, and (b) suggests that certain actions should be taken regarding textbooks currently in use in Greek primary and secondary education, in order to improve microorganisms teaching and learning. A conceptual shift to a representation of microorganisms that integrates their diversity and beneficial contributions seems to be an essential goal for science education, according to Simard (2021), who argued that education should emphasize the importance that microorganisms have in human lives, as well as their role in the ecosystem and industry. This may be much more crucial and difficult to accomplish today, i.e. after the SARS-CoV-2 virus spread in early 2020 which caused a global pandemic. Students have seen disproportionately negative effects from COVID-19 in terms of their wellbeing, educational possibilities, and physical and mental health (Valadez et al., 2020). Their daily routines have been severely disrupted, and the limitations they have been subjected to, along with the changes to their families' routines and practices, have had a significant psychological impact on them (Mukherjee, 2021). This could result in even stronger representations of microorganisms as "enemies" (Bonoti et al., 2022) or "always bad" (Ergazaki et al., 2010). Thus, the COVID-19 post-pandemic period will be important for the conceptualization of microorganisms. Due to a certain consolidation of pre-existing unfavourable perceptions, a considerable educational barrier must be anticipated; as a result, a knowledgeable education, based on textbooks promoting a more balanced understanding of microorganisms, will be necessary to prevent the crystallization of negative ideas of microorganisms in students (Simard, 2021).

References

Abd-El-Khalick, F., Waters, M., & Le, A.-P. (2008). Representations of nature of science in high school chemistry textbooks over the past four decades. *Journal of Research in Science Teaching, 45*(7), 835–855. https://doi.org/10.1002/tea.20226

Adamantiadou, S., Georgatou, M., Giapitzakis, C., Lakka, L., Notaras, Δ., Florentin, N., Chatzigeorgiou, G., & Chantikonti, O. (2013). *Biology for the 2nd and 3rd grade of General Lyceum.* CTI Diophantus.

Aleporou-Marinou, V., Argyrokastritis, A., Komitopoulou, A., Pialoglou, P., & Sgouritsa, V. (2013). *Biology for the 3rd grade of General Lyceum-Health Studies Specialization* (Vol. 2). CTI Diophantus.

Apostolakis, E., Panagopoulou, E., Savvas, S., Tsagliotis, N., Pantazis, G., Sotiriou, S., Tolias, V., Tsagkogeorga, A., & Kalkanis, G. (2015a). *Natural sciences for the 5th grade of primary school.* CTI Diophantus.

Apostolakis, E., Panagopoulou, E., Savvas, S., Tsagliotis, N., Pantazis, G., Sotiriou, S., Tolias, V., Tsagkogeorga, A., & Kalkanis, G. (2015b). *Natural sciences for the 6th grade of primary school.* CTI Diophantus.

Bandiera, M. (2007). Micro-organisms: Everyday knowledge predates and contrasts with school knowledge. In R. Pintó & D. Couso (Eds.), *Contributions from science education research* (pp. 213–224). Springer Netherlands). https://doi.org/10.1007/978-1-4020-5032-9_16

Bonoti, F., Christidou, V., & Papadopoulou, P. (2022). Children's conceptions of coronavirus. *Public Understanding of Science, 31*(1), 35–52. https://doi.org/10.1177/09636625211049643

Byrne, J. (2011). Models of micro-organisms: Children's knowledge and understanding of micro-organisms from 7 to 14 years old. *International Journal of Science Education, 33*(14), 1927–1961. https://doi.org/10.1080/09500693.2010.536999

Byrne, J., & Grace, M. (2010). Using a Concept Mapping Tool with a Photograph Association Technique (CoMPAT) to elicit children's ideas about microbial activity. *International Journal of Science Education, 32*(4), 479–500. https://doi.org/10.1080/09500690802688071

Byrne, J., Grace, M., & Hanley, P. (2009). Children's anthropomorphic and anthropocentric ideas about micro-organisms. *Journal of Biological Education, 44*(1), 37–43. https://doi.org/10.1080/00219266.2009.9656190

Ergazaki, M., Saltapida, K., & Zogza, V. (2010). From young children's ideas about germs to ideas shaping a learning environment. *Research in Science Education, 40*(5), 699–715. https://doi.org/10.1007/s11165-009-9140-2

Evans, B. R., & Leighton, F. A. (2014). A history of one health. *Revue Scientifique et Technique, 33*(2), 413–420.

Gibbs, E. P. J. (2014). The evolution of one health: A decade of progress and challenges for the future. *Veterinary Record, 174*(4), 85–91. https://doi.org/10.1136/vr.g143

Jones, M. G., & Rua, M. J. (2006). Conceptions of germs: Expert to novice understandings of microorganisms. *Electronic Journal for Research in Science & Mathematics Education, 10*(3).

Kapsalis, A., Bourmpouchakis, I. E., Peraki, V., & Salamastrakis, S. (2013). Biology for the 2nd grade of general lyceum-general education. CTI Diophantus.

Karadon, H. D., & Şahin, N. (2010). Primary school students' basic knowledge, opinions and risk perceptions about microorganisms. *Procedia – Social and Behavioral Sciences, 2*(2), 4398–4401. https://doi.org/10.1016/j.sbspro.2010.03.700

Kastorinis, A., Kostaki-Apostolopoulou, M., Mparona-Mamali, F., Peraki, V., & Pialoglou, P. (2011). *Biology for the 1st grade of Lyceum.* CTI Diophantus.

Liu, Y., & Khine, M. S. (2016). Content analysis of the diagrammatic representations of primary science textbooks. *Eurasia Journal of Mathematics, Science and Technology Education, 12*(8), 1937–1951. https://doi.org/10.12973/eurasia.2016.1288a

Mafra, P., & Lima, N. (2009). The microorganisms in the Portuguese national curriculum and primary school textbooks. In A. Mendez-Vilas (Ed.), *Current research topics in applied microbiology and microbial biotechnology* (pp. 625–629). World Scientific.

Mavrikaki, E., Gouvra, M., & Kampouri, A. (2017a). *Biology for the 1st grade of Gymnasium.* CTI Diophantus.

Mavrikaki, E., Gouvra, M., & Kampouri, A. (2017b). *Biology for the 2nd and 3rd grade of Gymnasium.* CTI Diophantus.

Mukherjee, U. (2021). Rainbows, teddy bears and 'others': The cultural politics of children's leisure amidst the COVID-19 pandemic. *Leisure Sciences, 43*(1–2), 24–30. https://doi.org/10.1080/01490400.2020.1773978

Simard, C. (2021). Microorganism education: Misconceptions and obstacles. *Journal of Biological Education, 0*(0), 1–9. https://doi.org/10.1080/00219266.2021.1909636

Simonneaux, L. (2000). A study of pupils' conceptions and reasoning in connection with "microbes", as a contribution to research in biotechnology education. *International Journal of Science Education, 22*(6), 619–644. https://doi.org/10.1080/095006900289705

Turnbaugh, P. J., Ley, R. E., Hamady, M., Fraser-Liggett, C. M., Knight, R., & Gordon, J. I. (2007). The human microbiome project. *Nature, 449*(7164), Article 7164. https://doi.org/10.1038/nature06244

Valadez, M. D. L. D., López-Aymes, G., Ruvalcaba, N. A., Flores, F., Ortíz, G., Rodríguez, C., & Borges, Á. (2020). Emotions and Reactions to the Confinement by COVID-19 of Children and Adolescents with High Abilities and Community Samples: A Mixed Methods Research Study. *Frontiers in Psychology, 11*. https://www.frontiersin.org/articles/10.3389/fpsyg.2020.585587

Weiss, I. R., Banilower, E. R., McMahon, K. C., & Smith, P. S. (2001). *Report of the 2000 national survey of science and mathematics education.* Horizon Research.

Part II
Students' Knowledge, Conceptions, Values, Attitudes and Motivation

Chapter 8
Investigative School Research Projects in Biology: Effects on Students

Wilton G. Lodge, Michael J. Reiss, and Richard Sheldrake

8.1 Context

School biology often entails undertaking practical work, which is generally intended to help students gain conceptual understanding, practical and wider skills, and understanding of how biologist work (Kampourakis & Reiss, 2018). However, the literature on practical work in school science indicates that it often achieves less than its proponents intend (Abrahams & Millar, 2008; Gatsby Charitable Foundation, 2017). Investigative school research projects are relatively uncommon, relative to other types of practical work, such as confirmatory practical activities (intended to produce the same result for all students every time), but it has been argued that they can give students a better understanding of what it is like to undertake authentic science. A systematic review found that investigative student science research projects could have a number of benefits for students including the learning of science ideas, affective responses to science, intentions to pursue careers involving science, and development of a range of skills, some specific to practical work and others, such as collaborative teamwork, more general (Bennett et al., 2018). Nevertheless, this same review concluded that further work is needed to enhance the quality of the available evidence and to explore more fully the potential longer-term benefits of participation in such projects at secondary school level.

W. G. Lodge (✉) · M. J. Reiss · R. Sheldrake
UCL's Faculty of Education and Society, London, UK
e-mail: wilton.lodge@ucl.ac.uk; m.reiss@ucl.ac.uk; r.sheldrake@ucl.ac.uk

© The Author(s) 2024
K. Korfiatis et al. (eds.), *Shaping the Future of Biological Education Research*,
Contributions from Biology Education Research,
https://doi.org/10.1007/978-3-031-44792-1_8

107

8.2   Conceptual Framing

We report on a study that is part of a larger project funded by the Institute for Research in Schools (IRIS) in the UK to explore school students' views related to science, their reasons for participating in IRIS projects, their views of these projects and their views about wider aspects of why students do or do not engage in IRIS projects, and what they have gained from their experiences. We focus on our analysis of qualitative data gathered from student interviews. Our specific research question is: 'How do secondary school students who are participating in a biology research project see both science and themselves in relation to it?'

Our research question was chosen because of an increasing recognition among science educators of the value of considering issues to do with students' science identity and their perceptions of science when researching students' responses to school science (e.g., Holmegaard & Archer, 2022). Aspirations and choices can involve identity, as ways to convey who someone is or who they want to become, and/or to undertake personally meaningful, interesting, and enjoyable activities (Holmegaard et al., 2015). Contemporary models of science identity highlight the relevance of someone recognising themselves and also being recognised by others as being a science person; someone may enact or embody a particular identity through undertaking particular practices or performances, which require specific knowledge and skills, and which are recognised by others (Carlone & Johnson, 2007). Wider analytical and theoretical frameworks also highlight the relevance of intersecting personal and socio-cultural aspects, such as finding science interesting and enjoyable (Avraamidou, 2020). Identities can change and develop over time; resources, contexts, and/or experiences can facilitate or limit some developments, and further experiences can be sought or avoided, through various trajectories (Gonsalves et al., 2021).

8.3   Materials, Methods and Analysis

The data used in this paper were taken from the qualitative phase of the larger study outlined above. Interviews were undertaken with participants between the ages of 12 and 18 from eight schools in England (seven state schools and one independent school). Thirteen of these students were undertaking biology projects and it is these thirteen interviews that are analysed here (nine from the state schools – three from one, two from another, one from each of four and none from one – and four from the independent school). All the participants undertook a scientific research project through IRIS and volunteered to take part in our study. We had no role in the design or undertaking of the projects. Examples of projects included mapping UK trees and investigating the impact of climate change, synthesising DNA, exploring how a reduction in diversity affects humans and other animals, and analysis of biological material using a scanning electron microscope. Information sheets (explaining the purpose of the research) and consent forms were sent to the schools to be distributed to potential participants and completed consent forms were obtained from the parents and the students whose data are used here. The participants' schools represented a wide coverage of geographic (urban and rural) and socio-economic background. One was in the north of England, two in the midlands, one in the northwest and four in the southeast.

Individual semi-structured interviews were conducted in the summer of 2021. Each interview lasted between 20 and 30 min and was directed by an interview agenda that was based on the research question. Initially, we had hoped to conduct face-to-face interviews. However, due to COVID-19 restrictions in place at the time, all the interviews were undertaken using the online platform Microsoft Teams. Although this approach provided reliable data on students' views about science and investigative research projects, it was not without obstacles. For example, in a few of the interviews the internet connection was unstable and consequently a small proportion of the respondents' replies were lost. In addition, in response to their school's safeguarding policy some of the participants chose not to turn their camera on during the interview. In such cases it was not possible to capture any non-verbal communication. All interviews, with participants' permissions, were audio-recorded and transcribed verbatim. Names of the participants cited in this article are all pseudonyms to protect participants' personal identities.

The interview responses were analysed following the approach to thematic analysis outlined by Braun and Clarke (2006, 2021). This approach has been widely recognized in qualitative research for its flexibility in not only developing significant themes from raw data but, as Boyatzis (1998, p.xiii) argues, in allowing for "meaning to be articulated or packaged in such a way, with reliability as consistency of judgment, that description of social 'facts' or observation seems to emerge". In this sense, thematic analysis echoes the general underpinnings of 'grounded theory' as articulated by Glaser and Strauss (1967), in which the central themes are developed in the actual data collected.

The students' responses were first read repeatedly and discussed by the first and second author to familiarise ourselves with the data, in line with Braun and Clarke's (2006) recommendation that researchers become intimately familiar with their entire data set since this lays the foundation for subsequent steps. Preliminary themes that we found relevant to our research question were identified by the first author. The relationships between these preliminary themes were given careful consideration by the first and second author. In this way, we were also able to identify features of the interview transcripts that required further probing and collapse some of the preliminary themes into new ones, thus reducing their overall number. As new themes were developed, participants' responses were re-read to check that the new themes validly represented the interview data. Towards the conclusion of the process, no new themes arose, which indicated that the most important ones had been identified.

We identified the following six themes, which are discussed in the findings section: motivation for participating in the project; benefits of participating in the project; views about science; views about school science; science aspiration; and family involvement. As is usual, we do not claim that another group of researchers would necessarily (or even probably) arrive at the same themes; there is an element of subjectivity in thematic analysis, though we tried to be as objective as possible.

8.4 Findings

8.4.1 Motivation for Participating in the Project

The findings suggest an association between the participants' motivations for engaging with the IRIS investigative research project and their science identities, with many of them articulating that participating in the project provided them with an "authentic" experience of what "real scientists" do. For example, John expressed that:

> For me, it's an ideal way to say you have actually had exposure to the type of equipment scientists use in the lab. This makes you feel like a real scientist, which you don't get when you do school practical.

Sarah, who, wasn't yet involved in an IRIS project, but intended to do one next year, when asked "so, what is motivating you to want to do the IRIS project?" replied:

> It is just because I am interested in it because I don't have a lot of chances in school or in my personal life to get very involved in any research projects. IRIS does provide a way to looking into researching and STEM and I find that very fascinating.

When Sarah was then asked "so, do you think the IRIS project will give you a better idea of what doing science or being a scientist is?", she went on to say:

> I think it does, especially the research and report and the presentation parts of it because you have to carry out your own research and you have to come up with really valid conclusions and I would think that's what real scientists do.

For a number of students, it seems that participation gave them an opportunity to understand themselves as scientists. They felt that by actively working with practising scientists they had a more representative experience of the processes of science and had become more socialised into the discursive practices [our phrase] of the science community. For example, Nara highlighted her work in using data to evidence the environmental changes in Antarctica. For her, the project not only informed her understanding of the extent to which human activities contribute to climate change but, as she put it, "provided real opportunities to engage in discussions and debates about scientific issues". These experiences reinforced Nara's conception of what a scientist does.

8.4.2 Benefits of Participating in the Project

Evidence from the interviews suggests that participating in the IRIS project increased the students' interests and attitudes towards science. For most of them the project was "engaging", "exciting", "quite interesting" or "refreshing". Participants' enthusiasm about participating in a project increased when they presented their findings at an IRIS conference and received feedback from peers and members of the science community. For example, Andy said:

> When I partook in the IRIS conference, I started to understand science and my world better, so obviously before I liked science but wasn't entirely sure how it applied to the world. I had a rough idea but obviously wasn't partaking in any research projects and when I partook in my first research project with IRIS I started to understand, and it really opened my eyes to how it can be different.

As highlighted above, self-efficacy can be seen as a powerful tool to comprehend and reinforce students' motivations to learn science. In other words, a student's self-efficacy will influence their attitude and tenacity to tackle problems they encounter in science classrooms. The interview data indicates that participating in the research project increased the students' self-efficacy. For example, Michael, who said that he had always struggled with biology, now said that he no longer found the subject difficult but was intrinsically motivated to succeed.

The participants valued the experience of working in a collaborative context, which they recognised as not just an important benefit of participating in the project but an important practice in the science community. Such collaboration was seen by the participants as a means for creating new forms of knowledge, negotiating meaning and developing new skills. Sarah, who hadn't started her own IRIS project yet but got involved in a friend's project on composting, said:

> Well as we saw from the pandemic working in a group is important for science. Scientists need to share their ideas with each other. So, working in a team with people I have never worked with before was nice to do, and although we had disagreements it really deepened our friendships. Some of the people in the group knew a lot more than me so I learnt a lot about science.

Michael said:

> Working as part of a team and discussing scientific ideas allowed me to develop my reasoning skills, and I've also learned how to take on feedback from people and also include the feedback in my work and use what peers have taught me.

One student, Pat, who conducted her own independent research, spoke enthusiastically about the skills she acquired from participating in the project. Planning and designing her research provided her with opportunities to develop reasoning and thinking skills; however, she lamented the lack of collaboration with peers, which she thought would have brought different perspectives to the process.

For some participants the project not only helped them experience some of the practices of the scientific community but gave them a deeper understanding of the nature of science – that science is tentative, and inquiry-based:

> … I think that IRIS actually helped me learn science. While doing the project I started to understand science and my world better so obviously before I liked science, but I wasn't entirely sure how it applied to the world. (Andy).

For some of the participants, engaging in investigative research projects afforded them opportunities not only to develop their scientific inquiry skills but also to acquire a deeper understanding of the epistemological dimensions of science and its connection with historical, cultural and social values. This greater appreciation and insight into the nature of science increased their science-related career aspirations, as illustrated by Nara:

> Doing the project gave me a better understanding of the subjective nature of science and how our personal view can influence the scientific process. It was exciting working with students from different backgrounds and this has reinforced my passion for science and biology especially.

Although there were varying views amongst the participants, there was broad agreement that the project gave them the opportunity to apply their scientific knowledge and engage in critical thinking, which they thought was an important characteristic of real science. Taken from this perspective, the IRIS project was generally felt to be quite different from school science. As George said:

> It's different from what we have done in the average science ... with science at school it's just we get told to do this and that and then we just have to try and work it out ... and like with the DNA, err the DNA project, we've like really had to think about how we're going to do this and how we're going to do that.

There was also broad consensus that participating in the IRIS project gave them a better idea of doing 'real' science or what it meant to be a scientist:

> So, I would say it's probably a reflection of real science more than it is the academic subjects. I think practical things are a massive part of science and something I don't think is reflected enough and working with samples and stuff like that, I think it's been nice to see what things are actually like ... (Michael).

Finally, some students talked about how undertaking a project helped them develop certain 'soft' skills, for example:

> It allows you to like work as a team, 'cos if someone's fallen behind and struggling you can go over get help off them or go over to someone get help or help someone yourself if you're further ahead. (Cameron).
>
> I've developed definitely, obviously, teamwork skills, and I've also learned how to take on feedback from people and also include the feedback in my work and use what peers have taught me and also my friends and things from the club and put it into my own work. (Nara).

8.4.3 Views About Science

Throughout the data, statements by the participants suggested that they held favourable views about the benefits of studying science mainly because of its perceived utility both at an individual and collective level. For many, the value of science lies in its ability to provide relevant real-world applications and a means of explaining and understanding the nature of the universe. For example, Paul said "Science can help us learn lots of information and people are still discovering information, for example, COVID-19. We need science to develop vaccines to help come up with statistics that help the community try to reduce the number of cases". In answer to the question "In what ways do you think science is beneficial for life or for the wider society?", Peter replied:

> It's kind of the answer to the future. I don't think that there is any other kind of special area which could, well, there isn't, progress humanity as a whole and propel our development.

So, I suppose you could say that science and innovation are, for the minute, the most prominent things that need improving. Science is how the world works and I think it can influence
our society in terms of the social aspects and it also plays a big role and science can change
so many different areas, like the environment, for example; that's the big thing at the
moment and science is the core of everything that people discuss about the environment.

For some participants, the perceived utility of science seems to operate at multiple
levels. Not only did they see the value of studying science as a way of understanding
the universe but also in relation to their own goals or career intentions. Michael
wanted to go into medicine but also said:

I think with medicine, you have to have a back-up plan … backup in biochemistry and that
is going to be my backup. I mean, that's like an ideal way to say you have actually had
exposure to the type of equipment and I know how to use this and look what I have studied
and look at the things that I got from and stuff like that.

For some of the participants, there seems to be an awareness of the importance of
scientific evidence in shaping public policy. For example, Peter said:

Science is how the world works and I think it can influence our society in terms of the social
aspects and it also plays a big role and science can change so many different areas, like the
environment for example; that's the big thing at the moment and science is the core of
everything that people discuss about the environment.

8.4.4 Views About School Science

Although most of the participants held favourable views about science, many of
them did not view school science as comparable to the science practised by scientists in the wider world. This was illustrated by Peter who expressed:

[School science] lacks sometimes the experience that you'd require for actual science work;
there is that odd occasion when we do a project that is quite similar to the work that a scientist actually does, so it is quite useful sometimes, but other than that, sometimes it kind of
lacks in that experience.

Some of the participants felt that school science was much too influenced by the
National Curriculum and consequently was much too prescriptive with, normally,
no opportunity to undertake investigative research projects. As Helen put it:

We are preparing for A levels, and we only learn what is in the GCSE specification. We
hardly do any practicals … we just spend most of the time remembering lots of information
for the exams.

A few of the participants, however, felt that school science was similar to 'real science' since it offered opportunities for students to develop a repertoire of scientific
practices. When Michael was asked "So, what do you enjoy about doing school
science?", he replied:

I just like the idea that there's always an answer, well most of the time, and I sort of like the
scientific way of working with things, I like working logically. I think a lot with non-
scientific subjects, it can be going all around the bend and you might not even get there. I

like the logicality and I like the systematic sort of process of everything and I think that's just the way my brain works.

Moreover, some students felt that school science allowed them to participate in scientific inquiry through which they could learn skills that scientists use in their day-to-day practice. When John was asked "Do you think that science at school is similar or different to what scientists do?", he answered:

I think that at this school, with the labs, yes. You get an opportunity to see what it's like to work in a real lab because it's all student-led. We have to take on our own roles and responsibilities so I think in our school, it's quite similar to a real lab.

At the same time, some of the participants acknowledged the importance of school science in providing a foundation to build on:

I think every scientist will have started by having a key understanding of science, that basic level of what we teach in school, and they then use those key skills to develop things in the wide world, so I think everyone needs the understanding of science from the core. (George).

8.4.5 Science Aspirations

Many of the participants during the interviews expressed that exposure to investigative research projects positively influenced their aspirations and career choices:

Yes, definitely … it's shown me what scientists do and it has shown me how fun it is and how enjoyable and how much I would like to actually do that as a job in the future and so definitely it has influenced that. (Jasmine).
… I didn't think about becoming a scientist before, I was pretty much focused on becoming a lawyer but now that I did the project, I started thinking well maybe I could become a scientist. It seems like a very cool field to work in. (Helen).
I think I have gained a better understanding and a better willingness to attempt new things. I am not going to just isolate myself into focusing into one area in science but I am going to look furthermore into the field of science. (Shane).

8.4.6 Family Involvement

Most of the participants felt that they were well supported by their family not just in their education and learning but in the active encouragement they received to do science. This, in part, may have accounted for their positive attitude towards science and motivation to participate in out-of-school related activities, such as IRIS:

… from a young age because my parents are math and science-oriented people, they encouraged me down the science and math path. (Sarah).
My mom thinks that I will be really good at science and encourages me to get involved in science activities to discover new stuff, especially in the environment. (Michael).

Moreover, many of the participants also reported that they regularly engage in family practices that support their interest in science. Such practices include debates around controversial scientific issues, watching science documentaries, and visits to outdoor science learning centres:

> My family is very interested in going to lots of science museums and the exhibitions, planetarium, and lots of science-related activities I was interested in. (Michael).

Notably, participants from families with 'high' science capital were more likely to be encouraged to continue studying science beyond the compulsory years and to pursue careers in science or science related fields. Nara talked about how all four of her grandparents are doctors "and they also helped me become interested in science along the way". She went on to say:

> My mom was really interested in science when she was at school and so was my dad and when they learned that I was really interested in science they were really happy and they encouraged me to join IRIS as well, so yeah, they really like it … the Science Museum is definitely somewhere we would go a lot and also I think I went to the Big Bang Fair, two years ago or something; we went there as a family, and it was really fun, in particular the planetarium … definitely, I'm interested in STEM and subjects surrounding that so I think a career in science would be really fun.

8.5 Discussion

It is clear, in response to our research question, 'How do secondary school students who are participating in a biology research project see both science and themselves in relation to it?', that many of the students felt that their participation in an IRIS project had changed how they saw both science and their relationship with science. Participants were strongly positive about the benefits of participating in such projects. Implicit in a number of the participants' statements above is a conceptualisation of learning science as an acculturation process in which learners develop science process skills by engaging in conversation about science. In this sense, as Lave and Wenger (1991) articulate, for learners to progress they much submerge themselves in the social practices of the community, guided by the expertise of more knowledgeable others who are skilled in their craft. Moreover, Brown et al. (1989) argue that learners are better able to acquire the knowledge and skills of science when they not only observe existing practices but become 'full participants' in the science community and its discursive practices.

Furthermore, an extensive body of critical scholarship has pointed to an increasing and worrying lack of interest among students in studying science at upper secondary school and aspiring to careers in science and related fields (see, for example, DeWitt & Archer, 2015). While we would argue that there are many factors that may relate to students' lack of interest (including home background, gender and ethnicity), recent research has identified an association between students' interest and engagement in science and pedagogical approaches that teachers' employ (Puslednik & Brennan, 2020). Both tacit and explicit statements from the participants seemed to reaffirm Puslednik and Brennan's (2020) views that exposure to investigative research projects positively influenced their aspirations and career choices, though it could be argued that since participation in the projects was voluntary, the interviewed students may have already had high aspirations towards science-related studies. Nevertheless, such findings illustrate the pedagogical value of investigative research projects in mediating science learners' orientation towards careers in science, especially those from underrepresented backgrounds.

At the same time, we acknowledge that we have no longitudinal data as to the long-term effects of participation in these projects. Further research on such long-term effects seems warranted, given the frequent enthusiasm exhibited by the interviewees in this study of their participation in investigative biology research projects.

References

Abrahams, I., & Millar, R. (2008). Does practical work really work? A study of the effectiveness of practical work as a teaching and learning method in school science. *International Journal of Science Education, 30*(14), 1945–1969.

Avraamidou, L. (2020). Science identity as a landscape of becoming: Rethinking recognition and emotions through an intersectionality lens. *Cultural Studies of Science Education, 15*, 323–345.

Bennett, J., Dunlop, L., Knox, K. J., Reiss, M. J., & Torrance-Jenkins, R. (2018). Practical independent research projects in science: A synthesis and evaluation of the evidence of impact on high school students. *International Journal of Science Education, 40*, 1755–1773.

Boyatzis, R. E. (1998). *Transforming qualitative information: Thematic analysis and code development*. Sage.

Braun, V., & Clarke, V. (2006). Using thematic analysis in psychology. *Qualitative Research in Psychology, 3*(2), 77–101.

Braun, V., & Clarke, V. (2021). One size fits all? What counts as quality practice in (reflexive) thematic analysis? *Qualitative Research in Psychology, 18*(3), 328–352.

Brown, J. S., Collins, A., & Duguid, P. (1989). Situated cognition and the culture of learning. *Educational Researcher, 18*(1), 32–41.

Carlone, H., & Johnson, A. (2007). Understanding the science experiences of successful women of color: Science identity as an analytic lens. *Journal of Research in Science Teaching, 44*(8), 1187–1218.

DeWitt, J., & Archer, L. (2015). Who aspires to a science career? A comparison of survey responses from primary and secondary school students. *International Journal of Science Education, 37*(13), 2170–2192.

Gatsby Charitable Foundation. (2017). *Good practical science*. Gatsby Charitable Foundation.

Glaser, B., & Strauss, A. L. (1967). *The discovery of grounded theory: Strategies for qualitative research*. Aldine Publishing Company.

Gonsalves, A., Cavalcante, A. S., Sprowls, E. D., & Iacono, H. (2021). "Anybody can do science if they're brave enough": Understanding the role of science capital in science majors' identity trajectories into and through postsecondary science. *Journal of Research in Science Teaching, 58*(8), 1117–1151.

Holmegaard, H. T., & Archer, L. (2022). *Science identities: Theory, method and research*. Springer.

Holmegaard, H. T., Ulriksen, L., & Madsen, L. M. (2015). A narrative approach to understand students' identities and choices. In E. K. Henriksen, J. Dillon, & J. Ryder (Eds.), *Understanding student participation and choice in science and technology education* (pp. 31–42). Springer.

Kampourakis, K., & Reiss, M. J. (Eds.). (2018). *Teaching biology in schools: Global research, issues, and trends*. Routledge.

Lave, J., & Wenger, E. (1991). *Situated learning: Legitimate peripheral participation*. Cambridge University Press.

Puslednik, L., & Brennan, P. C. (2020). An Australian-based authentic science research programme transforms the 21st century learning of rural high school students. *Australian Journal of Education, 64*(2), 98–112.

Chapter 9
Investigating Relationships Between Epistemological Beliefs and Personal Beliefs in Biological Evolution

Andreani Baytelman, Theonitsa Loizou, and Salomi Chadjiconstantinou

9.1 Introduction

Evolution is widely seen as the central, unifying and overarching theory in biology. The field of biology is made up of many broad topics threaded and held together by the theory of biological evolution. For example, content related to evolutionary theory can include anything that refers to organisms' adaptation to their environment and/or ability to survive and create offspring. It includes DNA, protein sequences, common ancestry, genetic variation of populations of organisms, fossils and plant and/or animal diversity. Therefore, educating students about biological evolution is vitally important because it is capable of explaining a large number of natural phenomena at different levels. In addition, an understanding of biological evolution is becoming increasingly relevant in practical contexts, including medicine, agriculture, and resource management (Dunk & Wiles, 2018; Fowler & Zeidler, 2016).

Despite the importance of biological evolution, it is still poorly understood by students throughout their time in education (Nehm & Reilly, 2007; Shtulman, 2006; Spindler & Doherty, 2009), science teachers, and the general public (Baytelman et al., 2023; Evans et al., 2011). This poor understanding has been attributed to diverse cognitive, epistemological, religious, and emotional factors (Rosengren et al., 2012) that evidently biological evolution education is generally not successfully coping with.

Previous research suggests a connection between students' acceptance and understanding of evolutionary theory and their epistemological beliefs toward

A. Baytelman (✉)
Learning in Science Group, University of Cyprus, Nicosia, Cyprus
e-mail: baytel@ucy.ac.cy

T. Loizou · S. Chadjiconstantinou
Cyprus Ministry of Education, Sport and Youth, Paralimni High School, Paralimni, Cyprus

© The Author(s) 2024   119
K. Korfiatis et al. (eds.), *Shaping the Future of Biological Education Research*,
Contributions from Biology Education Research,
https://doi.org/10.1007/978-3-031-44792-1_9

science. In particular, previous research has shown that there are relationships between students' sophisticated epistemological beliefs toward science and their acceptance and understanding of evolutionary theory (Borgerding et al., 2017; Mazur, 2005; Sinatra et al., 2003). On the other hand, it has been argued that a firm grasp of epistemological beliefs allows students to compare knowledge frameworks, in order to understand how and why knowledge produced through science is different from their religious beliefs. Additionally, students' (and other individuals') personal beliefs define how they view the world, which in turn can influence their learning, views of science and academic performance. Numerous studies have identified difficulties in learning about biological evolution throughout education, and there is evidence that some of these difficulties stem from epistemological beliefs, personal beliefs and cognitive biases (Cavallo & McCall, 2008; Harms & Reiss, 2019; Shtulman & Calabi, 2012; Shtulman & Schulz, 2008; Sinatra et al., 2003). The possible relationship between 12th grade students' epistemological beliefs toward science and their personal beliefs in biological evolution could be of interest to researchers, educators and biology teachers in the field, but has not yet been enough investigated.

In the present study, we address this gap in the literature, namely whether and to what extent 12th grade students' epistemological beliefs toward science can predict their personal beliefs in biological evolution, before biological evolution instruction. Based on previous research (Sinatra et al., 2003; Sinclair & Baldwin, 1996), we hypothesized that there would be a relation between students' epistemological beliefs toward science and their personal beliefs in biological evolution, before instruction. By doing this, we hope to gain a better understanding of the contribution of students' epistemological beliefs toward science to their beliefs in biological evolution, before instruction, and contribute to the development of a theoretical framework that will describe learning about biological evolution throughout education. Then, additional research could benefit from this study's findings to measure the possible interaction of these two concepts with students' understanding and acceptance of biological evolution.

In particular, we set out to answer the following research question: What are the relationships between 12th grade students' epistemological beliefs toward science and their personal beliefs in plant evolution, animal evolution and human evolution, before biological evolution instruction?

9.1.1 Conceptualization of Epistemological Beliefs

According to Kitchener (2002), epistemology is a theory of knowledge and how it develops, while personal epistemology is a personal theory about how individuals develop knowledge.

Researchers who study epistemology are interested in "*how individuals come to know, the theories and beliefs they hold about knowing, and the manner in which such epistemological premises are part of and have an influence on the cognitive processes of thinking and reasoning"* (Hofer & Pintrich, 1997, p. 88).

Epistemological beliefs refer to individuals' beliefs about the nature of knowledge and the process through which knowledge develops (Hofer & Pintrich, 1997). Different models have been proposed on how to conceptualize and examine epistemological beliefs. From these, two overarching kinds of models can be identified: (a) models that examine epistemological beliefs from a developmental perspective (Perry, 1970) and (b) models that explore epistemological beliefs from a multidimensional perspective (Hofer & Pintrich, 1997; Schommer, 1990).

Research on epistemological beliefs was initiated by Perry (1970). He found that students do have strong beliefs about knowing and knowledge, but they can change over time. Perry argued that students entering college perceive knowledge to be simple, certain, and provided by the instructor; however, upon graduation, the same students often held more sophisticated beliefs, viewing knowledge as complex, tentative, and derived from reason and observation. Perry proposed a developmental model that described nine levels in epistemological beliefs, ranging from the belief that knowledge is objective, to the belief that knowledge is radically subjective, and finally to the belief that knowledge has objective and subjective aspects.

Since Perry's research, perhaps one of the most influential studies in epistemological beliefs was conducted by Schommer (1990). Schommer suggested that students' epistemological beliefs consist of a collection of more or less independent beliefs (epistemological dimensions). Schommer proposed a multidimensional model and suggested five theoretical dimensions of epistemological beliefs: (a) the structure of knowledge (from the simple to the complex nature of knowledge), (b) the stability of knowledge (from the factual to the constantly changing nature of knowledge), (c) the source of knowledge (from the omniscient source to the empirically evidenced-based nature of knowledge), (d) the speed of learning (from the quick to the gradual nature of learning), and (e) the ability to learn (from the fixed or innate to the incremental nature of ability) (Cho et al., 2011).

While the dimensions of structure, stability, and source in Schommer's conceptualization fall under the more generally accepted definition of epistemological beliefs (Hofer & Pintrich, 1997), the speed and ability dimensions are controversial because they mainly concern beliefs about learning (speed) and intelligence (ability). Hofer and Pintrich (1997) argued that epistemological beliefs should be defined more purely, with two dimensions concerning the nature of knowledge (what one believes knowledge is) and two dimensions concerning the nature or process of knowing (how one comes to know).

According to Hofer and Pintrich (1997), the two dimensions concerning the nature of knowledge are: (a) Simplicity of knowledge (related to the structure of knowledge), ranging from the belief that knowledge consists of an accumulation of more or less isolated facts to the belief that knowledge consists of highly interrelated concepts, and (b) Certainty of knowledge (related to the stability of knowledge), ranging from the belief that knowledge is absolute and unchanging to the belief that knowledge is tentative and evolving. The two dimensions concerning the nature of knowing are: (c) Source of knowledge, ranging from the conception that knowledge originates outside the self and resides in external authority, from which it may be transmitted, to the conception that knowledge is actively constructed by the person in interaction with others, and (d) Justification for knowing, ranging from justification of knowledge claims through observation and authority, or on the basis of what feels right, to the use of rules of inquiry and the evaluation and integration of different sources (Hofer & Pintrich, 1997). Accordingly, Hofer and Pintrich's model differs from Schommer's by omitting the nature of learning factors and adding another nature of knowing factor: Justification.

Additionally, Conley et al. (2004) suggested a new dimension of epistemological beliefs, i.e., the Development of knowledge, which is related to the nature of the development of knowledge. Researchers in the field of epistemology, educational psychology and science education have proposed a variety of instruments for the examination of epistemological beliefs (e.g., Baytelman, 2015; Baytelman et al., 2020a, b, 2022; Baytelman & Constantinou, 2016a, b; Conley et al., 2004; Kuhn et al., 2000; Schommer, 1990; Schommer et al., 1992; Schommer-Aikins, 2004). The Dimensions of Epistemological Beliefs toward Science (DEBS) Instrument (Baytelman et al., 2020a, b, 2022; Baytelman & Constantinou, 2016a, b) is based on the multidimensional perspective of epistemological beliefs and captures five dimensions: three dimensions related to the nature of knowledge (Certainty, Simplicity and Development of Knowledge), and two dimensions related to the nature of knowing (Source and Justification of Knowledge). The DEBS Instrument is suitable for high school and university undergraduate students and was used for this study.

Despite the differences between the developmental and the multidimensional models, *"the fairly well-established trend is that individuals move from some more objectivist perspective through a relativistic one, to a more balanced and reasoned perspective on the objectivist–relativistic continuum, with this latter position reflecting a more sophisticated manner of thinking"* (Pintrich, 2002, p. 400).

According to Muis et al. (2015), since the multidimensional model of epistemological beliefs is a system of more or less independent epistemological dimensions which are not necessarily developing in synchrony with each other, it is important to make efforts to foster all dimensions of students' epistemological beliefs, using a variety of didactical approaches. Some recommended didactical approaches to promote students' epistemological beliefs are inquiry-based teaching and learning (Shi et al., 2020), teaching and learning using history of science (Matthews, 1992, 1994) dialogic argumentative activities (Baytelman, 2015; Baytelman et al., 2020a; Iordanou & Constantinou, 2014) and reflective judgment through socioscientific issues (Zeidler et al., 2009). However, the recommended didactical approaches are synergistic, built upon one another, and provide opportunities for fostering students' epistemological beliefs.

Researchers have argued that epistemological beliefs are related to learning and academic performance, comprehension, views of science, innate learning and choosing science as a career, self-efficacy beliefs, students' motivation and higher levels of self-concept and self-efficacy (Baytelman et al., 2023; Chen, 2012).

Additionally, studies argue that students' epistemological beliefs have a direct impact on the selection of learning strategies or approaches, the process of shaping conceptions, and problem-solving (Chan et al., 2011) and an individual's ability to generate alternative arguments, counterarguments and rebuttals (Baytelman et al., 2020a).

9.1.2   Personal Beliefs in Biological Evolution

Personal beliefs in a construct (e.g., biological evolution) are considered to be personal truths or personal views of the world. These personal truths are not held to the same epistemological criteria as knowledge; instead, personal beliefs are understood to be extra-rational. In other words, they are not based on the evaluation of evidence, they are subjective, and they are often intertwined with affect (Sinatra et al., 2003).

Personal beliefs in biological evolution are based on personal convictions, opinions, and degree of congruence with other belief systems, and are very resistant to change, despite instruction. Students' worldviews are sculpted mainly by culture, religion, politics and education (Mazur, 2005). Many times these beliefs influence students to place themselves in an either/or position in regard to evolution (Sinclair & Baldwin, 1996). These positions seem to fall into one of two camps: evolutionist or creationist. Evolutionists tend to believe that evolution is a process of change that is independent of the influence of any supreme design, while creationists tend to believe that there is some supreme force directing the development of life. These differing beliefs can affect how students approach learning evolution (Cavallo & McCall, 2008). One of the most influential factors regarding one's beliefs appears to be religion. Religious beliefs seem to contribute to the variation in student beliefs in biological evolution. Religion is a very personal aspect of one's life, and beliefs in general are a very personal aspect of viewing the world. Therefore, it stands to reason that religion can be an influence on beliefs about controversial topics such as biological evolution (Cavallo & McCall, 2008). In general, personal beliefs are shown to interfere with the students' ability to examine scientific evidence objectively, and the interference can be even stronger when learned religious ideas are against the information being taught (Cavallo & McCall, 2008; Sinclair & Baldwin, 1996).

Students have likely been exposed to some opinions about evolution from parents, religious leaders, or the media before entering the classroom. This exposure has most likely helped form ideas and beliefs in evolution prior to formal biological evolution instruction (Shtulman & Schulz, 2008; Woods & Scharmann, 2001). This suggests that biology teachers need to explore their students' worldviews and personal ideas in biological evolution before instruction, and explore how their personal beliefs may be impacted by science teaching and learning. Blackwell et al. (2003) highlight that evolutionary theory remains a topic that will often require penetration into a person's belief system prior to acceptance.

In an explanation of the role of emotions and epistemology when students learn biological evolution, Scharmann (1990) has suggested that students need to be aware that consideration of biological evolution does not require that they turn away from their firmly entrenched religious beliefs and culturally-based understandings. Yet, he has suggested that a diversified strategy that targets not only constructs related to biological evolution, but also focuses on students' understanding of the nature of scientific knowledge, allows for students to consider scientific concepts without forcing them to turn from their religious and cultural beliefs.

Additionally, Sinatra et al. (2003) have argued that a firm grasp of epistemological beliefs allows students to compare knowledge frameworks, to understand how and why knowledge produced through science is different from their religious beliefs. Yet, Cherif et al. (2001) found a strong relationship between beliefs and understanding of biological evolution. However, biology education currently has an incomplete understanding of potential relationships between students' epistemological beliefs toward science and personal beliefs in biological evolution.

9.2 Research Design and Method

9.2.1 Study Design

The present study examines relationships between 12th grade students' epistemological beliefs toward science and their personal beliefs in plant evolution, animal evolution and human evolution, before biological evolution instruction. The aim of the study is a deeper understanding of the contribution of students' epistemological beliefs toward science to their beliefs in biological evolution, before instruction, and the development of a theoretical framework that will describe teaching and learning about biological evolution throughout education. In particular, we seek to answer the following research question: What are the relationships between 12th grade students' epistemological beliefs toward science and their personal beliefs in plant evolution, animal evolution and human evolution, before biological evolution instruction?

To answer our research question, we asked 12th grade students to respond to instruments (questionnaires and semi-structured interviews) that assess their epistemological beliefs toward science and their personal beliefs in plant evolution, animal evolution and human evolution, before biological evolution instruction.

The quantitative and qualitative data were collected in three stages: (a) First stage: Assessment of 51 participants' epistemological beliefs, using a questionnaire based on the multidimensional perspective of epistemological beliefs; (b) Second stage: Assessment of 51 participants' personal beliefs in animal, plant and human evolution, using a specific questionnaire; (c) Third stage: Conducting semi-structured interviews with five participants. The interview guidelines made specific reference to the questionnaire in order to investigate further 12th grade students' epistemological beliefs and personal beliefs in biological evolution and obtain a more comprehensive understanding of them. The major purpose of using quantitative and qualitative approaches and methods of data collection was to increase their validity and credibility (Greene, 2007).

9.2.2 Participants

In this study, participants included 51 12th grade students at a public secondary school in Cyprus, (female 31, male 20, with a mean age of 17.5 years). In Cyprus, 12th grade students have biology as an elective course. The unit on evolution is taught at the end of high school, and, according to the Cyprus National Curriculum, students do not have any lessons on biological evolution before grade 12. The participants were Caucasian native speakers of Cyprus and shared a homogeneous middle-class social background and the Greek language. They were of Christian affiliation, with the majority being Christian Orthodox. They attended the same school and came from the same geographical area of Cyprus. All instruments that were used for this study were in the Greek language, and all data were treated anonymously and confidentially.

9.2.3 Data Collection

Instruments In order to answer the research question (What are the relationships between 12th grade students' epistemological beliefs toward science and their personal beliefs in plant evolution, animal evolution and human evolution, before biological evolution instruction?), we used two different questionnaires and semi-structured interviews.

Participants' epistemological beliefs were assessed using the Dimensions of Epistemological Beliefs toward Science Instrument (DEBS) (Baytelman, 2015; Baytelman et al., 2020a, b; Baytelman & Constantinou, 2016a, b), which is based on the multidimensional perspective of epistemological beliefs and has been validated in the particular culture in which the research was conducted. This instrument contained 30 Likert-scale items designed to assess three dimensions concerning knowledge (i.e., Certainty, Simplicity and Development of knowledge), and two dimensions concerning knowing (i.e., Source and Justification of knowledge). Each dimension consisted of six items. Scoring of the DEBS was done by rating the 30 items on a four-point Likert scale, ranging from strongly disagree to strongly agree (strongly disagree = 1, disagree = 2, agree = 3, and strongly agree = 4). High scores on this measure represent more sophisticated epistemological beliefs, while low scores represent less sophisticated beliefs. The DEBS Instrument is suitable for high school and university undergraduate students (Baytelman, 2015; Baytelman et al., 2023).

Table 9.1 Main questions used in the semi-structured interviews

A/A	Main questions used in the semi-structured interviews
1	Do you believe that scientific knowledge and theories are reliable and unchanging? Please explain why or why not
2	Do you believe that in order to gain real insight into scientific issues it is necessary to form a personal opinion about what one reads/listens to, or to accept this information as reliable? Please explain why or why not
3	Do you believe that the plants that we know today have evolved from earlier species? Please explain why or why not
4	Do you believe that the animals that we know today have evolved from earlier species? Please explain why or why not
5	Do you think that human beings have evolved from earlier species? Please explain why or why not

To assess participants' personal beliefs in biological evolution, we used a specific instrument, namely the Personal Beliefs in Biological Evolution Instrument (PBBE), which was developed specifically for this study. This instrument contained four items designed to assess students' beliefs in plant evolution, animal evolution, human evolution, and human creation by God. Similarly to epistemological beliefs, each item was rated on a four-point Likert scale, ranging from strongly disagree to strongly agree (strongly disagree = 1, disagree = 2, agree = 3, and strongly agree = 4).

Interview Guidelines In order to triangulate and verify the findings of the data collected by the DEBS and PBBE Instruments, interviews were conducted. Interviews were semi-structured and conducted individually. In this part of the study, five 12th grade students (female 3; male 2) volunteered to participate in interviews, six days after they completed the DEBS and PBBE questionnaires, and before biological evolution instruction. In particular, five main questions were posed, supported by a number of sub-questions to help students elaborate on the topic if necessary. Interviews lasted 20 min each. The participants were all asked the same questions, but, in some cases, the manner in which they were asked varied, in order to obtain in-depth information (Bryman, 2008). Table 9.1 illustrates the main questions that were used in the semi-structured interviews.

9.2.4   Data Analysis

The quantitative data from the four-point Likert-scale questionnaires were analyzed statistically with the help of a computer-based statistical program: SPSS 20. First, the means, standard deviations, and the minimum and maximum scores of all variables of this study were calculated. Then, to investigate whether the variables of the study were positively or negatively and significantly correlated among them, Pearson correlations were calculated.

To answer whether the 12th grade students' epistemological beliefs can predict their personal beliefs in plant evolution, animal evolution and human evolution, before biological evolution instruction, multiple regression analyses were carried out with epistemological beliefs (epistemological dimensions) as predictors. This approach enabled us to examine a relationship between dependent variables (personal beliefs in plant evolution, animal evolution and human evolution), and multiple independent variables (dimensions of epistemological beliefs), before biological evolution instruction.

Qualitative data from semi-structured interviews were analyzed through content analysis, using both inductive and deductive qualitative content analysis in order to develop coding categories (Mayring, 2000). The semi-structured interviews were audio recorded and transcribed. The content analysis of interview transcripts was conducted by two researchers who were familiar with epistemological beliefs and biological evolution. The content analysis was undertaken through a manual method of analysis. Coding categories emerged from students' data through repeated examination, comparison, and interpretation.

In the case of disagreements regarding content analysis of semi-structured interviews, the two coders discussed all discrepancies. Inter-rater reliability for the main questions of the semi-structured interviews was estimated using Cohen's Kappa, with $k = .91$.

9.3   Results

Table 9.2 displays the means, standard deviations, and the minimum and maximum scores of all variables of this study.

As seen in Table 9.2, participants' scores on the epistemological beliefs toward science measure suggested relatively sophisticated beliefs about the nature of knowing (source and justification of knowledge), and less sophisticated beliefs about the nature of knowledge (certainty [stability of knowledge], simplicity [structure of knowledge] and development of knowledge). Yet, participants' scores on their

Table 9.2 Descriptive statistics for all variables of the current study ($N = 51$)

Variable	M	SD	Min.	Max.
Epistemological beliefs				
Certainty of knowledge	2.59	0.39	1.33	3.16
Simplicity of knowledge	2.55	0.37	1.66	3.33
Development of knowledge	2.56	0.27	1.83	3.00
Source of knowledge	2.80	0.43	2.00	3.66
Justification of knowledge	2.90	0.34	2.16	4.00
Personal beliefs in biological evolution				
Beliefs in plant evolution	3.23	0.73	1.00	4.00
Beliefs in animal evolution	3.30	0.74	1.00	4.00
Beliefs in human evolution	2.65	0.79	1.00	4.00
Beliefs in human creation by God	3.41	0.92	1.00	4.00

personal beliefs in plant and animal biological evolution measure were higher than their scores on beliefs in human evolution. However, participants' scores on beliefs in human creation by God were the highest.

Table 9.3 displays the Pearson correlations between all variables of this study.

As seen in Table 9.3, the Pearson correlations indicated significant positive correlation (Cohen, 1988, 1992) between source epistemological beliefs before evolution intervention and personal beliefs in plant evolution and animal evolution ($r = .33$, $p < .05$; $r = .35$, $p < .05$), suggesting that more sophisticated epistemological beliefs about the source of knowledge were correlated with high personal beliefs scores on plant and animal biological evolution. The results of the Pearson correlations indicated no significant correlation between epistemological beliefs and personal beliefs in human biological evolution or human creation by God.

Table 9.4 displays the unstandardized regression coefficients (B) and intercept, the standardized regression coefficients (β), R^2, and adjusted R^2 after entry of all independent variables (IVs).

As seen in Table 9.4, the regression analyses revealed a similar pattern to the Pearson correlations. Using the personal beliefs in plant, animal, human evolution and human creation by God measures as dependent variables and the measures of epistemological beliefs (dimensions) as predictors in separate analyses revealed that there was a weak significant predictive relation between the epistemological beliefs dimension of source of knowledge and personal beliefs in plant evolution ($R^2 = 0.11$, Finc(1, 49) =6.09, p = 0.02) and animal evolution ($R^2 = 0.12$, Finc(1,49) = 6.68, p = 0.01).

The results of the regression analyses indicated no predictive relation between epistemological beliefs and personal beliefs in human evolution ($R^2 = 0.49$, Finc(5, 45) = 0.47, p = 0.8), and beliefs in Human Creation by God ($R^2 = 0.11$, Finc(5, 45) = 1.12, p = 0.36).

Table 9.3 Pearson correlations for all variables of the current study (N = 51)

Variable	1.	2.	3.	4.	5.	6.	7.	8.	9.
Epistemological beliefs									
1. Certainty of knowledge	–								
2. Simplicity of knowledge	0.08	–							
3. Development of knowledge	0.30*	0.28*	–						
4. Source of knowledge	0.30*	0.08	0.37**	–					
5. Justification of knowledge	0.13	0.09	0.23	0.36**	–				
Personal beliefs in biological evolution									
6. Beliefs in plant evolution	0.14	0.00	0.18	0.33*	0.00	–			
7. Beliefs in animal evolution	0.16	0.30	0.21	0.35*	0.00	0.96***	–		
8. Beliefs in human evolution	0.12	0.05	0.06	0.12	0.17	0.18	0.18	–	
9. Beliefs in human creation by God	−0.9	0.08	−0.02	0.13	−0.10	−0.23	−0.23	−0.05	–

Note: ***p < .001, **p < .01, two-tailed; *p < .05, two-tailed

Table 9.4 Results of regression analyses for epistemological beliefs (dimensions) variables predicting personal beliefs in biological evolution (N = 51)

Predictor variables	Beliefs in plant biological evolution B(SE)	β	Beliefs in animal biological evolution B(SE)	β	Beliefs in human Biological evolution B(SE)	β	Beliefs in human creation by God B(SE)	β
Epistemological dimensions								
Certainty of knowledge	0.05(0.21)	0.03	0.09(0.28)	0.05	0.17(0.36)	0.08	−0.31(0.36)	−0.13
Simplicity of knowledge	0.03(0.30)	0.02	0.10(0.30)	0.05	−0.18(0.34)	−0.08	0.38(0.39)	0.15
Development of knowledge	0.21(0.44)	0.08	0.21(0.40)	0.09	−0.16(0.50)	−0.04	−0.27(0.56)	0.08
Source of knowledge	0.61(0.29)	0.35*	0.64(0.29)	0.37*	0.26(0.35)	0.03	0.60(0.37)	0.28
Justification of knowledge	−0.30 (0.32)	−0.14	−0.31(0.33)	−0.19	0.29(0.37)	0.80	−0.48(0.42)	−0.18

Note: ***p < .001, **p < .01, two-tailed; *p < .05, two-tailed
Note: For personal beliefs in Plant Evolution: R = 0.37, R^2 = 0.13, Adjusted R^2 = 0.04
For personal beliefs in Animal Evolution: R = 0.39, R^2 = 0.16, Adjusted R^2 = 0.06
For personal beliefs in Human Evolution: R = 0.22, R^2 = 0.49, Adjusted R^2 = −0.06
For personal beliefs in Human Creation by God: R = 0.23, R^2 = 0.11, Adjusted R^2 = 0.01

The semi-structured interviews results indicated a similar pattern to that of the Pearson correlations and the regression analyses. All interviewed students expressed relatively sophisticated epistemological beliefs toward scientific theories and nature of knowledge and knowing, indicating that knowledge is tentative and evolving. For example, two students commented that "… scientific theories are reliable and well established, but sometimes they can change because of new evidence, new instruments or new interpretations …" Additionally, all five interviewees mentioned that it is necessary to form a personal opinion about what one reads/listens to. For example, three students commented that "… some scientists can make errors that harm people's health and the environment …".

All interviewed students mentioned that they believe in plant and animal biological evolution, but only two students mentioned that they believe in human evolution. The other students mentioned that God created human beings. For example, one student stated that "… animals and plants have evolved from other organisms, but the human being is God's creation. Another interviewed student used data from the Bible to explain the creation of Earth and Life and then expressed the idea that God guided evolution. One student mentioned that "…according our religion, humans were created by God, and I believe this…".

However, students who expressed highly sophisticated epistemological beliefs toward science were more likely to believe in human evolution compared to students with less strong / (OR weaker) and less sophisticated epistemological beliefs. Additionally, the results from the semi-structured interviews indicated that students' personal beliefs in human evolution were more related to their degree of religious commitment and not to their epistemological beliefs.

9.4   Discussion and Conclusions

The present research extends the current literature examining relationships between 12th grade students' epistemological beliefs toward science and their personal beliefs in plant evolution, animal evolution and human evolution, before biological evolution instruction. The findings indicate that students with relatively sophisticated epistemological beliefs, particularly beliefs about the source of knowledge, believe more in animal and plant evolution than students with less sophisticated epistemological beliefs. In other words, students who view science and scientific knowledge as a tentative and a dynamic process, and the result of the logical processing of facts and evidence with coherence, are also more likely to believe in plant and animal biological evolution. On the contrary, the more the students view science and scientific knowledge as fixed and authoritative, the more likely they are to not believe in evolutionary theory. On the other hand, our data showed no relationship between epistemological beliefs and beliefs in human evolution. Instead, our interviews findings showed high beliefs in human creation by God.

The finding that 12th grade students' epistemological source beliefs predicted their personal beliefs in plant and animal evolution, before biological evolution instruction, constitutes a novel contribution of the present study. Furthermore, our interview results show that religious belief is an important influential factor in determining students' personal beliefs in human biological evolution. This finding is consistent with previous findings reported in the literature (Cavallo & McCall, 2008; Gould, 1997; Winslow et al., 2011; Woods & Scharmann, 2001). According to Gould (1997), although epistemologically, religion and science can be considered as non-overlapping magisteria, pedagogically these two magisteria could potentially overlap each other in a student's mind; thus feeding their opposition to biological evolution and promoting beliefs in human creation by God.

The finding that human beings are not automatically considered as animal organisms by the 12th grade students highlights the need for biology teachers to address students' misconception in order to foster their conceptual understanding of the biological evolution of all living organisms. Researchers of conceptual change have explored the impact of different factors on students' understanding of the process of conceptual understanding (Pintrich, 1999; Sinatra & Pintrich, 2003). They argue that affective constructs, epistemological beliefs and religious factors can be brought intentionally to bear on the process of conceptual change and learning (Pintrich, 1999; Reiss & Harms, 2019; Rosengren et al., 2012; Sinatra & Pintrich, 2003). This means that the way in which students understand the nature of knowledge and knowing, their religious beliefs and their personal views of the world may impact their conceptual change mechanism and learning about biological evolution.

Our research has important educational implications, showing that one goal of biology education should be to teach students to inquire about the world around them in an objective manner, taking into consideration that science and religion are distinct systems, and being aware that questioning what they know and think does not necessitate that they change their faith. Yet, biology teachers should address students' misconceptions about human biological evolution (among others) and foster students' epistemological beliefs and their familiarity with the methodological principles of scientific knowledge that – by their very nature – set the boundaries on what science can address.

Additionally, our research points to the need to invest in efforts to foster students' source epistemological beliefs. When students understand that knowledge is actively constructed by the person in interaction with others, and they view their current knowledge about the world as something that will change with new knowledge, perhaps they will become open to continued inquiry and questioning in every aspect of their lives.

In summary, the present study extends the current literature by examining relationships between epistemological beliefs toward science and personal beliefs in plant evolution, animal evolution and human evolution, before biological evolution instruction. Our results show that 12th grade students' epistemological beliefs predict their personal beliefs in plant evolution and animal evolution, but not in human evolution. Our findings suggest the need to design educational programs to support the development of students' epistemological beliefs toward science, supporting students: (a) to understand and be able to practice the processes of science, to experience the tentative and evolving nature of science, and to logically and thoughtfully analyze scientific evidence, making their own logical arguments that justify their personal beliefs; (b) to understand how and why knowledge produced through science is different from their religious, social and cultural beliefs; (c) to understand the nature of knowledge and knowing and the methodological principles of scientific knowledge which may impact students' conceptual change mechanism and learning about the theory of biological evolution, with an emphasis on human evolution.

Inquiry-based teaching and learning (Shi et al., 2020), learning by using history of science (Matthews, 1994), dialogic argumentative activities (Iordanou & Constantinou, 2014) and reflective judgment through socioscientific issues (Zeidler et al., 2009) are some of the recommended didactical approaches to foster students' epistemological beliefs. However, the recommended didactical approaches are synergistic, build upon one another, and provide opportunities to foster students' epistemological beliefs (Baytelman et al., 2023).

There are some limitations to the current study that may provide impetus to further work in this area. The first limitation concerns the small size of our sample, consisting of 51 students. The second limitation concerns the impact of the school unit and the residence of the participants. All of the participants came from the same school unit and from the same region: They shared a homogeneous middle-class social background, and the same language and religion. The third limitation concerns the small size of semi-structured interviews, consisting of five interviews. The last limitation concerns the Pearson coefficients, as well as the R^2 in the multiple regression analyses in the present study, which were low. Further research is required to replicate these findings, which will be of interest to researchers, educators and biology teachers in the field.

References

Baytelman, A. (2015). *The effects of epistemological beliefs and prior knowledge on pre-service teachers' informal reasoning regarding socio-scientific issues*. University of Cyprus, Faculty of Social Sciences and Education.

Baytelman, A., & Constantinou, C. P. (2016a). Development and validation of an instrument to measure epistemological beliefs in science. In J. Lavonen, K. Juuti, J. Lampiselkä, A. Uitto, & K. Hahl (Eds.), *Proceedings of the ESERA 2015 conference. Science education research: Engaging learners for a sustainable future, part 11 (coed. Jens Dolin and per kind)* (pp. 1047–1058). University of Helsinki.

Baytelman, A., & Constantinou, C. P. (2016b). Development and validation of an instrument to measure student beliefs on the nature of knowledge and learning. *Themes of Science and Technology in Education, 9*(3), 151–172.

Baytelman, A., Iordanou, K., & Constantinou, C. P. (2020a). Epistemic beliefs and prior knowledge as predictors of the construction of different types of arguments on socioscientific issues. *Journal of Research in Science Teaching, 57*(8), 1199–1227. https://doi.org/10.1002/tea.21627

Baytelman, A., Iordanou, K., & Constantinou, C. P. (2020b). *Dimensions of epistemological beliefs toward science instrument (DEBS) [Database record]*. APA PsycTests.

Baytelman, A., Iordanou, K., & Constantinou, C. P. (2022). Prior knowledge, epistemic beliefs and socio-scientific topic context as predictors of the diversity of arguments on socio-scientific issues. In K. Korfiatis & M. Grace (Eds.), *Current research in biology education* (pp. 45–57). Springer.

Baytelman, A., Loizou, T., & Chatziconstantinou, S. (2023). Relationships between epistemological beliefs and conceptual understanding of evolution by natural selection. *Center of Education Policy Studies Journal, 13*(1), 63–93. https://doi.org/10.26529/cepsj.1484

Blackwell, W. H., Powell, M. J., & Dukes, G. H. (2003). The problem of student acceptance of evolution. *Journal of Biological Education, 37*, 58–67.

Borgerding, L. A., Deniz, H., & Anderson, E. S. (2017). Evolution acceptance and epistemological beliefs of college biology students. *Journal of Research in Science Teaching, 54*(4), 493–519.

Bryman, A. (2008). *Social research methods* (3rd ed.). Oxford University Press.

Cavallo, A. M. L., & McCall, D. (2008). Seeing may not mean believing: Examining Students' Understandings & Beliefs in evolution. *The American Biology Teacher, 70*(9), 522–530.

Chan, N.-M. T., Ho, I. T., & Ku, K. Y. (2011). Epistemic beliefs and critical thinking of Chinese students. *Learning and Individual Differences, 21*(1), 67–77.

Chen, J. A. (2012). Implicit theories, epistemic beliefs, and science motivation: A person-centred approach. *Learning and Individual Differences, 22*(6), 724–735.

Cherif, A., Adams, G., & Loehr, J. (2001). What on earth is evolution: The geological perspective of teaching evolutionary biology effectively. *The American Biology Teacher, 63*(8), 569–591.

Cho, M., Lankford, D., & Wescott, D. (2011). Exploring the relationships among epistemological beliefs, nature of science, and conceptual change in the learning of evolutionary theory. *Evo Edu Outreach, 4*, 313–322.

Cohen, J. (1988). *Statistical power analysis for the behavioral sciences* (2nd ed.). Erlbaum.

Cohen, J. (1992). A power primer. *Psychological Bulletin, 112*(1), 155–159.

Conley, M., Pintrich, P., Vekiri, I., & Harrison, D. (2004). Changes in epistemological beliefs in elementary science students. *Contemporary Educational Psychology, 29*(2), 186–204.

Dunk, R. D. P., & Wiles, J. R. (2018). *Changes in acceptance of evolution and associated factors during a year of introductory biology: The shifting impacts of biology knowledge* (p. 280479). Demographics, and Understandings of the Nature of Science. BioRxiv.

Evans, E. M., Legare, C., & Rosengren, K. (2011). Engaging multiple epistemologies: Implications for science education. In M. Ferrari & R. Taylor (Eds.), *Epistemology and science education: Understanding the evolution vs. intelligent design controversy* (pp. 111–139). Routledge.

Fowler, S. R., & Zeidler, D. L. (2016). Lack of evolution acceptance inhibits students' negotiation of biology-based Socioscientific issues. *Journal of Biological Education, 50*(4), 407–424.

Gould, S. J. (1997). Nonoverlapping magisteria. *Natural History, 106*, 16–22.

Greene, J. C. (2007). *Mixed methods in social inquiry*. Jossey-Bass.

Harms, U., & Reiss, M. J. (2019). *Evolution education re-considered: Understanding what works*. Springer.

Hofer, B. K., & Pintrich, P. R. (1997). The development of epistemological theories: Beliefs about knowledge and knowing their relation to learning. *Review of Educational Research, 67*(2), 88–140.

Iordanou, K., & Constantinou, C. P. (2014). Developing pre-service teachers' evidence-based argumentation skills on socio-scientific issues. *Learning and Instruction, 34*, 42–57.

Kitchener, R. (2002). Folk epistemology: An introduction. *New Ideas in Psychology, 20*, 89–105.

Kuhn, D., Cheney, R., & Weinstock, M. (2000). The development of epistemological understanding. *Cognitive Development, 15*(3), 309–328.

Matthews, M. R. (1992). History, philosophy and science teaching: The present rapprochement. *Science & Education, 1*(1), 11–47.

Matthews, M. R. (1994). *Science teaching: The role of history and philosophy of science*. Routledge.

Mayring, P. (2000). Qualitative content analysis. *Forum: Qualitative Social Research, 1*(2), 20.

Mazur, A. (2005). Believers and disbelievers in evolution. *Political Life Science, 23*(2), 55–61.

Muis, K., Pekrun, R., Sinatra, G., Azevedo, R., Trevors, G., Meier, E., & Heddy, B. (2015). The curious case of climate change: Testing a theoretical model of epistemic beliefs, epistemic emotions, and complex learning. *Learning and Instruction, 39*(2), 168–183.

Nehm, R. H., & Reilly, R. (2007). Biology majors' knowledge and misconceptions of natural selection. *Bioscience, 57*(3), 263–272.

Perry, W. (1970). *Forms of intellectual and ethical development in the college years: A scheme.* Holt, Rinehart & Winston.

Pintrich, P. R. (1999). Motivational beliefs as resources for and constraints on conceptual change. In W. Schnotz, S. Vosniadou, & M. Carretero (Eds.), *New perspectives on conceptual change* (pp. 33–50). Pergamon.

Pintrich, P. R. (2002). Future challenges and direction for theory and research on personal epistemology. In B. K. Hofer & P. R. Pintrich (Eds.), *Personal epistemology: The psychology of beliefs about knowledge and knowing* (pp. 389–414). Lawrence Erlbaum Associates, Inc..

Reiss, M., & Harms, U. (2019). What now for evolution education? In U. Harms & M. Reiss (Eds.), *Evolution education: Re-considered* (pp. 331–343).

Rosengren, K. L., Brem, S. K., Evans, E. M., & Sinatra, G. M. (Eds.). (2012). *Evolution challenges integrating research and practice in teaching and learning about evolution.* Oxford University Press.

Scharmann, L. C. (1990). Enhancing an understanding of the premises of evolutionary theory: The influence of a diversified instructional strategy. *School Science and Mathematics, 90*(2), 91–100.

Schommer, M. (1990). Effects of beliefs about the nature of knowledge on comprehension. *Journal of Educational Psychology, 82*(3), 498–504.

Schommer, M., Crouse, A., & Rhodes, N. (1992). Epistemological beliefs and mathematical text comprehension: Believing it is simple does not make it so. *Journal of Educational Psychology, 84*, 435–443.

Schommer-Aikins, M. (2004). Explaining the epistemological belief system: Introducing the embedded systemic model and coordinated research approach. *Educational Psychologist, 39*(1), 19–29.

Shi, W., Ma, L., & Wang, J. (2020). Effects of inquiry-based teaching on chinese university students' epistemologies about experimental physics and learning performance. *Journal of Baltic Science Education, 19*(2), 289–297.

Shtulman, A. (2006). Qualitative differences between naïve and scientific theories of evolution. *Cognitive Psychology, 52*(2), 170–194.

Shtulman, A., & Calabi, P. (2012). Cognitive constraints on the understanding and acceptance of evolution. In K. S. Rosengren, S. K. Brem, E. M. Evans, & G. M. Sinatra (Eds.), *Evolution challenges* (pp. 47–65). Oxford University Press.

Shtulman, A., & Schulz, L. (2008). The relation between essentialist beliefs and evolutionary reasoning. *Cognitive Science, 32*(6), 1049–1062.

Sinatra, G. M., & Pintrich, P. R. (2003). *Intentional conceptual change.* Lawrence Erlbaum Associates.

Sinatra, G. M., Southerland, S. A., McConaughy, F., & Demastes, J. W. (2003). Intentions and beliefs in students' understanding and acceptance of biological evolution. *Journal of Research in Science Teaching, 40*(5), 510–528.

Sinclair, A., & Baldwin, B. (1996). The relationship between college zoology students' religious beliefs and their ability to objectively view the scientific evidence supporting evolutionary theory. In *Paper presented at the annual meeting of the American Educational Research Association.*

Spindler, L., & Doherty, J. (2009). Assessment of the teaching of evolution by natural selection through a hands-on simulation. *Teaching Issues and Experiments in Ecology, 6.*

Winslow, M., Staver, J., & Scharmann, L. (2011). Evolution and personal religious belief: Christian university biology-related majors' search for reconciliation. *Journal of Research in Science Teaching, 48*, 1026–1049.

Woods, C. S., & Scharmann, L. C. (2001). High school students' perceptions of evolutionary theory. *Electronic Journal of Science Education, 6*(2).

Zeidler, D. L., Sadler, T. D., Applebaum, S., & Callahan, B. E. (2009). Advancing reflective judgment through socioscientific issues. *Journal of Research in Science Teaching, 46*(1), 74–101.

Chapter 10
Plant Blindness Intensity Throughout the School and University Years: A Cross-Age Study

Alexandros Amprazis and Penelope Papadopoulou

10.1 Introduction

In the mid-1980s, James Wandersee (1986) recorded for the first time in the literature a phenomenon according to which people show more interest in animals than plants. Since then, he and other researchers (Bebbington, 2005; Fančovičová & Prokop, 2011; Schussler & Olzak, 2008) have identified the phenomenon known as 'Plant Blindness', which mainly refers to: (a) the inability of humans to see, notice or focus their attention to plants in their daily life, (b) the inability to recognize the importance of plants' aesthetics and the uniqueness of their biological characteristics, (c) the tendency of humans to classify plants as inferior to animals and (d) the lack of basic knowledge regarding plant organisms (Amprazis et al., 2019; Strgar, 2007; Wandersee & Schussler, 2001). The phenomenon seems to have social and environmental impacts, as links to sustainable development are documented (Amprazis & Papadopoulou, 2020; Thomas et al., 2022). It is noteworthy that lately some researchers have questioned the validity of plant blindness as a 'whole world phenomenon' (Balding & Williams, 2016). This notion is based upon the fact that the vast majority of plant blindness research derives from the Western world, as well as upon some ethnobotany studies which outline how indigenous communities can be closer to flora than social groups in Western civilisation (Katz, 1989).

Regarding the causes of the phenomenon, the first researchers focused on the lack of intense plant movement, the types of animal activities (feeding, communication) and the external, morphological relevance of mammals to the human species (Hoekstra, 2000). Additionally, the human brain processes only a portion of the millions of information units (bits) that are continuously being sent from the eyes (Wandersee & Schussler, 2001). Plants as a common part of the visual background

A. Amprazis (✉) · P. Papadopoulou
University of Western Macedonia, Florina, Greece
e-mail: aamprazis@uowm.gr; ppapadopoulou@uowm.gr

137

K. Korfiatis et al. (eds.), *Shaping the Future of Biological Education Research*,
Contributions from Biology Education Research,
https://doi.org/10.1007/978-3-031-44792-1_10

and without visible movement are not usually contained in the information that the brain chooses to process. The phenomenon appears to be even more complex and intricate, as there is a kind of 'zoocentrism' in the school context. Hershey (1996) was the researcher who first referred to this 'zoo-centric approach' in school curriculum. This notion was reinforced by later research, which recorded more visual or written references to animals than to plants in textbooks (Link-Perez et al., 2009). Regarding school teachers, they also seem to contribute to zoocentrism when, for example, they want to elaborate upon the concept of life. Here, they are more likely to refer to an animal than a plant (Hershey, 2005).

Regarding the restriction of plant blindness, in the literature one can find research mainly in the educational context. Several researchers report that specifically designed educational interventions can stimulate students' interest in plants and contribute to the phenomenon's restriction. Fančovičová and Prokop (2011) implemented and evaluated educational projects in Slovakia, involving pupils aged 10–11 years old. The children visited areas with rich vegetation and the researchers recorded a statistically significant improvement in pupils' attitudes towards plants after the educational activities. Similarly, in Lindemann-Matthies' (2005) research, students from 146 secondary schools in Switzerland increased their interest in non-common plant species through contact and guided observation of plants. Educational activities outside of school were also chosen by Borsos (2019) to enhance students' interest towards plants. Based on the results of her research, the children who participated in the study significantly improved their relative knowledge. Çil (2016) evaluated an intervention to enhance knowledge and interest in plant organisms, which was based on the integration of elements of visual arts and chemistry in the botany lesson. According to the results, 25 children aged 10 to 12 years old improved their attitudes towards plants as they became more aware of the importance of plant organisms and increased their interest in them. In Stagg and Verde's (2018) interdisciplinary approach, English primary school pupils attended or participated in an interactive drama performance that aimed to teach basic, plant-based functions. Afterwards, an increase in knowledge and an improvement of children's attitudes towards plants was recorded. Kissi and Dreesmann's (2018) study implemented an educational intervention in a botanical garden using mobile phones. Children were involved in the whole process by observing and collecting information about different plant species in the botanical garden. This led to improved knowledge and more positive attitudes towards plants.

In consideration of all that has been mentioned above, the research questions we address here are as follows:

1. How does the intensity of plant blindness change as students move from primary school to university?
2. How does the correlation between the core elements of the phenomenon alter as students move from primary school to university?

10.2   Research Design and Method

Data were collected from 1237 students from primary school, junior high school, senior high school and university. More specifically, these were 333 students from the sixth grade of primary school, 301 from the third grade of junior high school, 305 from the third grade of senior high school and 298 from the fourth year of university. Regarding the latter participant group, they were students from primary education pedagogical departments of Greek universities. This choice was made for students from such departments who have a broader knowledge, similar to those of the general population, without having expertise in plants or any other particular field. Moreover, they are potentially compulsory education teachers who will later be called upon to teach and act as role models for the children. It has to be mentioned that only graduates from each school level were selected, in order for the participants to reflect all the background knowledge offered by every educational grade. Since primary school in Greece lasts 6 years, junior and senior high school last 3 years each, and pedagogical departments in tertiary education last 4 years, participants age range was 12, 15, 18 and 22 years old accordingly.

As there were different age groups, a cross-sectional study was chosen (Abdolmohammadi & Reeves, 2000). In general, cross-sectional studies are widely used in the field of education to examine students' attitudes and learning patterns (Prochaska et al., 2003).

The research instrument was a questionnaire. Participants' attitudes towards plants and animals were assessed through five-point Likert-type scale questions. The same type of questions was also used to assess the amount of knowledge about plants that is being offered to students in school. For example, 'how often do your teachers talk about plants?' In order to evaluate students' knowledge about flora, 'right or wrong' questions were used. More specifically, students were given specific statements and asked to rate them as right, wrong, or to admit that they could not give an opinion on the sentence correctness. Examples of such statements for primary school were 'There are plants that can grow without a root' and 'Plants make their own food', while examples for senior high school were 'Plants participate directly or indirectly in the production of all food consumed by humans' and 'All plants contain phloem and xylem to move nutrients'.

The content of these statements was determined by the knowledge offered at each school level, as this is formed by the official curriculum. During the analysis, students' wrong answers and the 'I don't know' option were consolidated and classified as lack of knowledge. The instrument also consisted of a particular question in which students were asked to freely complete a list of five living things that they could think of. That was included in order to examine participants' spontaneous recall of plants as living things (Anderson et al., 2002). Finally, a question was included in which students had to distinguish and mark out plant-derived products among 12 common daily products. To construct the research instrument, specific steps were followed, as these are determined by the literature of quantitative methodology (Creswell, 2012; Little, 2013; Teo, 2013). Initially, the primary school questionnaire was created and afterwards it was adapted to the other age groups. After determining the theoretical framework, several exploratory semi-structured interviews were conducted with students to gain insight into their conceptions and make a first assessment of the existence of plant blindness aspects. The first version of the research instrument was created and assessed by in-service teachers and students regarding its comprehension. New versions of the research instrument were created and tested through three pilot implementations (Krosnick et al., 2018). The evaluation of the questionnaire's validity was established by a group of experts in biology education and, regarding the internal consistency, the value of Cronbach's Alpha was above 0.8 for every age group. It is also important to note that by following the methodology of cross-sectional studies, wording changes were applied to the instrument in order to reflect the participants' age at each level.

In the research instrument there were six factors of the questionnaire to reflect the core elements of the definition of plant blindness. These factors were (a) Interest in plants, (b) Interest in animals, (c) Assessment of students' knowledge about plants, (d) Identification of plant-derived products, (e) Recall of plants as living things and (f) Amount of knowledge about plants offered by school.

The Spearman's rank correlation coefficient test was conducted to examine correlations between these factors. To graphically present these correlations, and hence the relationships between the phenomenon's core elements, a network analysis was also conducted (Borgatti et al., 2009). Only the statistically significant positive correlations were used for the network analysis. In order to classify the statistically significant positive correlations between the factors in the questionnaire as strong or weak, the following scale was used: .00–.19 'very weak', .20–.39 'weak', .40–.59 'moderate', .60–.79 'strong' and .80–1.0 'very strong' (Moore, 2004). To illustrate the results of the Spearman's test, the open-source software called Gephi (Bastian et al., 2009) was used. Gephi generated specific diagrams in which the width of the lines represents the level of positive correlation between the factors – the wider the line, the stronger the correlation.

10.3   Results

A Wilcoxon signed-ranks test was conducted and indicated that students' scores regarding preference for animals were statistically significantly higher than their scores regarding preference for plants, $Z = -32.51$, $p < .000$. The effect size of this statistical test was found to be large, $r = .86$. The same test was conducted for each grade individually and produced the same result every time.

A Kruskal-Wallis H test was conducted to clarify whether there is a statistically significant difference in the 'Preference in Flora' factor among the four school level groups of our participants (Table 10.1). The test was significant [$X^2(3, n = 1248) = 95.69$, $p < .05$] and follow-up Mann-Whitney U tests indicated that the high school students recorded the most negative answers regarding preference for flora. No difference was recorded among primary school and university students.

Table 10.1 Kruskal-Wallis H test results regarding students' level of schooling effect on their preference for plants (N = 1248)

Students grade	N	Mean ranks	Mean	x^2	df	Asymp. sig
Primary school (12 years old)	333	3.35	731	95.697	3	.000
Junior high school (15 years old)	301	2.99	553.2			
Senior high school (18 years old)	305	2.94	540			
University (22 years old)	298	3.42	767.8			

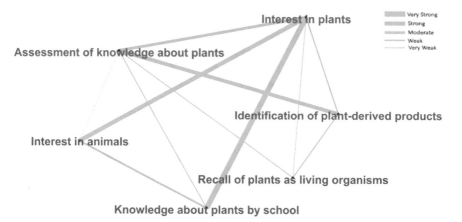

Fig. 10.1 Plant blindness' core elements' network analysis for primary school students

Different correlations have been recorded for each grade by conducting the Spearman's rank correlation coefficient tests. As shown by the follow-up network analysis, in primary school all of the phenomenon's core elements appear to be interconnected (Fig. 10.1). More specifically, regarding strong or moderate correlations, wide lines connect interest in plants to (a) interest in animals, (b) knowledge about plants and (c) assessment of knowledge about plants. Another strong correlation that is recorded is between assessment of knowledge about plants and identifications of plant-derived products. It is interesting that the strong correlations that have been recorded in primary school were also recorded in the junior high school (Fig. 10.2). By examining the correlations recorded in Fig. 10.2, one can conclude that the more junior high school students are interested in animals, the more they are interested in plants. Respectively, the more they are interested in plants, the more knowledge they have about plants. Regarding the latter correlation (interest in plants – knowledge about plants offered by school), as we move to senior high school, it is not so strong anymore (Fig. 10.3). The same applies to the correlation between interest in plants and assessment of knowledge about plants. On the contrary, the correlation between interest in plants and interest in animals remains strong, as the more senior high school students are interested in animals, the more they are interested in plants and vice versa.

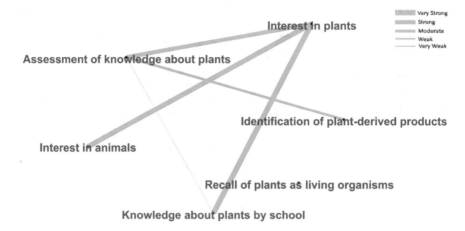

Fig. 10.2 Plant blindness' core elements' network analysis for junior high school students

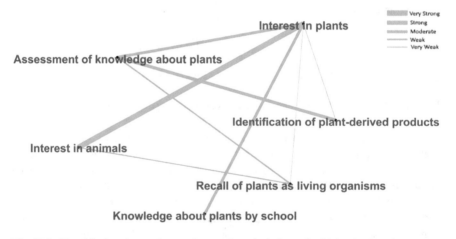

Fig. 10.3 Plant blindness' core elements' network analysis for senior high school students

Finally, by examining the network analysis of the university group (Fig. 10.4), one can draw interesting conclusions. Firstly, an all-pervasive lack of statistically significant positive correlations is recorded. Plant blindness core elements do not seem to be so connected when the phenomenon is examined among university students. The only strong correlation that one can identify is between interest in plants and knowledge about plants. Interest in plants and interest in animals are still connected, but not so intensely anymore.

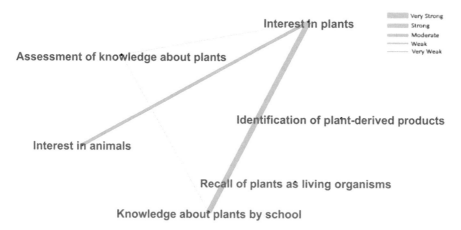

Fig. 10.4 Plant blindness' core elements' network analysis for university students

10.4   Discussion and Conclusions

Plant blindness seems to be a complex phenomenon that should be examined with caution. It is a multidimensional issue which can be incorporated, but cannot be described in its totality by just the concept of 'humans' interaction with plants. Moreover, plant blindness is an important challenge that the academic and educational community must face, especially because of its link to sustainable development. This link confirms emphatically the need for connectedness to nature as a prerequisite to achieve sustainability (Jordan & Kristjánsson, 2017).

The main conclusion regarding the cross-sectional approach is that plant blindness seems to intensify as students grow older and move from primary to secondary and higher education. Elaborating more upon this finding, one can interpret it by taking adolescence into consideration: This is a period during which people rarely show interest in issues such as the environment, as they focus more on themselves (Prochaska et al., 2003).

The alterations that have been recorded during the network analysis of each age group in this research bring to the limelight the need for a different didactic approach in each educational grade. Inside a holistic and inclusive context as educators, we may have to be flexible and adjust the educational interventions to the distinctive characteristics of each school level. Regarding primary school, the strong positive correlation that has been recorded between interest in plants and interest in animals possibly integrates the whole plant blindness issue into the biodiversity awareness context. By promoting an educational approach of fostering all living organisms on earth, human's appreciation for flora can be a collateral benefit. Moreover, the strong positive correlation between interest in plants and knowledge about plants offered by school is an indication that this cognitive background should remain solid, if not be enhanced even more. This enhancement can be achieved through a wide perspective that, besides botany, will also highlight the importance of plants for human welfare and the life phenomenon in general. Regarding junior high school, the main strong correlations remained mostly the same as in primary school. Therefore, learning about the entire life spectrum's importance and enhancing plants' meaningful and sufficient presence in the curriculum, are once again recommended Moving to senior high school, the aspect of knowledge seems to withdraw a bit, and the only strong correlation recorded is between interest in plants and interest in animals As we now have to do with older students, the biodiversity education in this grade can be based more on transformative learning and critical thinking, since the goal is to provide students with a new positive perspective of all other living components of the ecosystem. Finally, regarding university students, knowledge seems to be the most important factor correlated to their interest in flora. Consequently, provision of a comprehensive and a specialised amount of plant knowledge probably creates a framework in which fostering plant awareness can be achieved more easily. Once again, it is important to have a broad perspective and use a cross-thematic approach beyond botany that will clarify plants' relation to human culture, history, economy, and even great social endeavours, such as sustainable development.

All the educational implications mentioned above can also be integrated in environmental education and education for sustainable development. These frameworks are highly appreciated for being able to alter students' attitudes and enhance children's appreciation of the natural environment. Environmental education can promote a comprehensive approach to the living world and enable the improvement of human's relationship with animals and plants. The way environmental education projects are organised and implemented provides multiple opportunities for experiential learning, interdisciplinarity and observation of plant and animal species. In addition, the long duration and scope of these projects within a school year allows for a deeper understanding of the developmental timelines, biological functions and aesthetic properties of plants. Accordingly, it is highly recommended that educators take advantage of school gardens and botanical gardens, which are immersive educational contexts that can maximise quantitative and qualitative contact with plant organisms. They can be used consistently throughout the school year in order to enhance people's interest in flora.

Concerning limitations, the educational implications for the university students' group should be examined with caution and cannot be generalised. The choice made to include students from pedagogical departments was exactly for the purpose of being able to simulate the characteristics of the general population, at least to a certain extent. However, the relative conclusions should be limited only to pre-service teachers and be assessed in relation to every educational department's course of study. Additionally, as mentioned in the methodology section of this contribution, convenience sampling was used during participants' selection to allow supervision of data collection. This can be considered as another limitation, for plant blindness is a phenomenon linked to the natural environment in which individuals live. Social groups of different geographical areas have distinctive habits and practices regarding their contact with plants and the contribution of plant organisms in meeting their daily needs. Thus, a study covering more geographical areas of the country (e.g. islands) could allow for a more reliable generalisation of the results.

As already mentioned, cultural diversity is important when one endeavours to provide a critical reflection of plant blindness; here lies a great and exciting challenge for researchers that focus on that subject. Examining the phenomenon in indigenous communities or countries that rely heavily on agriculture may lead to interesting conclusions. Research is also needed for the collection of data regarding plant blindness in age groups above 22 years old. For the most part, in the literature one can find data only about school and university students and, therefore, we do not have a clear picture of other age groups. In total, all the above future research directions can enrich the relevant literature and restrict the intensity of plant blindness even more over the forthcoming years. This is a goal that should be a priority for both developed and developing countries and may be integrated in the sustainability context.

References

Abdolmohammadi, M. J., & Reeves, M. F. (2000). Effects of education and intervention on business students' ethical cognition: A cross sectional and longitudinal study. *Teaching Business Ethics, 4*(3), 269–284.

Amprazis, A., & Papadopoulou, P. (2020). Plant blindness: A faddish research interest or a substantive impediment to achieve sustainable development goals? *Environmental Education Research, 3*(11), 238–256.

Amprazis, A., Papadopoulou, P., & Malandrakis, G. (2019). Plant blindness and children's recognition of plants as living things: A research in the primary schools context. *Journal of Biological Education, 55*(2), 139–156.

Anderson, D., Piscitelli, B., Weier, K., Everett, M., & Tayler, C. (2002). Children's museum experiences: Identifying powerful mediators of learning. *Curator: The Museum Journal, 45*(3), 213–231.

Balding, M., & Williams, K. J. (2016). Plant blindness and the implications for plant conservation. *Conservation Biology, 30*(6), 1192–1199.

Bastian, M., Heymann, S., & Jacomy, M. (2009). Gephi: An open source software for exploring and manipulating networks. *Proceedings of the International AAAI Conference on Web and Social Media, 3*(1), 361–362. https://doi.org/10.1609/icwsm.v3i1.1393

Bebbington, A. (2005). The ability of A-level students to name plants. *Journal of Biological Education, 39*(2), 63–67.

Borgatti, S. P., Mehra, A., Brass, D. J., & Labianca, G. (2009). Network analysis in the social sciences. *Science, 323*(5916), 892–895.

Borsos, E. (2019). The gamification of elementary school biology: A case study on increasing understanding of plants. *Journal of Biological Education, 53*(5), 492–505.

Çil, E. (2016). Instructional integration of disciplines for promoting children's positive attitudes towards plants. *Journal of Biological Education, 50*(4), 366–383.

Creswell, J. W. (2012). *Educational research: Planning, conducting, and evaluating quantitative and qualitative research* (4th ed.). Pearson.

Fančovičová, J., & Prokop, P. (2011). Plants have a chance: Outdoor educational programmes alter students' knowledge and attitudes towards plants. *Environmental Education Research, 17*(4), 537–551.

Hershey, D. R. (1996). A historical perspective on problems in botany teaching. *The American Biology Teacher, 58*(6), 340–347.

Hershey, D. R. (2005). *Plant content in the national science education standards*. Retrieved from https://files.eric.ed.gov/fulltext/ED501357.pdf

Hoekstra, B. (2000). Plant blindness – The ultimate challenge to botanists. *The American Biology Teacher, 62*(2), 82–83.

Jordan, K., & Kristjánsson, K. (2017). Sustainability, virtue ethics, and the virtue of harmony with nature. *Environmental Education Research, 23*(9), 1205–1229.

Katz, C. R. (1989). Herders, gatherers and foragers: The emerging botanies of children in rural Sudan. *Children's Environments Quarterly, 1*, 46–53.

Kissi, L., & Dreesmann, D. (2018). Plant visibility through mobile learning? Implementation and evaluation of an interactive Flower Hunt in a botanic garden. *Journal of Biological Education, 52*(4), 344–363.

Krosnick, S. E., Baker, J. C., & Moore, K. R. (2018). The pet plant project: Treating plant blindness by making plants personal. *The American Biology Teacher, 80*(5), 339–345.

Lindemann-Matthies, P. (2005). 'Loveable' mammals and 'lifeless' plants: How children's interest in common local organisms can be enhanced through observation of nature. *International Journal of Science Education, 27*(6), 655–677.

Link-Perez, M. A., Dollo, V. H., Weber, K. M., & Schussler, E. E. (2009). What's in a name: Differential labeling of plant and animal photographs in two nationally syndicated elementary science textbook series. *International Journal of Science Education, 32*(9), 1227–1242.

Little, T. D. (Ed.). (2013). *The Oxford handbook of quantitative methods, volume 1: Foundations*. Oxford University Press.

Moore, D. S. (2004). *The basic practice of statistics*. W.H. Freeman.

Prochaska, J. J., Sallis, J. F., Slymen, D. J., & McKenzie, T. L. (2003). A longitudinal study of children's enjoyment of physical education. *Pediatric Exercise Science, 15*(2), 170–178.

Schussler, E., & Olzak, L. (2008). It's not easy being green: Student recall of plant and animal images. *Journal of Biological Education, 42*(3), 112–118.

Stagg, B. C., & Verde, M. F. (2018). Story of a seed: Educational theatre improves students' comprehension of plant reproduction and attitudes to plants in primary science education. *Research in Science & Technological Education, 37*(1), 15–35.

Strgar, J. (2007). Increasing the interest of students in plants. *Journal of Biological Education, 42*(1), 19–23.

Teo, T. (Ed.). (2013). *Handbook of quantitative methods for educational research*. Sense Publishers.

Thomas, H., Ougham, H., & Sanders, D. (2022). Plant blindness and sustainability. *International Journal of Sustainability in Higher Education, 23*(1), 41–57.

Wandersee, J. (1986). Plants or animals – Which do junior high school students prefer to study? *Journal of Research in Science Teaching, 23*(5), 415–426.

Wandersee, J., & Schussler, E. (2001). Toward a theory of plant blindness. *Plant Science Bulletin, 47*(1), 2–9.

Chapter 11
Where Do Plants Get Their Mass From? Using Drawings to Assess Adolescent Students' Modelling Skills and Their Ideas About Plant Growth

Eliza Rybska, Joanna Wojtkowiak, Zofia Chyleńska, Pantelitsa Karnaou, and Costas P. Constantinou

11.1 Introduction

11.1.1 Modelling

Evidence-informed modelling is a vital element of scientific reasoning, and the competencies nurtured through modelling-based learning are part of the foundations of science education (Constantinou et al., 2019). Modelling broadens students' views of the world and teaches them to engage in scientific thinking about observable phenomena (Zimmerman, 2007). By developing and using models, students can learn to represent how scientific phenomena operate, test hypotheses and evaluate predictions. The construction and use of models enable students to develop an understanding of the mechanisms underlying phenomena. A scientific model also demonstrates a predictive function, enabling the user to predict changes and map out future behaviour in various aspects of a phenomenon (Nicolaou & Constantinou, 2014).

Drawing, as a mode of expression of students' ideas, enables the development of different representations of the same scientific concept, which in turn leads to creative reasoning (Ainsworth et al., 2011). The application of drawing-driven modelling can be used to support students' reasoning. Drawing-based modelling is particularly useful in school biology education (Brooks, 2009).

E. Rybska (✉) · J. Wojtkowiak · Z. Chyleńska
Faculty of Biology, Laboratory of Nature Education and Conservation,
Adam Mickiewicz University, Poznań, Poland

P. Karnaou · C. P. Constantinou (✉)
Learning in Science Group, Department of Educational Sciences, University of Cyprus,
Nicosia, Cyprus

© The Author(s) 2024   147
K. Korfiatis et al. (eds.), *Shaping the Future of Biological Education Research*,
Contributions from Biology Education Research,
https://doi.org/10.1007/978-3-031-44792-1_11

Drawing can play an essential role in helping students develop a deeper understanding of biological concepts, such as photosynthesis. By creating visual representations of complex ideas, students can more easily grasp key concepts and develop an understanding of their interconnections. Through the act of drawing, students can also identify gaps in their understanding and negotiate misconceptions. Additionally, drawing allows students to communicate their understanding to others, and can serve as a valuable assessment tool for teachers. Overall, incorporating drawing into the learning process can enhance students' comprehension of biological concepts and promote a deeper appreciation for the mechanisms supporting phenomena in the natural world.

11.1.2   Photosynthesis

Photosynthesis is a process that is crucial for our planet, and very important in understanding the biology of plants and the processes that are taking place in sustaining ecosystems. Understanding photosynthesis involves developing a robust sense of how the process operates at both the micro- and macro- level of the organisation of biological matter and the facility to express those understandings at a symbolic level. Photosynthesis has been identified as one of the most challenging topics in biology education and one in which students encounter diverse challenges in their efforts to develop a robust understanding.

11.1.3   Students' Conceptions of Photosynthesis

Science education research studies of school students' understanding of the process of photosynthesis have been reviewed by Messig and Groß (2018) and by Russell et al. (2004), whose findings can be summarised as follows:

(a) students find photosynthesis to be conceptually difficult, a fact that often leads to a lack of interest and the emergence of misconceptions.
(b) students find it particularly difficult to visualise the process, or relate it to things they can observe, especially when the topic is presented purely as a molecular process or too many levels of representation are introduced at the same time.
(c) there are limitations to the practical demonstration of photosynthesis because the equipment is either unreliable and antiquated, or prohibitively expensive.

This study departed from the idea of invoking drawing as a medium for modelling-based learning to explore a potentially productive approach to support the development of adolescent students' understanding of the process of photosynthesis. We focussed on three age groups with the aim to explore the extent to which we could use students' drawings to identify how their modelling skills evolve across this age range and how their understanding of the mechanism of photosynthesis progresses.

11.2 Objectives

The aim of this study was to explore the extent to which we could use student-constructed drawings to identify how modelling skills evolve with age and how they relate to students' understanding of the mechanism of photosynthesis. To achieve this aim, we chose to work with students in grades 5, 7, and 10, i.e. from upper elementary, middle, and high school.

Our research questions were formulated as follows:

- Do students take into consideration the three functions of a model in their efforts to use drawing to express their understanding of the process of photosynthesis?
- How do age, interest, prior experience and knowledge about plants influence students' ability to model photosynthesis?
- What conceptions about photosynthesis do school students reveal in the process of modelling by drawing?

11.3 Research Design and Methodology

A total of 75 students took part in the study: 17 students from grade 5, 20 students from grade 7, and 38 students (2 classes) from grade 10. The participating students from grades 5 and 7 had just completed the curriculum-prescribed topic of photosynthesis. However, grade 10 students completed the tasks before participating in the teaching unit on photosynthesis that is offered for their grade level. None of the students had received prior teaching about modelling.

The research was carried out in three different schools covering the whole of the school education spectrum in Poland. The survey was anonymous, the students worked individually and were only asked to record their age, gender and class name. The research tool consisted of two parts, one with five questions, including two closed and three open-ended questions, and a task requiring students to make a drawing of the process of photosynthesis. In the first question, the students used a five-point Likert scale (from strongly dislike to strongly like) to express how much they like biology. In the second question, the students were asked whether they were growing plants at home. The next three questions were open-ended, and students were asked to provide short answers to:

- What is the role of plants in maintaining life on Earth?
- What life processes do plants carry out?
- What do plants do at night?

The open-ended questions were used to gain insight into students' personal knowledge about plants. Students were asked to work independently when answering the questions on provided paper with printed questions. Students' responses were coded using an iterative categorisation process of identifying suitable codes and checking for consistency and coherence in applying them to the whole dataset (Papadouris & Constantinou, 2010). The same responses were also checked for accuracy with respect to established scientific knowledge and points were given for appropriate answers.

Between responding to the open-ended questions and before the drawing activity, students were exposed to an introductory framing activity. The researcher presented them with two photographs, one of an acorn of a *Quercus robur*, and another of a *Quercus robur* tree. The following questions were addressed in a whole-group discussion: In what ways has this tree changed over the timespan of a few years? Can you identify as many changes as possible? What do you see? What might you assume? Where does a tree get its mass from? What do you think plants need to live? What might happen to stop plants from performing the processes that support life?

This framing activity served to introduce a problem situation and the discussion facilitated the creation of a common shared formulation of the problem. After the framing activity, students were asked to work individually to develop a drawing as an answer to the question: Where do trees get their mass from?

There were three aspects of assessment in this task, thanks to which it was possible to determine whether a given product meets the criteria for a scientific model or not.

11.3.1 Representative Function of Students' Drawings

The first assessed aspect of the drawings were elements that correspond to their *representative function*. The features of this function are divided into two groups: objects presented in the model and process variables. Drawings were coded for both features. For each depicted object and variable/process, students received one point. If the drawing depicted an object/variable connected correctly with other objects/ variables, then it received two points. The detailed coding scheme used to analyse student drawings for this aspect is presented in Table 11.1.

11.3.2 Explanatory Function of Students' Drawings

The second evaluated aspect of the students' drawings was their explanatory function, i.e. their facility to provide an interpretation of how photosynthesis takes place. Table 11.2 presents the coding scheme and the anticipated exemplary answer, i.e. our reference framework, alongside the allocation of points (Table 11.2).

For each interpretive aspect (Table 11.2, 2nd row) that could be identified in each drawing, the student's construction received one point. Within explanatory functions, the main focus was to track students' ability to provide some kind of a mechanism that can be adapted to drawing as modeling. Thus, here we did not focus on

Table 11.1 Categories used to evaluate drawings in terms of the representative function of a model

Objects depicted in the model	Trait scoring
Plant	0 points: not included in the model
Soil	1 point: object included in the model with some incorrect connections with other objects
Water	
Sun	2 points: included in the model with correct interactions with other objects
Chlorophyll	
Glucose/sugar/starch	
Oxygen	
Carbon dioxide	
Nutrients	
Variables/processes	**Trait scoring**
Energy transfer	0 points: variable not identified
Transport of water/nutrients/starch	1 point: variable identified with some incorrect effects
Growth – Size of the plant	2 points: variable identified correctly with appropriate influences on other variables or objects
Chemical reaction producing glucose	
Role of sunlight in sustaining the chemical reaction	
Gas exchange	

biological correctness or language. We scored responses based on whether they indicated student awareness of a mechanism that can somehow explain the phenomenon. We ignored issues of terminology or language (for example using the term "duct" in place of "vascular tissue") and we refrained from assessing understanding of concepts that were not directly relevant to an underlying mechanism that the student was seeking to describe.

11.3.3 Predictive Function of Students' Drawings

Finally, students' drawings were also coded for their facility to serve a predictive function, the third defining feature of a scientific model. Table 11.3 presents examples of predictions that we would anticipate as possible aspects of students' drawings. In coding for this third aspect, we looked out for any information about a future event or any form of anticipated change and the conditions under which it might emerge. Very few drawings included information that alluded to predictions or the facility to use the drawing/model to make predictions. For this reason, in our coding, we simply recorded the predictive features, where present. For predictive functions, the main focus was to track students' ability to provide some kind of a prediction, hypothesis or educated guess that can be adapted to drawing as a modeling medium. Thus, here we did not focus on biological correctness or language. We scored responses based on whether they indicated some (potential) use of the model to formulate one or more predictions related to future changes in the phenomenon.

Table 11.2 Exemplary version of student response and grading points

Exemplary student response with some misconceptions
Water is transported from soil to the ducts of the root and from the ducts of the root shoots to the leaves, where a chemical reaction takes place with CO_2 absorbed from the atmosphere. This reaction, facilitated by the sunlight that binds to the chlorophyll in the leaves, produces glucose, needed for plant growth, and oxygen, which is released into the environment. Glucose is converted to starch to be stored. At night, when photosynthesis stops, glucose is transferred to other parts of the plant through the ducts of the shoot. If the amount of water is too high, then the leaves will produce excessive amounts of starch and the plant will overheat. Conversely, inadequate amounts of water lead to the production of insufficient glucose and the plant becomes undernourished. The increase in the mass of the plant is due to the carbon left over from the chemical reactions. Mineral nutrients from the soil are dissolved in the water absorbed by roots and transported through ducts to those parts of the plant (trunks, branches, leaves and roots) where the nutrient plays a role in cellular processes and plant growth
Elements of the mechanistic story – explanatory functions (with points allocation):
1. Water absorption: water is transported from soil to the ducts of the root (1)
2. Water transport: water is transported from the ducts of the root shoots to the leaves (1)
3. A chemical reaction takes place in the leaves; water and carbon dioxide are involved (2)
4. CO_2 is absorbed from the atmosphere (1)
5. This reaction is facilitated by sunlight (1)
6. Sunlight binds to chlorophyll in the leaves (1)
7. The reaction in the leaf produces glucose, needed for plant growth, and also releases oxygen into the atmosphere (2)
8. Glucose is converted to starch for storage (2)
9. Glucose transfer to other parts of the plant (1)
10. Glucose transfer takes place at night, when photosynthesis stops, through the ducts of the shoot (2)
11. The increase in the mass of the plant is due to carbon left over from the chemical reactions (2)
12. Nutrients that are in the soil are absorbed by roots and transported through ducts to all parts of the plant (2)
13. Glucose/starch provide the energy needed by the plant. Nutrients play a role in the biological processes inside the plant cells (2)

Table 11.3 Examples of possible predictions that we recorded students alluding to in their drawings

Prediction	Points allocation
If excessive water is available, leaves will produce too much starch and the plant will overheat	2
If inadequate amounts of water are available, the plant will not produce enough glucose and will become undernourished, i.e. it will not have enough energy to grow or support the cellular functions	2
Plants that are facing the sun for more hours are more likely to grow faster	1

Even if a response included biologically incorrect or inaccurate claims – for example "If excessive water is available, leaves will produce too much starch and the plant will overheat", we focused on assigning points for the display of predictive power of the modeling ability, not for biological knowledge. For this particular example, the response did not receive points for representative power (since it is biologically incorrect), but we did deem that the response was indicative of the student's understanding that modeling practice includes the formulation of predictions.

11.4   Results

11.4.1   Analysis of the Students' Conceptions Based on the Questionnaire Data

In the open-ended part of the questionnaire, the first question was dedicated to the explanation of the role of plants in maintaining life on Earth. Most students (61%) received 1 point in this task. Sample answers were *Plants give food; Plants produce oxygen; They clean the air; They decorate rooms; They absorb carbon dioxide.* Two points were obtained by 23% of the respondents. One student who received the most points, i.e. 5, wrote down the following answers: *They produce oxygen; They take in carbon dioxide; They purify the air; They give fruits and vegetables; Some plants have a healing effect.* Three students, i.e. 4% of the respondents, did not receive any points in this task.

In the next question, the students were asked to list plant processes that support life. Furthermore, in this task, students received one point for each written answer. There were two students (3%) who did not provide any answer to this question. 25% of participating students received 4 points. One student got the maximum number of points (8) for vital functions, such as respiration, nutrition, reproduction, growth, excretion, development, movement and receiving stimuli. Students most often mentioned that plants grow (54 responses, 72% of participants). The least frequent answer was that plants react to stimuli: only 17 students (23%) included the response to stimuli in their answers.

The last task in the first part of the study was for students to describe what they thought plants do at night. 49% (36) of the respondents scored 1 point and 23 students (31%) scored 2 points. Two pupils (3% of participants) received 4 points in this task. The most frequent correct answer was that plants grow at night; such an answer was found in 20% (15) of the responses. From the students' responses, the largest group of answers indicates a misconception that plants produce oxygen at night (over 20% of respondents). Only 7% of respondents pointed to breathing and 15 (20%) to growth. 13% (9) of respondents indicated that plants sleep at night.

The coding of students' responses also showed that younger students tended to focus more on phenomenological features of plants and less on life processes. Younger students were more likely to only mention water as a pre-requisite to plant growth. Some also mentioned soil and in fewer cases alluded to its 'richness'. More sophisticated responses became more prevalent in the higher age groups and were more likely to identify water, carbon dioxide and light as necessary factors for plant growth. In grade 7, the process of photosynthesis was identified far more commonly than the process of respiration. Only a minority of the older, grade 10, students could associate plant growth with the dual processes of photosynthesis and respiration and identify the reactants, environmental conditions, and the products for both processes.

11.4.2 Analysis of the Students' Models

Through their drawings, most students demonstrated a good facility to visualize the process of photosynthesis, at least in its most basic elements. Every drawing was based on the student's own knowledge and experience. The students' drawings included rich information representing aspects of the process of photosynthesis, including participating objects, relevant variables and processes. Many drawings included some interpretive information about the workings of the process of photosynthesis and in a small number of cases we could identify connections with the process of cellular respiration and with plant growth. Predicting changes to the process of photosynthesis following specific stimuli, e.g. changes in the environmental conditions (Constantinou et al., 2019), or even changes to plant growth were much rarer. This is a limitation of this study, which we attribute to the preparatory framing activity that preceded the drawing/modelling activity: no reference was made in that activity to the process of making predictions, and their significance in science was not referred to either. In a follow-up study we intend to remedy this, in which case we expect that we would have more evidence to work with in terms of students' ability to use models to formulate predictions. Below we present findings for each of the three aspects of a scientific model, as revealed through our analysis and coding of students' drawings.

The students' drawings revealed the richest information and obtained the highest number of points in the category related to the representative function of models. This indicates that participating students found it easy to represent features of photosynthesis and plant growth. A variety of elements were represented in the students' drawings, such as plants, water, sun, soil, oxygen, carbon dioxide, and nutrients (Fig. 11.1). In addition, the rich phenomenological information represented in students' drawings may reveal that the act of drawing is more conducive to representing aspects of the phenomenon than to expressing explanatory features (Tsivitanidou et al., 2018). An example of a student produced drawing is presented in Fig. 11.1.

She is very fond of biology and grows plants at home. Evaluation of the model: representative function – 12 points (objects: plant, soil, water, sun; variables/processes: plant growth); explanatory function – 1 point (arrows suggesting plant growth); prediction – 0 points.

80% of students' drawings contained explanations about changes in weight expressed in kilograms or arrows, suggesting plant growth is a consequence of the process of photosynthesis. A rare explanation captured in student models was information about reactants and products of the photosynthesis process and their role in

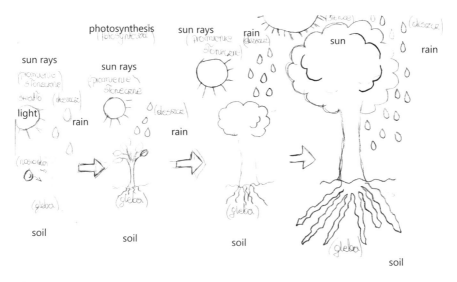

Fig. 11.1 Drawing produced by a student from Grade 7

Table 11.4 List of interpretations given by students, which were used to score the explanatory function of drawings

No.	Explanation
1	Plant size changes as a consequence of photosynthesis (e.g. arrows, weight in kg)
2	Indication of the reactants of photosynthesis (carbon dioxide, water), with information that these are needed for glucose production during photosynthesis
3	Providing the names of the products of the process of photosynthesis (food, glucose, oxygen)
4	Additional explanations: * Uptake of water from the soil through the roots * Uptake of nutrients (ions) from the soil through the roots (Morgan & Connolly, 2013) * Energy from the sun's rays (Papadouris & Constantinou, 2011) * The effect of temperature on the process of photosynthesis
5	Respiration as the process of breaking down glucose/food to release energy that is a necessary pre-requisite for plant growth. Where and when this process takes place. Release of carbon dioxide

supporting plant life. This explanation was found in 10 drawings (approx. 13% of the participating students). This is consistent with the prevalence of these processes in the students' responses to the open-ended questions, an example of triangulation that strengthens the credibility of this finding.

Table 11.4 presents typical categories of actual students' explanations that were scored based on the coding scheme of Table 11.2. The categories are presented in order of increasing complexity.

We identified only two elements that could indicate predictive functions in the student constructed models. These were: (1) Drawings that included the rings of annual increments (growing ring or tree-ring dating) and explaining that a new wood ring arises every year, from which we can know the age of the tree. These students explicitly identified that the older a tree is, the more wood rings it will have. (2) Drawings that included some indication of time (e.g. a clock), and the explanation that the tree increases its mass over time; so, the more time passes, the bigger/heavier the tree will be.

In summary, students' drawings were relatively lacking in the interpretive and especially the predictive functions. The most common interpretation that was identifiable in the drawings (approx. 80%) was the explanation of the change in mass, commonly illustrated with arrows indicating plant growth or expressing mass in kilograms. Only four drawings contained elements that could be identified with the function of prediction. These were comments about the fact that new tree rings grow every year and after cutting down the tree, you can count how old it is. In Fig. 11.2, this idea is illustrated with a drawing of a representation of a cross-section of a tree trunk, illustrating that, over time, the tree grows in weight and the more time passes, the bigger the tree will be (Fig. 11.2). A similar idea was presented in Fig. 11.3, where the student notes that after years, the mass of the tree increases (Fig. 11.3).

She is very fond of biology and she grows plants at home. Evaluation of the model: representative function – 16 points (objects: plant, water, sun, glucose, oxygen, carbon dioxide, nutrients; variables: transport of materials, plant growth, exchange of gases); explanatory function – 4 points (arrows, reactants, products, additional descriptions); prediction – 1 point (information about the annual increment of a new wood growth ring and the possibility of counting how old the tree is).

She is moderately fond of biology and grows plants at home. Evaluation of the drawing: total representative function –10 points (objects: plant, soil, water, sun; variables: plant growth); interpretive function – 1 point (after years, the mass of the tree increases through the help of light); prediction – 0 points.

11.4.3 The Effect of Age and Personal Knowledge on Students' Drawings

In the statistical analysis of the data, a linear regression model (GLM) was developed. The dependent variable used was the sum of the points attributed to each student drawing for all three features of its function as a model. The regression model indicates a moderate relationship between the dependent variable and the predictors, i.e. age of students and the class they attend. Other variables turned out to be statistically insignificant: whether students grow any plants at home: $p = 0.2$; whether students like biology: $p = 0.3$. In addition, the students' gender had no impact on the results obtained in the survey ($p = 0.4$). The coefficient of specificity (r^2) calculated for this study was 0.246. This shows that the regression model

Fig. 11.2 Drawing model made by a student from Grade 10

stipulates that 24% of the variance of the dependent variable is explained by the independent variable at a significance level of $p < 0.01$. This is consistent with the sample size and the exploratory nature of this study.

Overall, our findings show that the richer the resources of personal knowledge revealed through the students' written responses, the more detailed and the more accurate their drawings were ($r = 0.86$, $p < 0.0001$). Based on these findings, we infer that the written text and visual representations complement each other concerning the knowledge and the thinking they reveal about the process of photosynthesis.

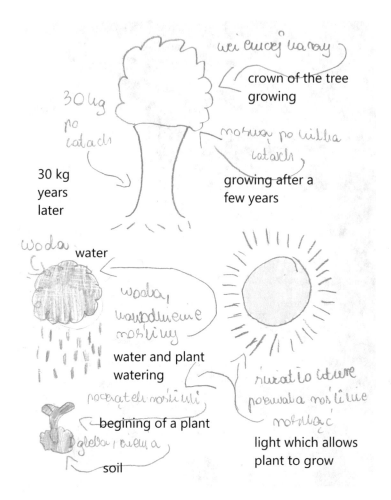

Fig. 11.3 Drawing by a girl in Grade 5

11.5   Discussion

Students' conceptions of the process of photosynthesis and their ability to construct models expressing the representative and interpretive aspects of their understanding of the phenomenon develop with age from upper primary through secondary education. The study confirms that the complexity of photosynthesis presents challenges for students in balancing the understanding of multiple processes and their role in a plant producing its source of energy. In addition, our findings reveal a strong dependence of the ability to model photosynthesis using constructed drawings on a student's knowledge about photosynthesis.

The presented research findings demonstrate that the richer the knowledge resources that were revealed in written verbal expression, the more accurate the students' drawings were, with more relevant details in representing the phenomenon of plant growth and interpreting the underlying mechanisms of photosynthesis and (more rarely) cellular respiration. It can therefore be argued that the verbal and visual communication complement each other. In other words, when students strive to develop and express their understandings, coherent information about their thought processes can be obtained by combining information from both text and graphics that are generated by the students (Lemke, 1998). In furthering our personal knowledge, we elaborate meanings, we manipulate connections between ideas and phenomena, and we express our understandings of these connections by resorting to the use of scientific concepts as epistemological tools. To acquire complete understanding of scientific concepts, we need to gradually develop the facility to move freely and consciously between the verbal channel and the visual, as well as between quantitative reasoning and symbolic mathematical logic, and between operational, localized sense-making and action (Aikens et al., 2021).

We note with interest that the resources revealed by the students verbally in this research study were also reflected in their graphical productions in ways that mostly cohere across age and maturation/educational level.

Moreover, the drawings which were more robust and accurate had more excessive written text in them. Those more complex drawings were rare, and students' explanations were basic and included alternative conceptions. Our findings demonstrate a correlation between students' age and robustness of drawings. This finding has also been observed in the past (Rybska, 2017).

Messig and Groß (2018) propose a novel way of teaching photosynthesis – one that starts the whole process from our anthropocentric point of view, in which we, as heterotrophic organisms, decompose the matter produced by plants so that the plants, from simple inorganic compounds contained in the soil and from the sun, can convert the sun's energy into the energy of the chemical bonds contained in the organic matter they produce. We acknowledge this proposal noting at the same time that, as illustrated in our work also, alternative conceptions tend to be persistent and do not disappear automatically with age or educational maturation. For example, students all the way to upper secondary education sometimes find the role of nutrients absorbed from the soil as difficult to grasp and tend to confound it with the process of 'feeding the plant or providing for its energy needs'. Morgan and Connolly (2013) address this issue and the need to differentiate between the outcomes of photosynthesis and its role in satisfying the need of the plant to have an energy fuel on the one hand, and the role of mineral nutrients absorbed through the soil in sustaining the biological processes that take place in plant cells, including in the release of energy by metabolising glucose.

Our study largely confirms that the act of drawing can serve as a resourceful part of scientific modelling activities. Scientific modelling includes three functions. In the presented research, students were able to fulfil the representational function to a greater extent. Objects and variables relating to the process of photosynthesis appeared in every one of the student-constructed drawings. Working with the inclusion of the other two functions of a scientific model turned out to be a more challenging task. As Upmeier Zu Belzen et al. (2019, p. 157) writes, tips and guidance from the teacher are essential to getting students through the modelling process, considering its three functions. Only with such support and experience can students develop an understanding of scientific models and modelling (e.g. Akerson et al., 2011; Gilbert & Justi, 2016; Schwarz et al., 2009; Windschitl & Thompson, 2006). In the presented research, students were not exposed to any special activities dedicated to developing the practice of scientific modelling, which probably influenced the appearance of a small number of drawings with a predictive level of modelling. Providing a scaffold in the form of a designed question turned out to be insufficient in most cases. As Karnaou et al. (2018) write, the process of constructing and interpreting models includes effective teaching practices that broaden students' understanding of science concepts and scientific modelling skills. Students' interest in working with models in the sciences assumes that models act as a link between theory and the discussed phenomena. In this work, we surmise that such a link was not obvious to many students. The findings demonstrate that to make this connection and to fully access the rich potential afforded by drawing activities as a context for modelling-based learning, a structured preparation would be necessary that would go substantially beyond our framing activity in explicitly developing ideas related to modelling and its emergence from applying theory to phenomena, and its utility in representing, interpreting and predicting aspects of the phenomena under study (Constantinou et al., 2019).

Fulfilling all three requirements (representation, explanation, and prediction) has also been shown in prior research to be a challenging task to achieve for most students (Karnaou et al., 2018; Cheng & Brown, 2015; Krell et al., 2014). Additionally, Krell and co-workers (2014) noticed that despite the global level of understanding of models, or aspect-dependent levels, important elements are students' understanding of aspects, such as the nature of models, multiplicity of models, the purpose of models, model evaluation processes, and changing models. In their research, they noticed that students seem to have a complex and at least partly inconsistent form of understanding of models. Students with high nonverbal intelligence and higher grades (indicating stronger personal knowledge) seem to have a more reliable and more expanded understanding of models and modelling than students with a less robust understanding of the phenomena under study. This study confirms these findings and also illustrates the need for further research before we can claim to fully understand how we can enculturate students into the scientific practice of modelling and have the assessment approaches to know when we do so effectively.

11.6   Conclusions

Our findings demonstrate that participating students have a certain amount of personal knowledge about plant life and the process of photosynthesis, and that modelling through drawing enables students to reveal this knowledge. During the modelling process, students focussed mainly on the representation of phenomeno-logical aspects, including relevant objects and variables. Explanatory details were also present, but more rarely. Where they did exist, they provided richer information about student thinking and understanding. Predictive functions received markedly less attention in the students' drawings, possibly because participants were not prompted to think about dynamic changes in the preparatory framing activity. The findings confirm that the participating students from all levels of the school educational system are not familiar with modelling principles and that there is a strong need for enriching the curriculum and instructional tools and approaches with structured modelling-based learning activities if we are to claim that scientific practices are nurtured in ways that promote learning, autonomous thinking and creativity.

References

Aikens, M. L., Eaton, C. D., & Highlander, H. C. (2021). The case for biocalculus: Improving student understanding of the utility value of mathematics to biology and affect toward mathematics. *CBE—Life Sciences Education, 20*(1), ar5. https://doi.org/10.1187/cbe.20-06-0124

Ainsworth, S., Prain, V., & Tytler, R. (2011). Drawing to learn in science. *Science, 333*(6046), 1096–1097.

Akerson, V. L., White, O., Colak, H., & Pongsanon, K. (2011). Relationships between elementary teachers' conceptions of scientific modeling and the nature of science. In *Models and modeling: Cognitive tools for scientific enquiry* (pp. 221–237). Springer.

Brooks, M. (2009). Drawing, visualisation and young children's exploration of "big ideas". *International Journal of Science Education, 31*(3), 319–341.

Cheng, M. F., & Brown, D. E. (2015). The role of scientific modeling criteria in advancing students' explanatory ideas of magnetism. *Journal of Research in Science Teaching, 52*(8), 1053–1081.

Constantinou, C. P., Nicolaou, C. T., & Papaevripidou, M. (2019). A framework for modeling-based learning, teaching, and assessment. In *Towards a competence-based view on models and modeling in science education* (pp. 39–58). Springer.

Gilbert, J. K., & Justi, R. (2016). *Modelling-based teaching in science education* (Vol. 9). Springer.

Karnaou, P., Tsivitanidou, O., Livitzis, M., Nicolaou, C., & Constantinou, C. (2018). *Practices and challenges in an undergraduate teachers' course: Modeling-based learning and peer assessment in science.* Nova Science Publishers.

Krell, M., Upmeier zu Belzen, A., & Krüger, D. (2014). Students' levels of understanding models and modelling in biology: Global or aspect-dependent? *Research in Science Education, 44*, 109–132.

Lemke, J. L. (1998). Teaching all the languages of science: Words, symbols, images, and actions.

Messig, D., & Groß, J. (2018). Understanding plant nutrition—The genesis of students' conceptions and the implications for teaching photosynthesis. *Education Sciences, 8*(3), 132.

Morgan, J. Á., & Connolly, E. Á. (2013). Plant-soil interactions: Nutrient uptake. *Nature Education Knowledge, 4*(8), 2.

Nicolaou, C. T., & Constantinou, C. P. (2014). Assessment of the modeling competence: A systematic review and synthesis of empirical research. *Educational Research Review, 13*, 52–73.

Papadouris, N., & Constantinou, C. P. (2010). Approaches employed by sixth-graders to compare rival solutions in socio-scientific decision-making tasks. *Learning and Instruction, 20*(3), 225–238.

Papadouris, N., & Constantinou, C. P. (2011). A philosophically informed teaching proposal on the topic of energy for students aged 11–14. *Science & Education, 20*(10), 961–979.

Russell, A. W., Netherwood, G. M. A., & Robinson, S. A. (2004). Photosynthesis in silico. Overcoming the challenges of photosynthesis education using a multimedia CD-ROM. *Bioscience Education, 3*(1), 1–14.

Rybska, E. (2017). *Przyroda w osobistych koncepcjach dziecięcych–implikacje dla jej nauczania z wykorzystaniem rysunku.* Kontekst.

Schwarz, C. V., Reiser, B. J., Davis, E. A., Kenyon, L., Achér, A., Fortus, D., et al. (2009). Developing a learning progression for scientific modeling: Making scientific modeling accessible and meaningful for learners. *Journal of Research in Science Teaching: The Official Journal of the National Association for Research in Science Teaching, 46*(6), 632–654.

Tsivitanidou, O. E., Constantinou, C. P., Labudde, P., Rönnebeck, S., & Ropohl, M. (2018). Reciprocal peer assessment as a learning tool for secondary school students in modeling-based learning. *European Journal of Psychology of Education, 33*, 51–73.

Upmeier zu Belzen, A., van Driel, J., & Krüger, D. (Eds.). (2019). Introducing a framework for modeling competence. In *Towards a competence-based view on models and modeling in science education* (pp. 3–19). Springer Nature.

Windschitl, M., & Thompson, J. (2006). Transcending simple forms of school science investigation: The impact of preservice instruction on teachers' understandings of model-based inquiry. *American Educational Research Journal, 43*(4), 783–835.

Zimmerman, C. (2007). The development of scientific thinking skills in elementary and middle school. *Developmental Review, 27*(2), 172–223.

Chapter 12
Diagnosing of Valuing and Decision-Making Competencies in Biology Lessons

Malte Ternieten and Doris Elster

12.1 Introduction (Problem Definition and Research Questions)

Valuing and decision-making competencies in the context of Education for Sustainable Development (ESD) play an essential role in the national educational standards (NBS) of biology. The students should be prepared to recognize biological facts, evaluate them, and justify their own or third-party judgement. On this basis, they should develop their point of view, considering individual and socially negotiable values. This implies multi-perspective thinking concerning the ESD dimensions: ecology, economy, and social issues (KMK, 2005). However, the diagnosis of these competencies is perceived by teachers as a challenge. Based on empirical data, German teachers perceive the openness of evaluation processes and the associated performance assessment as a difficulty (Alfs, 2012). Therefore, the goal of this study is to develop, test and evaluate a diagnosis grid for valuing and decision-making competencies. In addition, a teaching unit to promote these competencies in heterogeneous classrooms should be promoted.

M. Ternieten (✉) · D. Elster
Institute for Science Education – Biology Education, University of Bremen, Bremen, Germany
e-mail: malte.ternieten@uni-bremen.de; doris.elster@uni-bremen.de

© The Author(s) 2024

K. Korfiatis et al. (eds.), *Shaping the Future of Biological Education Research*, Contributions from Biology Education Research, https://doi.org/10.1007/978-3-031-44792-1_12

The overarching research question (RQ) of this Design-Based-Research (DBR) project is:

(RQ.DBR): *How must a lesson be designed to comply with the conditions of the cooperating school and address the school-specific requirements?*

Explanation: The focus of the paper will be to describe how the diagnostic grid was developed, tested, and evaluated. In this sense, "designing a lesson" means mainly the aspect addressed by the diagnostic grid.

In addition, cycle-specific questions are defined as:

(RQ. Cycle.1): *Which aspects of the design prototype show the most significant potential for development (practical/theoretical)?* (RQ. Cycle.2): *Which criteria must be modified to improve the selected methods for promotion and diagnosis of the decision-making competence (cf. material-based writing/competence grid)?* (RQ. Cycle.3): *How suitable is the modified teaching method (cf. material-based writing) to promote decision-making competencies? How suitable is the developed diagnostic grid (PARS model; short for: "In Partnership with Competences Diagnosing") for diagnosing decision-making competence?*

Explanation: The research question for the first cycle was intentionally formulated in an open-ended way, since at that time (2016–2017) it was not yet clear whether the promotion (and diagnosis) of decision-making competence would be the focus of the research project. From cycle.2 onwards, the methods for promotion were defined, but it was also clear that these would have to be modified to meet the needs of the target group. In cycle.3, the focus was on the important testing of the modified methods in practice.

12.2 Theoretical Framework

12.2.1 Design-Based Research (DBR)

Didactic development research or design research starts from a concrete problem in educational practice. A solution is developed and validated by a theory developed in the context of this problem. A solution usually involves the further development of a method, selection of materials or medium that addresses the problem. To verify this solution, evaluation methods must then be selected and adapted to measure the effectiveness of the solution. DBR does not offer a systematic repertoire of methods, but falls back on the evaluation methods that are common in subject didactics. They can be qualitative or quantitative, depending on the needs of the research design. DBR works in iterative cycles, which is why data are collected and

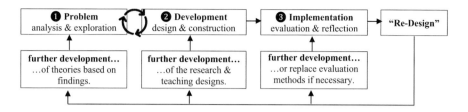

Fig. 12.1 Exemplary DBR cycle based on the Model of McKenney and Reeves (2012). The numbers ❶ → ❸ refer to the chronological order in which a design should be developed, implemented, evaluated, and later revised

evaluated several times and the design is further developed based on the knowledge gained, so that an optimized prototype is available at the end of the development (see Fig. 12.1) (McKenney & Reeves, 2012).

12.2.2 Material-Based Writing

The ability of students to write their own judgements based on existing content knowledge is a competence that is required in many contexts today (Schüler, 2017). Different sources are used for writing these texts. These can be continuous texts, statistics, explanatory graphs, graphs, or images (Abraham et al., 2015). The method of material-supported writing transfers the usual combination of reception and production phases into a new task type (target text) and is thus a demanding form of reading and writing promotion that serves the purpose of (subject matter) learning (see Fig. 12.2) (Phillip, 2017).

Regarding the relevance of material-based writing for education, Abraham et al. (2015) state that the skills required of students to meet the objectives of this method are central to school, extracurricular and future professional activities. However, this alone does not yet legitimize the use of the method in biology lessons from a didactic perspective. Previous studies on the handling of situational writing tasks in biology lessons lead to the conclusion that the writing of subject-specific texts is largely outsourced from the classroom in the form of homework (Thürmann et al., 2015). It could be observed that in biology lessons themselves, only about 6% of the class time is spent on written work phases. In addition, the writing tasks ultimately used are mainly reproductive forms of writing. This also applies to subject-specific text types, such as experimental protocols (Thürmann et al., 2015). From this, it can be concluded that although writing activities are a prerequisite for teaching science lessons, they are hardly ever integrated into subject-specific work (Gebhard et al., 2017). Many of the special topics of biology lessons defined in the curriculum touch on bioethical issues of society, which are of relevance to the students. Material-based writing offers a methodical approach that enables the students to deal with

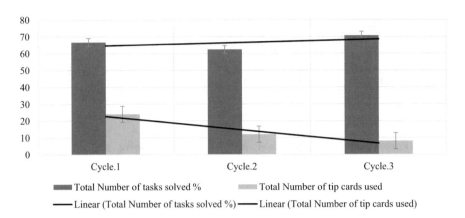

Fig. 12.2 Comparative representation of the total number of tasks solved and tip cards used over the course of cycles 1–3. Since the total number of possible tasks to be solved decreased from cycle 1 → 2, the figures were converted into percentages for the purpose of normalization. This does not apply to the "tip card selection", where the real numbers were taken over unchanged. "Linear" represents the linear trend developments based on the numbers

these topics in depth and independently. This has the potential that students learn to position themselves on controversial socio-scientific issues in a controlled way. In this way, they are gradually introduced to a scientific language and discourse culture (Schüler, 2017).

12.3 Methodology for Data Sources, Collection, and Analysis

The participants of this study are ninth graders of two biology classes and two German language classes. In the biology classes, the course lasted 360 min (for the design see Table. 12.1 – topic 1–3) and in the German language classes 180 min (only topic.3). The procedure in topic.3 always includes the reception phase (collecting arguments), the production phase (writing a statement) and a written survey. In addition, in the biology classes, the subjects were sensitized to the topic beforehand. A total of 181 students took part in the study (Ø age: 14,5). The students' performance was defined by their last report card grade in biology.

The basis for this paper was the analysis of students' written judgements, as well as teacher interviews (Table 12.2). The transcription and subsequent analysis of the collected writing products were carried out by F4 analysis (Version 2.5) based on the method of content-structuring qualitative content analysis according to Mayring (2015). Cohen's Kappa was calculated using the online tool: ReCal2. The Ø coefficient is 0.79. The competence levels of the written judgements of the students were analysed using the diagnostic grid (see Table 12.3).

All judgements were evaluated using the PARS model (see Table 12.3 again). When working with the model, it quickly became apparent that only four out of five categories (cf. **Perspective**, **Scope**, **Knowledge**, and **Solutions**) were suitable for

Table 12.1 Overview of the lesson design (post-cycle.3)

	Ecosystem relationships	90 min
Topic.1	The introduction is problem-oriented with the question of why bog bodies are so well preserved. Starting with the further question of who is responsible for the decomposition of dead organisms, the energy flow, as well as the relationship between consumers, producers and decomposers are worked out by the students. This knowledge will then be used to create a food network	
	The ecosystem swamp	**90 min**
Topic.2	**Topic.1** is taken up again by asking the problem-oriented question of why the decomposers do not "work" in the swamp. This question is then clarified through learning at learning stations. The students learn about biotic and abiotic factors, pH value, and the importance of peat and peat moss. The students present the results of the individual learning stations	
	Man's relationship with the swamp	**90 min**
Topic.3	At the end of the unit, **topic.2** is taken up again. The students should now develop an idea of the influence humans have on the swamp. Here the method of **material-based writing** is used. Based on the question "*Should the swamp be protected?*", the students work out arguments for or against the preservation of the bog from various information materials and then write a well-founded judgement about their personal decision	

Table 12.2 Overview of the data used in this paper

	Learning products	Interviews	Other data
Cycle.1		Teacher – interview (1)	Process log (1)
Cycle.2	Judgements (28)	Teacher – interview (2)	Process log (2)
Cycle.3	Judgements (88)	Teacher – interview (4)	Process log (2)

Notice: The process log is a basic quantitative method in the form of a table documenting how many tasks were completed per student, and how often the tip cards were used

evaluation, as the category **Values** was not explicitly part of the task, and was therefore not designated in the judgement. In this way, the use of this category was only possible if the evaluator had knowledge of the authors' value system. The teachers were involved in the evaluation through the interviews (in cycle.3). They were asked to evaluate two previously selected judgements with the PARS model before the interviews. Care was taken to ensure that the two judgements differed in quality to create a contrast. This was to ensure that the differences in quality were also independently confirmed. The actual interviews then dealt with the evaluation as well as the work with the model and possible suggestions for improvement.

12.3.1 Diagnostic Grid (PARS-Model)

Based on the results of the cycle.2, it has become clear that there is a lack of judgement competence grids that are practicable in the everyday work of a teacher and context-specific concerning the learning products.

Table 12.3 PARS-model – diagnostic grid – characterization of the quality of argumentation structures in the context of ESD science discussions in biology lessons (post-cycle.3)

	Level.0 Unstructured response	Level.1 Simple answer (Use of multiple terms)	Level.2 Multi-layered answer (Linking terms to a concept)	Level.3 Coherent response (Link multiple concepts)	Level.4 Abstract response (Meta-reflection of concepts)
Perspective					
Are the situation or the views of the involved parties considered from all different ESD perspectives?	...perceives no difference in perspective.	...**describes** the situation from one perspective.	...**links** two perspectives.	...**links** all three perspectives together.	...**reflects** controversies in the problem and puts the various assertions into perspective from the perspective of the actors.
Scope					
Are changes/ consequences in spatial or temporal dimensions recalled?	...mentions no consequences.	...**describes** only one consequence in time and/or space.	...**weighs** different consequences in time and/or space.	...**links** different consequences and dimensions and questions the sustainability of decisions.	...**reflects** spatial and temporal interactions of varying degrees.
Knowledge					
How is the handling of reference knowledge structured?	... does not refer to the assignment.	...**uses** superficial knowledge from the materials.	...**links** almost completely the acquired knowledge with own previous experiences.	...**discusses** differences between personal and acquired knowledge.	...**reflects** on and weighs the importance of the different forms of knowledge.
Value					
Is there an awareness of the underlying individual and collective values?	...does not consider values.	...**is aware of** the presence of values.	...**distinguishes** between different values (own and foreign values).	...**describes** possible conflicts between different values.	...**reflects** the meaning of the values employed from a personal and collective perspective.
Solutions					
Does the solution consider the relationship between special and collective interests?	...gives no solution.	...**describes** a pre-existing solution for one party.	...**describes** an already existing solution for two different parties.	...**develops** a new solution for two different parties.	...**reflects** and weighs up different solutions between the actors.

First published in Ternieten and Elster (2020)

In 2014, Christenson and Chang Rundgren (2015) presented a framework for evaluating argumentation structures in the context of socio-scientific issues, which aims to capture both the content of a learning product and the argumentation structure. The aim is to provide teachers with a tool to identify quality-producing and defining features of arguments in the context of problem-oriented judgement. Against this background, a grid was developed that additionally includes further sub-elements from other models to take the ESD perspective relevant to biology lessons more strongly into account. The basis of this grid is the SEE-SEP model, which relates the thematic complexes: sociology/culture (sociology/culture; S), environment (environment; E), economy (economy; E), science (science, S) ethics/morality (ethics/morality; E) and politics (politics; P) to the personal aspects of knowledge, values, and personal experiences. In addition, the SOLO classification was used to differentiate the levels (Biggs & Collis, 1982). The claim of complexity reduction is expressed above all in the approach postulated by Christenson and Chang Rundgren (2015) of a concrete judgement or evaluation and the associated justification, whereby both pro and contra arguments are considered. The arguments can consist of value-based, as well as technical, arguments. Value-based arguments indicate how something should or may be, i.e., they are normative. The technical arguments are mainly descriptive. They are therefore descriptive in character. In the PARS model, the value-based arguments are considered, but unlike the SEE-SEP model, they are not equally weighted and are only represented in one category (see Table 12.3 – Values). The technical arguments are taken into account to a greater extent than if an entire category is dedicated to the ESD aspect mentioned above (cf. perspective ESD) and, according to the Göttingen model of evaluation competence, spatial dimensions [local/global], as well as the time dimension (short- /long-term consequences), are considered in a separate category (cf.) In addition, the category of solution finding was taken over from the Göttingen model. The latter two categories had to be modified because the original grid did not take as its basis the specific characteristics of the construct of judgement competence described in the German educational standards for biology (KMK, 2005). However, since the approach to socio-scientific issues is comparable, if not identical, to judgement competence, only slight modifications were necessary (Hostenbach et al., 2011). For application in everyday school practice, an additional operationalization according to the curriculum of the federal state of Bremen was carried out for the secondary schools, considering the increasing level of difficulty (cf. Level.0 → Level.4) of operators with a higher level of difficulty (see Table 12.3 again).

12.4 Results and Discussion

The following three sub-chapters are structured based on the questions mentioned at the beginning, with the three cycle-specific questions being answered first and the core question in the conclusion.

12.4.1 Results: Cycle.1

> (RQ.): Which aspects of the design prototype show the most significant potential for development (practical/theoretical)?

The results show that the first prototype of the lesson design only partially worked in the given framework conditions of the cooperation school and that there is a lot of potential for improvement.

> "(...) And that discouraged and that's why it didn't work and because the methods built on each other the next one didn't work either and that's why the discussion was heavily moderated. (...) The students told me in feedback to the (.) about these methods, the reason why they didn't adopt them is that it would have required text work, so the implementation."
> **Source:** Interview.1_Teacher.1_Cycle.1_Paragraph.341 (Translation: German → English)

A more general criticism that applied to the entire prototype was that the information texts used were simply too extensive and complex to be processed in the time provided. The biggest problem, however, was the third part of the lesson, since the promotion of decision-making competence was simply not successful. The central problem was that the transfer of information (arguments) from the used educational film to the list of arguments did not take place, which is why the subsequent steps of classification into pro/con protection and the discussion in plenary did not work. Regarding the research design, there was also the problem that theoretically only the argument lists were available as a data source and that, for example, a video of the discussion did not take place. There was also a lack of a suitable tool to analyze the quality of this discussion in relation to ESD.

On the practical side, accessibility must be increased through simpler texts and more explanatory illustrations, and the difficulty of the lesson must be significantly reduced. On the theoretical side, the promotion of decision-making competence in the context of the chosen topic shows the greatest potential for promotion. This is also confirmed by the teachers interviewed in the first cycle.

> "(...) So, in relation to the ecosystem, I think decision-making competence is very, very important, but evaluation is not possible without content. Basically, you must have worked out something properly to present the problem, (...) you can only love something that you also know. (...)"
> **Source:** Interview.1_Teacher.1_Cycle.1_Paragraph.114 (Translation: German → English)

12.4.2   Results: Cycle.2

(RQ.): Which criteria must be modified to improve the selected methods for promotion and diagnosis of the decision-making competence (cf. material-based writing/competence grid)?

Compared to the first cycle, the revised lesson worked better, which is reflected, among other things, by the fact that the number of tip cards used has decreased (see Fig. 12.2). In particular, the reduction of the amount of text and the accompanying reduction of technical terms, as well as the replacement of text content with explanatory illustrations, seems to have improved the processing behavior. Here, however, much more needs to be reduced or simplified for low-performing students with German as a second language. The tip cards worked better after the revision in the sense that there were fewer comprehension problems. However, they were used less often, which is an indication that the need for this form of assistance was not quite as high.

The promotion of decision-making skills through the method of material-based writing in the third part of the lesson was successful. The students worked out arguments from information material and arranged them in a list (of arguments). In a reasoned judgement on the question of whether the swamp should be protected, the use of these arguments works. However, it was problematic that the task posed enormous problems, especially for the low-performing students, as the formulation of the judgements requires linguistic skills that were not yet present in this target group. A further problem is the focus of the diagnostic grid used on precisely these linguistic skills.

Regarding its use in biology lessons, the focus of the grid in cycle.3 should be shifted to the subject-specific ESD aspects and the task for formulating a judgement should be better supported.

12.4.3   Results: Cycle.3

(RQ.): How suitable is the modified teaching method (cf. material-based writing) to promote decision-making competence? How suitable is the developed competence grid (PARS model) for diagnosing decision-making competence?

The learning effectiveness of the design could be further improved compared to cycle.3. The triple differentiated working materials (differentiated according to the reduced amount of text, shorter sentences, fewer technical terms, and more

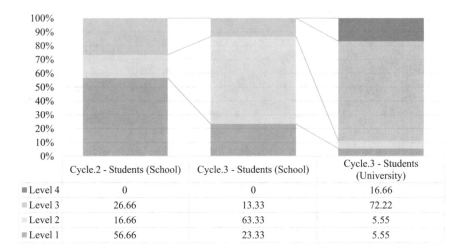

	Cycle.2 - Students (School)	Cycle.3 - Students (School)	Cycle.3 - Students (University)
■ Level 4	0	0	16.66
■ Level 3	26.66	13.33	72.22
Level 2	16.66	63.33	5.55
■ Level 1	56.66	23.33	5.55

Fig. 12.3 Comparative presentation of the levels achieved between the students of cycle.2 and cycle.3. Contrast this with the levels achieved by the expectably higher-performing university students. Since the number of participants differed between the various groups, the values were converted into percentages for normalization. Levels 1–4 are derived from the average scores of the four categories (cf. **perspective, scope, knowledge,** and **solutions**) achieved by each student

explanatory illustrations), as well as the possibility for the students to choose them independently, have further increased the processing behavior, especially of the students who tend to be low performing. The tip cards were used even less frequently compared to the previous cycles, which is seen as a positive sign that the need for such support has decreased, even though it is still there (see Fig. 12.2 again).

The third part of the lesson works better with the revised methods of material-based writing compared to the second cycle. Through the support of judgement formation by a guideline, as well as the use of partner work, even low-performing students can be led to write a reasoned judgement in the ESD context. The results show that the students always argue in favor of swamp protection, whereby the ecological and economic aspects are particularly important to them. The social aspect only plays a minor role, and its relevance seems to be linked to the students' personal experiences. In about half of the cases, the approaches to solutions are compensatory: the students mostly try to offer compromises between the different parties of interest. Most of the students think in very long-term time frames and global, spatial dimensions. They perceive swamp protection as a task for society (Fig. 12.3).

12.5 Conclusions

(RQ.): How must a lesson be designed to comply with the conditions of the cooperating school and address the school-specific requirements?

To measure the (learning) efficiency over the course of the 3 cycles using the designated designs, five markers have been defined (repetitive data acquisition in each cycle at predefined points). These markers are (I) the process log (a competency grid to record the progress of individual students), (II) the complexity of the food web, (III) the quality of the decision, (IV) the lines of argumentation, and (V) the revision of the PARS-model. The overall results show that the advancement of students in highly heterogeneous groups can be successful if the education material is differentiated on multiple levels. These materials must include suitable language (simplified language with reduced use of specific terminology) and a higher percentage of descriptive illustrations than complex text passages. Further support for students can be provided by taking on preparatory tasks, considering time as an additional crucial differentiation factor. The study shows that low-performing students can successfully produce a qualified opinion that meets the requirements of the ESD context by using supporting guidelines and/or partner work. The results also show that the students always argue in favour of swamp protection, whereby they describe ecological and economic aspects as the most important ones. However, students with personal experiences regarding swamps assign social aspects as another critical factor, unlike students without personal experiences. Approximately half of the students try to use compromises to negotiate between the interested parties. Most of the students consider extended periods and global dimensions and classify swamp protection as a challenge to be addressed by the whole society. To further improve the developed materials and methods of the education units, it is recommended that the use of specific terminology in the evaluation process of the opinions be implemented. Furthermore, the linguistically improved materials should be integrated into a more extended teaching lesson on an ESD-approved topic. This should guarantee a continuous improvement of the pre-existing lesson. The continuous improvement of how lessons are taught is of significant importance regarding differentiated educational materials for Oberschulen (high schools) in Bremen, in the field of biology and science. For future projects to further improve the PARS model, it is recommended using additional empirical studies with a higher number of test persons and different school types, such as grammar schools. What remains is the problem described at the beginning with the category: **values**, which requires an understanding of the students' values and can therefore only be assessed by teachers who have knowledge of these values. Level.4 of the PARS model was only achieved by university students. This was not surprising given the requirements, but for the purposes of clarity, it should be mentioned that the task for ninth graders did not require them to reflect on the topic, and level.4 thereby set requirements that the students were not supposed to achieve at all. A further development of the PARS model could therefore be to reduce the requirements at this level or use the model only with level.3 or to revise the assignment for the students regarding the ability to reflect. Overall, the exchange within a community of practice, including science educators, to create a lesson that fulfils the school-specific requirements of an integrated comprehensive school, was a success.

References

Abraham, U., Baurmann, J., & Feilke, H. (2015). Materialgestütztes Schreiben. *Praxis Deutsch – Zeitschrift für den Deutschunterricht, 251*(42), 4–11.

Alfs, N. (2012). *Ethisches Bewerten fördern. Eine qualitative Untersuchung zum fachdidaktischen Wissen von Biologielehrkräften zum Kompetenzbereich „Bewerten".* Hamburg: Verlag Dr. Kovač.

Biggs, J., & Collis, K. (1982). *Evaluating the quality of learning the SOLO taxonomy.* Academic Press.

Christenson, N., & Chang Rundgren, S. N. (2015). A framework for teachers' assessment of socio-scientific argumentation: An example using the GMO issue. *Journal of Biological Education, 2*(49), 204–212.

Gebhard, U., Höttecke, D., & Rehm, M. (2017). *Pädagogik der Naturwissenschaften. Ein Studienbuch.* Springer VS.

Hostenbach, J., Fischer, H. E., Kauertz, A., Mayer, J., Sumfleth, E., & Walpuski, M. (2011). Modellierung der Bewertungskompetenz in den Naturwissenschaften zur Evaluation der Nationalen Bildungsstandards. *Zeitschrift für Didaktik der Naturwissenschaften, 17*(1), 261–288.

Kultusministerkonferenz – KMK. (2005). *Bildungsstandards im Fach Biologie für den Mittleren Schulabschluss.* Wolter Kluwer.

Mayring, P. (2015). *Qualitative Inhaltsanalyse. Grundlagen und Techniken.* Beltz.

McKenney, S., & Reeves, T. (2012). *Conducting educational design research.* Routledge.

Phillip, M. (2017). *Materialgestütztes Schreiben.* Beltz Juventa.

Schüler, L. (2017). *Materialgestütztes Schreiben argumentierender Texte. Untersuchungen zu einem neuen wissenschaftspropädeutischen Aufgabentyp in der Oberstufe.* Schneider Hohengehren.

Ternieten, M., & Elster, D. (2020). Diagnosis and promotion of decision-making competence of students with methods of internal differentiation in biology lessons. In L. G. Chova, A. López Martínez, & I. Candel Torres (Eds.), *EDULEARN20 proceedings. 12th international conference on education and new learning technologies, Palma, Spain. 6–7 July 2020* (pp. 1413–1422). IATED Academy. https://doi.org/10.21125/edulearn.2020

Thürmann, E., Pertzel, E., & Schütte, A. U. (2015). Der schlafende Riese: Versuch eines Weckrufs zum Schreiben im Fachunterricht. In S. Schmölzer-Eibinger & E. Thürmann (Eds.), *Schreiben als Medium des Lernens: Kompetenzentwicklung durch Schreiben im Fachunterricht* (pp. 17–45). Waxmann Verlag.

Chapter 13
Factors Influencing the Intention of Students in Regard to Stem Cell Donation for Leukemia Patients: A Comparison of Non-intenders and Intenders

Julia Holzer and Doris Elster

13.1 Theoretical Background

Leukemia includes cancers of the hematopoietic and lymphatic systems. The only curative treatment for leukemia is stem cell transplantation. However, there is still a high demand for new and fitting donors. A healthy and tissue-compatible donor is required for stem cell donation. Despite around 80% of patients being able to find a close match, the demand for new donors is still very high (Hochhaus et al., 2004; ZKRD, 2018, 2021).

In order to explain why some people decide themselves to become donors or not, the Theory of Planned Behavior (TPB) can be used (Ajzen, 2005). TPB assumes that the proximal cause of behavior is one's intention to adopt the action in question. Intention can be defined as an indicator of how hard a person is willing to try in order to perform a given behavior (Frey et al., 2001). Therefore, intention includes motivational factors. Intention is influenced by three determinants (Fig. 13.1), which are *attitudes* towards the behavior, *subjective norms* and *perceived behavioral control* (Eagly & Chaiken, 1993). The attitudes are defined as affective evaluations which show how likely or unlikely an individual is to perform the behavior in question. Subjective norms represent a person's belief that significant others (people whose opinion is valued as important) think that they should adopt a certain behavior. It represents the social pressures on individuals regarding the performance of a given behavior. Finally, perceived behavioral control can be defined as the individual's perception of how easy or difficult the performance of behavior is (Ajzen, 1991; Ajzen & Fishbein, 2002). According to the TPB, it can be assumed that the

J. Holzer (✉) · D. Elster
Institute for Science Education – Biology Education, University of Bremen,
Bremen, Germany
e-mail: doris.elster@uni-bremen.de

© The Author(s) 2024

K. Korfiatis et al. (eds.), *Shaping the Future of Biological Education Research*,
Contributions from Biology Education Research,
https://doi.org/10.1007/978-3-031-44792-1_13

177

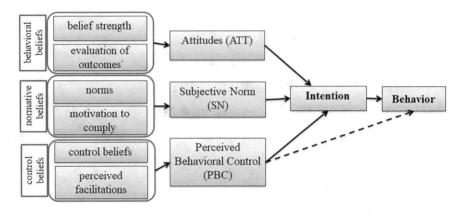

Fig. 13.1 TPB-model. (Based on Ajzen, 2005)

more positive the attitudes, subjective norm and perceived behavioral control, the stronger the intention should be and thus the more likely the conduct of the behavior (Ajzen, 2005). These three determinants of intention can be traced to specific sets of beliefs, referred to as *behavioral beliefs, normative beliefs* and *control beliefs* (Fig. 13.1): *"These beliefs provide us with insight into the underlying cognitive foundation of the behavior, i.e., they tell us why people hold certain attitudes, subjective norms, and perceptions of behavioral control and, therefore, why they intend to perform the behavior in question"* (Ajzen, 2005, p. 137).

According to the TPB, changing behavior requires changing or modifying these underlying beliefs using the fact that beliefs are the fundamental determinants of any behavior.

Ajzen (2011) has admitted the potential for other constructs to be added to the TPB to enhance its prediction of intention or behavior. As an example, studies on blood donation have identified that three variables are particularly influential in intention prediction, namely *knowledge* (Holdershaw et al., 2007), *moral obligation* (Schwartz & Howard, 1981) and *self-identity* (Conner & Armitage, 1998). These findings have been partially confirmed by a further empirical investigation in the context of this study: In order to identify influential predictors of intention to register as a stem cell donor, a regression analysis was performed. It became clear that above all, *knowledge* and *moral obligation* are factors that determine intention (for more detailed information, see Holzer & Elster, 2021).

Furthermore, in this study two further variables, namely *empathy* and *moral reasoning*, are investigated, as they are positively related to pro-social behavior (see e.g. studies of Bierhoff, 2010; Emler & Rushton, 1974; Krebs & Rosenwald, 1977; Demir & Kumkale, 2013). These five variables are presented in more detail below.

Moral obligation can be described as an individual's perception of the moral correctness or incorrectness of performing a behavior and represented as *"feelings of personal responsibility or duty to perform a behavior"* (Ajzen, 1991, p. 199;

Schwartz & Howard, 1981). *Self-identity* is a construct that reflects the different identities of a person. In the context of stem cell donation in this study, this construct reflects the extent to which a person considers themselves as a *helper* (self-identity as a helper) (Stryker, 1968, 1997). *Moral reasoning* involves cognitive processes in which action decisions are made based on moral considerations (Richardson, 2018). Such decisions are not under the influence of individual moral standards (what individuals prefer to do), but are governed by what is expected of them in the face of moral standards. *Empathy* has been extensively studied in social psychology. The perception of someone else's need or suffering tends to create emotional distress in the viewer. The experience of this arousal is called empathy, and it is believed that it comes from a learned or acquired or genetic competence to look at events from the perspective of those to whom one feels similarly (Schwartz, 1977).

The present study not only expands the norm component by additionally considering *moral obligation* and the variable *self-identity*, but also examines *knowledge* regarding stem cell donation and leukemia. For example, Feeley (2007) found from the analysis of about 27 studies that students have an insufficient level of knowledge regarding tissue and organ donation. However, research shows that knowledge is positively related to other key factors, such as attitudes and intention (Bilgel et al., 2006; Horton & Horton, 1991; Morgan & Miller, 2002; Rubens et al., 1998). Kaiser and Frick (2002) and Kaiser and Fuhrer (2003) distinguish three forms of knowledge, namely *system knowledge*, *action-related knowledge* and *effectiveness knowledge*. Even if the differentiation of the three dimensions of knowledge could not always be confirmed empirically (cf. Kaiser & Frick, 2002), the differentiation on the theoretical level nevertheless appears to be meaningful. *System knowledge* is comparable to the constructs of conceptual and situational knowledge (Zeyer, 2012). *Action-related knowledge* comprises knowledge about possible options for action and provides an assessment of whether corresponding actions can be carried out and with which "costs" they are associated (Frick et al., 2004). The third form of knowledge is *effectiveness knowledge*, which indicates how effective an action option can be, and whether it is ultimately profitable to bear the costs associated with performing the action (Kaiser & Fuhrer, 2003). Therefore, *effectiveness knowledge* describes the potential of a certain action or the relative potential of different actions (Kaiser & Frick, 2002). It has been empirically shown that *system knowledge* predicts *action-related knowledge* and *effectiveness knowledge*, whereby *action-related knowledge* predicts *effectiveness knowledge* and the last two types of knowledge finally predict behavior (Frick et al., 2004).

Since the presented constructs (*moral obligation, moral reasoning, knowledge, self-identity, empathy*) turned out to be useful extensions of the TPB, especially in the context of blood and organ donation, according to the results of the literature research, they are also examined in this study alongside the classic constructs of the TPB. Within this study, such an extended TPB is named as the TPB+ model.

13.2   Research Design and Methodology

According to TPB, anything that changes beliefs in the appropriate direction will increase the likelihood of behavioral change. However, also focusing on predictive factors of intention regarding stem cell donation can affect behavioral change positively (Ajzen, 2006). For this purpose, the overall objective of this study is to investigate changes of factors of the extended version of TPB (including the model-external variables *moral obligation, self-identity as helper, knowledge, empathy, moral reasoning*) among a sample of students.

A questionnaire was distributed before and after the teaching unit "Wake up" to the student sample (age ⌀: 17.14; n = 48♀/n = 46 ♂). The sample was split by the median of "intention", which was 4.0 on the 7-point Likert scale. All students with "intention" equal or lower than 4.0 were defined as *non-intenders* (persons with low level of intention) and all students with higher values than 4.0 in their intention were defined as *intenders* (persons with high level of intention to act as a donor). All of them participated as students in the teaching unit "Wake up". T-tests were performed to assess post-intervention changes in the measured constructs. This is a so-called "paired t-test", since the changes were tested among *intenders* and *non-intenders*.

The same questionnaire was used in the pre- and post-test. Measurement of TPB (Table 13.1) follows closely the guidelines recommended by Ajzen (2005). Thereby, the TPB constructs were formulated by the authors, but the correspondence principle (degree of specificity of the items) according to Ajzen (2005) was observed in the formulation of the items, as well as the 7-point Likert scales. When looking at the beliefs, Ajzen (2005) distinguishes between the expectation component, i.e., the expectation or probability that a certain belief can become true, and the evaluation component, i.e., the evaluation of this belief. Both components have to be multiplied together and product terms have to be formed and interpreted in the course of the data analysis (Ajzen, 2005). Therefore, product terms are also obtained in this study. Thus, in Table 13.1 belief-based-constructs of TPB are product terms (expectation component × evaluation component). Regarding behavioral beliefs, the present study distinguishes between *belief-based positive attitudes* (positive behavioral beliefs) and *belief-based negative attitudes* (negative behavioral beliefs) based on the empirical investigation.

Table 13.1 provides an overview of the TPB variables, model external variables, Cronbach's alpha values of scales, and examples of items. The construct knowledge includes two forms of knowledge, namely *system knowledge* and *action-related knowledge*. Due to time limitations, *effectiveness knowledge* was not assessed in the questionnaire. *System knowledge* comprises knowledge about normal blood formation on the one hand, and knowledge about its malfunction, which is discussed in this study in the context of leukemia, on the other hand. *Action-related knowledge* includes relevant knowledge that is important for the action or registration as a stem cell donor or for stem cell donation. It should be emphasized that *system knowledge* and *action-related knowledge* are not considered separately, but under one construct, since the constructs were not empirically confirmed as separate constructs in

Table 13.1 Factors and examples of TPB, and model external variables and the reliability of scales

Factor (number of Items)	Items (examples only for expectation component)	Mean (post)	SD (post)	post- α
Belief-based (positive) attitudes (4)	I could possibly save a life	40.53[a]	8.01[a]	.750
Belief-based (negative) attitudes (4)	I could harm my health through a possible donation	27.79[a]	10.99[a]	.647
Belief-based subjective norms (5)	Do your parents think that you should register as a stem cell donor?	18.79[a]	8.72[a]	.892
Belief-based perceived behavioral control (6)	I will have a very bad conscience for a very long time if I do not register as a stem cell donor	28.64[a]	8.53[a]	.707
Intention (2)	How likely do you think you will register as a stem cell donor for leukemia patients in the foreseeable future?	4.47	1.55	.922
Moral obligation (4)	I feel morally responsible to support the leukemia patients by registering as a potential stem cell donor	4.93	1.29	.859
Moral reasoning (5)	I choose a course of action that considers the rights of all people involved	5.08	0.86	.795
Self-identity helper (5)	I see myself as someone who likes to help other people	5.57	0.96	.863
Empathy (5)	I sometimes find it difficult to see things from the "other person's" point of view	5.27	0.77	.567
Content knowledge1 (9) (dichotomous scale)	Stem cells are ordinary body cells. Body cells are cells that do not pass on their genetic material to the next generation	18.48	1.98	.618
Content knowledge2 (5) (open questions)	What is a stem cell transplant? [Cohen's Kappa (Ø):,935]	7.16	2.74	–

Here, for better clarity, only the reliability of the expectation component of the belief-based TPB constructs in post-test is presented. However, the evaluation component showed similarly high reliability

[a]The mean values and standard deviations of the multiplication terms (expectation component × evaluation component) are shown here

this study. In Table 13.1, the knowledge that was collected using open and closed questions is considered separately in order to clearly present the calculated quality criteria. However, closed and open knowledge questions measure both types of knowledge, namely *system knowledge* and *action-related knowledge*. With regard to the open and closed knowledge questions, it should be mentioned that while the closed knowledge questions were coded with 0 (if wrong answer) and 1 (if right answer), the open knowledge questions were analyzed using scaling structuring based on the qualitative content analysis (Mayring, 2015). Different points were assigned depending on the different characteristics of an answer. To ensure objectivity, 25% of the data was coded by another coder and a Cohen's Kappa was calculated (Table 13.1).

13.3   Teaching Unit "Wake Up: Sensitization for Stem Cell Donation"

During the 5-h-long intervention "Wake up", the students gathered information and participated in discussions relevant to leukemia and stem cell donation. Table 13.2 summarizes the schedule and topics of the teaching intervention "Wake up", as well as all promoted constructs of TPB+.

Table 13.2 Teaching unit "wake up"

Lecture phase		Topics	TPB+ constructs
Introductory phase (approx. 40 min)	Interactive introduction: "Stem cells are special cells!" Jonas (sick with leukemia) tells his story	Blood formation Stem cells The role of stem cells in medicine and scientific research	Cognitive level of TPB (beliefs) Knowledge Empathy
Elaboration phase: learning at different stations (approx. 160 min)	**Station 1:** What is leukemia?	Causes and consequences of leukemia: Analysis and evaluation of blood Analysis and evaluation of genetic aberration	Cognitive level of TPB (beliefs) Knowledge Empathy Moral reasoning Moral obligation
	Station 2: Stem cell donation – what is behind it?	Accessible sources of adult stem cells in humans Types of transplants Side effects and risks	
	Station 3: From HLA-typing to transplantation	Understanding of HLA-typing Important steps before stem cell transplantation	
	Station 4: Reports of one's experiences	Reading and reflection	Empathy Moral reasoning Moral obligation Self-identity as a helper
Evaluation phase (approx. 45 min.)	Work in groups: discussion	Different opinions about stem cell donation	Cognitive level of TPB (beliefs) Knowledge Empathy Moral reasoning Moral obligation Self-identity as a helper

The teaching unit "Wake up" comprises three teaching phases, namely the introductory phase (confrontation with the problem), the elaboration phase and the evaluation phase (Table 13.2). The "problem" – personified by the figure of the adolescent, Jonas, who is diagnosed with leukemia – is presented immediately after the short introduction "Stem cells are special cells". The students are then asked to write down possible questions that a 17-year-old boy might ask himself. In the following elaboration phase, Jonas's questions are answered. The elaboration phase is divided into four stations, in which the clinical picture of leukemia is treated (station 1) and the aspects of stem cell donation are elaborated (station 2–4). In the evaluation phase, the central results of the teaching unit are discussed, as well as different points of view regarding stem cell donation. In the course of the intervention, not only are many misconceptions and beliefs regarding stem cell donation and leukemia discussed, but also normative and moral aspects are reflected (Table 13.2). In addition, it should be noted that the intention to become a stem cell donor is not directly promoted by the "Wake-up" teaching unit, but, above all, is indirectly influenced by reflective discussions in the final evaluation phase. The goal of the teaching is to raise awareness about the problems experienced by people with leukemia. It is important to emphasize that there is no intention to exert any influence on the participants in either direction.

13.4   Findings

In order to analyze the changes in all assessed TPB+ factors after the intervention, especially among *intender* and *non–intenders*, t-tests were performed. While *positive attitude-related beliefs* and *normative beliefs* remain stable on the level of "whole sample" in the post-test (Table 13.3), *negative attitude-related beliefs* decrease and *control beliefs* increase significantly in the post-test. In the following, the results regarding *belief-based negative attitude* and *belief-based perceived behavioral control* are reported by taking a closer look at beliefs among *intenders* and *non-intenders*. Negative belief-based attitudes include beliefs such as fear of pain and damage to health when stem cells are donated. In addition, there is a belief regarding the state of health, which reflects the perception of whether this is considered suitable for carrying out a stem cell donation. The last one expressed the level of trust in the medical system in the field of stem cell donation (Table 13.4). It can be concluded that all *negative attitude-related beliefs* significantly decrease in the post-test, especially with higher effect sizes among *non-intenders* (Table 13.4).

After the intervention, it was observed that among the control beliefs, three out of the six assessed beliefs underwent changes (see Table 13.1). The belief "perception of effort" describes if students perceive effort regarding stem cell donation. "Awareness of opportunity" shows if participants see an opportunity in their daily life to register as a stem cell donor and donate their cells. The last belief "time consuming" expresses if students perceive the process of registration and the donation

Table 13.3 Results of the t-test: changes in TPB-factors in whole sample (n = 94)

TPB-constructs		M	SD	T	df	p (2-sided)	Cohens d
Belief-based positive attitude (n = 94)	Pre	40.45	7.57	−.141	93	.889	–
	Post	40.53	8.02				
Belief-based negative attitude (n = 92)	Pre	19.55	9.23	−7.390	91	.000	.807
	Post	27.79	10.99				
Belief-based subjective norm (n = 75)	Pre	18.08	7.89	−.943	74	.349	–
	Post	18.79	8.72				
Belief-based perceived behavioral control (n = 88)	Pre	25.97	6.45	−4.124	87	.000	.339
	Post	28.64	8.53				
Intention (n = 93)	Pre	3.68	1.55	−6036	92	.000	.506
	Post	4.47	1.54				

Belief-based attitude, belief-based subjective norm and belief-based perceived behavioral control are averaged from all measured behavioral, normative, and control beliefs. Due to the inverting of the polarity of the negative items during data processing, the increase in belief-based negative attitude are to be interpreted as it decreases

Table 13.4 Results of the t-tests: changes in negative attitude-related beliefs in subgroups

Subgroup			M	SD	T	df	p (2- sided)	Cohens d
No fear of pain	Non-intender (n = 57)	Pre	14.61	11.51	−3.816	56	.000	.622
		Post	22.49	13.69				
	Intender (n = 36)	Pre	16.78	11.31	−3.640	35	.001	.522
		Post	23.44	13.74				
No harm to health	Non-intender (n = 57)	Pre	18.32	11.67	−5.834	56	.000	.986
		Post	31.25	14.36				
	Intender (n = 35)	Pre	22.09	12.46	−3.563	34	.001	.614
		Post	30.66	15.10				
Perceived adequate health status	Non-intender (n = 57)	Pre	19.54	10.78	−3.188	56	.002	.494
		Post	25.14	11.83				
	Intender (n = 36)	Pre	22.39	11.94	−3.575	35	.001	.586
		Post	29.92	13.59				
Trust in the medical system	Non-intender (n = 57)	Pre	21.04	11.09	−5.022	56	.000	.720
		Post	30.18	13.91				
	Intender (n = 35)	Pre	25.51	13.18	−4.379	34	.000	.600
		Post	33.23	12.45				

Due to the inverting of the polarity of the negative items during data processing, the increases in beliefs are to be interpreted as they are decreased

process as time consuming. There are increases, for example, among *non-intenders* in some control beliefs: "no perception of effort", "awareness of opportunity", as well as in "not time consuming" (Table 13.5).

The results of the t-test (n = 93) demonstrate that the intention after the intervention increases significantly (M = 3.68_{pre} /4.47_{post}; Table 13.3). More specifically, the intention in the *non-intenders* group (M = 2.68_{pre} /3.87_{post}; p = .000, d = .904)

Table 13.5 Results of the t-tests: changes in control beliefs in subgroups

Subgroup			M	SD	T	df	p (2-sided)	Cohens d
No perception of effort	Non-intender (n = 55)	Pre	19.69	7.70	−2.689	54	.010	0.417
		Post	23.82	11.40				
	Intender (n = 36)	Pre	29.50	8.69	−2.093	35	.044	0.371
		Post	33.22	11.01				
Awareness of opportunity	Non-intender (n = 55)	Pre	18.71	7.49	−2.961	54	.005	0.442
		Post	22.80	10.49				
	Intender (n = 36)	Pre	29.75	8.75	−1.785	35	.083	–
		Post	32.86	11.70				
Not time consuming	Non-intender (n = 56)	Pre	15.26	6.38	−2.963	55	.004	0.456
		Post	18.21	6.57				
	Intender (n = 36)	Pre	19.97	8.07	−2.497	35	.017	0.421
		Post	23.33	7.88				

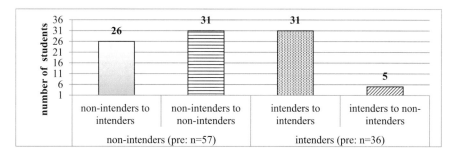

Fig. 13.2 Numbers of intenders and non-intenders after Wake-up-intervention

increases significantly in the post-test, while it remains stable in the *intenders* group (M = 5.28~pre~/5.42~post~; p = .388). There are also changes in the number of *non-intenders* and *intenders* after the intervention (Fig. 13.2): 26 *non-intenders* show higher values after the intervention (>4.0) in their intention than in the pre-test; therefore they become *intenders*. Only 5 *intenders* move after intervention to *non-intenders*, which means that their value of intention decreases (1.00 to ≤4.0).

Focusing on post-changes, it can be summarized that in the *non-intenders* subgroup, the *moral obligation* (while among *intenders* there were increases in *moral reasoning* instead of in *moral obligation*) and *self-identity helper* increase significantly in the post-test (Table 13.6). There is also a significant increase in *knowledge* in both groups (Table 13.7). The effect sizes range from $r = 0.545$ to 0.609, which are interpreted as strong effects according to Cohen (1988) (Table 13.7). There is no significant change in *empathy*, but it shows high values already in the pre-test (Table 13.6).

Table 13.6 Results of the t-tests: changes in model-external factors among subgroups

Subgroup			M	SD	T	df	p (2-sided)	d
Moral obligation	Non-intender (n = 56)	Pre	4.11	1.21	−2.801	55	.007	.367
		Post	4.58	1.33				
	Intender (n = 36)	Pre	5.26	1.11	−1.633	35	.112	–
		Post	5.48	1.05				
Self-identity helper	Non-intender (n = 57)	Pre	5.12	0.94	−2.311	56	.025	.183
		Post	5.30	0.98				
	Intender (n = 36)	Pre	5.79	0.76	−2.71	35	.01	.228
		Post	5.97	0.79				
Moral reasoning	Non-intender (n = 57)	Pre	4.76	0.80	−0.949	56	.347	–
		Post	4.83	0.85				
	Intender (n = 36)	Pre	5.17	0.86	−2.165	35	.037	.330
		Post	5.44	0.77				
Empathy	Non-intender (n = 57)	Pre	5.12	0.93	−1.114	56	.270	–
		Post	5.20	0.77				
	Intender (n = 36)	Pre	5.32	0.73	−1.149	35	.259	–
		Post	5.42	0.73				

Table 13.7 Wilcoxon-test: changes in *knowledge* (K) among subgroups

Skala	Subgruppen		M	SD	z	Asympt. significance (2-sided)	r
K1	Non-intender (n = 57)	Pre	15.69	2.35	−5814	.000	.545
		Post	18.42	2.14			
	Intender (n = 36)	Pre	15.26	2.62	−4888	.000	.576
		Post	18.61	1.73			
K2	Non-intender (n = 57)	Pre	2.70	1.46	−6084	.000	.570
		Post	6.75	3.00			
	Intender (n = 36)	Pre	2.92	1.84	−5164	.000	.609
		Post	7.83	2.16			

K1 knowledge was assessed with open questions, *K2* knowledge was assessed with closed questions

13.5   Discussion and Conclusion

While positive attitudes do not change significantly after the intervention in the whole sample – as they are already very high – negative attitudes show significant decreases. The negative attitudes show significant decreases among *intenders* and *non-intenders*, although these increases occur slightly more among *non-intenders* (Table 13.4). It can be concluded that "fear of pain" decreases significantly in the post-test. Furthermore, the students think that the donation process is "less harmful to health" and they perceive "health status" as more appropriate for stem cell donation after the intervention. They also show a higher degree of "trust in the medical system". These results clearly show that the intervention has a positive effect, especially on negative attitudes, but to varying degrees depending on the subgroup.

Above all, negative behavioral beliefs, such as fears, are reduced during the elaboration phase of the intervention (Table 13.2) by specific tasks and the acquisition of knowledge.

It can be summed up that there is a significant change in some control beliefs in the post-test: The students (*intenders/non-intenders*) perceive "less effort in performance of behavior", they are "more aware of opportunities for registration as a stem cell donor", and they are more aware that the registration and donation processes do "not consume a lot of time". It can be summarized that the significant increases within the subgroups after the intervention (Table 13.5) can be attributed to the effectiveness of the intervention. The control beliefs were also specifically addressed in the elaboration phase (Table 13.2), primarily through text work, as well as video material. Behavioral beliefs and control beliefs were also discussed and doubts clarified in the plenary session during the evaluation phase.

Students, especially *non-intenders*, show a significantly higher *intention* with regard to the registration as a stem cell donor after the intervention. When looking at the two subgroups, it becomes clear that the intenders' intention is already high in the pre-test and remains stable in the post-test. 26 students, which were characterized as *non-intenders* before intervention, change their status to *intenders*, because of a significant increase in their intention regarding stem cell donation, which reflects a positive effect of the teaching unit.

Taking a closer look at model-external factors, such as *moral obligation, self-identity, moral reasoning, empathy*, as well as *knowledge*, which especially in the context of blood donation and in the context of prosocial behavior proved to be influential factors of intention, it can be summarized that some of them show lower values before the teaching unit. However, this changes after the intervention. After the intervention, the constructs *self-identity helper* and *knowledge* increase significantly in both groups (Tables 13.6 and 13.7). While a significant increase in *moral obligation* is observed among *non-intenders*, a significant increase in the *moral reasoning* construct is observed only in the *intenders* group. These findings demonstrate impressively that the intervention addresses both groups (albeit in slightly different ways) regarding internal and external model factors. Furthermore, these results show that it is important to take into account and promote influential factors of intention regarding stem cell donation. While *knowledge* (regarding stem cell donation und leukemia) and *empathy* were promoted throughout all phases of the teaching intervention, through text material and specific tasks (Table 13.2), *moral reasoning, moral obligation,* and *self-identity* were targeted mainly in the elaboration phase and in the evaluation phase (Table 13.2).

While *moral obligation, moral reasoning, knowledge* regarding leukemia and stem cell donation and *self-identity* change after the intervention, *empathy* remains stable. This finding could possibly be explained by the fact that *empathy* already had a very high level in the pre-test (mean value <5.0 measured on the 7-point Likert scale) and therefore did not increase after the intervention. In order to promote *empathy*, activities focusing on perspective taking are used. In further studies, other activities and methods, like "role playing" can be tested and reflected regarding their effectiveness in promoting empathy.

Significant changes in TPB+ factors can be attributed to the positive effectiveness of the "Wake up" intervention, as already indicated. However, the significant increases in the measured constructs could also be attributed to sources other than the intervention. It is therefore advisable for further studies to survey a control group (without intervention) in addition to the intervention group.

Negative beliefs, as well as misconceptions regarding the topic, could be reduced by the "Wake up" teaching unit and most constructs of TPB could be fostered and promoted. Thus, the TPB+ model proves to be a successful decision-making model in the context of health education and should be tested in other contexts, such as organ donation.

What about the actual behavior regarding the registration as a stem cell donor? This question can be answered only in terms of measured *intention*, as behavior or actual registration as a stem cell donor was not captured in this study. Since many model-internal as well as model-external factors of the TPB+ increase after the intervention, as well as the intention, it can be concluded with regard to behavior that this becomes more likely or that the gap between intention and behavior is reduced. In other words, the higher the intention, the more probable the actual performance of behavior (Ajzen, 2005). However, further studies are needed, which also measure behavior in order to verify this theoretical assumption.

Regarding TPB-based intervention, it can be concluded that it can act as a possible instructional design tool to assist teachers in structuring and planning their teaching units. Biology teachers should be the key players in the promotion of cancer education (Barros et al., 2016), but their intention to teach such a complex interdisciplinary health topic is low and based on the emotional nature of topic (Carey, 1992). Furthermore, Heuckmann et al. (2020) pointed out that teachers' fear of the emotional reactions of students strongly contributed to the perceived burden of teaching the subject of cancer. Therefore, TPB-based interventions in the classroom are necessary in order to make a complex subject easier to talk about. In this way, the teaching unit "Wake up" provides teachers with support on how to deal with the topic of cancer in the classroom, using the example of leukemia and stem cell donation.

References

Ajzen, I. (1991). The theory of planned behavior. *Organizational Behavior and Human Decision Processes, 50*(2), 179–211.

Ajzen, I. (2005). *Attitudes, personality and behavior* (2nd ed.). Open University Press.

Ajzen, I. (2006). *Behavioral interventions based on the theory of planned behavior*. University of Massachusetts Amherst. Download 10 Mar 2019 https://people.umass.edu/aizen/pdf/tpb.intervention.pdf

Ajzen, I. (2011). The theory of planned behavior: Reactions and reflections. *Psychology & Health, 26*(9), 1113–1127.

Ajzen, I., & Fishbein, M. (2002). *Understanding attitudes and predicting social behavior*. Prentice-Hall.

Barros, A., Santos, H., Moreira, L., Ribeiro, N., Silva, L., & Santos-Silva, F. (2016). The cancer, educate to prevent model: The potential of school environment to primary prevention of cancer. *Journal of Cancer Education, 31*(4), 646–651.

Bierhoff, H.-W. (2010). *Psychologie prosozialen Verhaltens. Warum wir anderen helfen* (2nd ed.). W. Kohlhammer.

Bilgel, H., Sadikoglu, G., & Bilgel, N. (2006). Knowledge and attitudes about organ donation among medical students. *Transplantation, 18*, 91–96.

Carey, P. (1992). Teachers' attitudes to cancer education: A discussion in the light of a recent English survey. *Journal of Cancer Education, 7*(2), 153–161.

Cohen, J. (1988). *Statistical power analysis for the behavioral sciences* (2nd ed.). Erlbaum.

Conner, M., & Armitage, C. J. (1998). Extending the theory of planned behavior: A review and avenues for further research. *Journal of Social and Psychology, 28*, 1429–1464.

Demir, B., & Kumkale, G. T. (2013). Individual differences in willingness to become an organdonor: A decision tree approach to reasoned action. *Personality and Individual Differences, 55*(1), 63–69.

Eagly, A. H., & Chaiken, S. (1993). *The psychology of attitudes.* Wadsworth Cengage Learning.

Emler, N. P., & Rushton, J. P. (1974). Cognitive-developmental factors in children's generosity. *British Journal of Social and Clinical Psychology, 13*, 277–281.

Feeley, T. H. (2007). College students' knowledge, attitudes, and behaviors regarding organ donation: An integrated review of the literature. *Journal of Applied Social Psychology, 37*(2), 243–271.

Frey, D., Stahlberg, D., & Gollwitzer, M. P. (2001). Einstellung und Verhalten: Die Theorie des überlegten Handelns und die Theorie des geplanten Verhaltens. In D. Frey & M. Irle (Eds.), *Kognitive Theorien-Theorien der Sozialpsychologie* (pp. 361–398). Hans Huber.

Frick, J., Kaiser, F. G., & Wilson, M. (2004). Environmental knowledge and conservation behavior: Exploring prevalence and structure in a representative sample. *Personality and Individual Differences, 37*(8), 1597–1613.

Heuckmann, B., Hammann, M., & Asshoff, R. (2020). Identifying predictors of teachers' intention and willingness to teach about cancer by using direct and belief-based measures in the context of the theory of planned behaviour. *International Journal of Science Education, 42*(4), 547–575.

Hochhaus, A., Berger, U., & Hehlmann R. (2004). *Chronische myeloische Leukämie. Empfehlungen zur Diagnostik und Therapie* (2. Aufl.). Uni-Med.

Holdershaw, J., Gendall, P., & Wright, M. (2007). *Factors influencing blood donation behavior.* Paper presented at the Australia and New Zealand Marketing Academy Conference Dunedin, New Zealand.

Holzer, J., & Elster, D. (2021). Einflussfaktoren für die Intention Jugendlicher zu einer Stammzellenspende. In S. Kapelari, A. Möller, & P. Schmiemann (Eds.), *Lehr- und Lernforschung in der Biologiedidaktik* (pp. 179–199). Innsbruck.

Horton, R. L., & Horton, R. L. (1991). A model of willingness to become a potential organ donor. *Social Science in Medicine, 33*, 1037–1051.

Kaiser, F. G., & Frick, J. (2002). Entwicklung eines Messinstrumentes zur Erfassung von Umweltwissen auf der Basis des MRCML-Modells. *Diagnostica, 48*, 181–189.

Kaiser, F. G., & Fuhrer, U. (2003). Ecological behavior's dependency on different forms of knowledge. *Applied Psychology, 52*(4), 598–613.

Krebs, D. L., & Rosenwald, A. (1977). Moral reasoning and moral behavior in conventional adults. *Merill-Palmer Quarterly of Behavior and Development, 23*, 77–87.

Mayring, P. (2015). *Qualitative Inhaltsanalyse. Grundlagen und Techniken.* Beltz.

Morgan, E. S., & Miller, K. J. (2002). Communicating about gifts of life: The effect of knowledge, attitudes, and altruism on behavior and behavioral intentions regarding organ donation. *Journal of Applied Communication Research, 30*(2), 163–178.

Richardson, H. S. (2018). Moral reasoning. In E. N. Zalta (Ed.), *The Stanford encyclopedia of philosophy* (Fall 2018 ed). Download 31 Oct 2019 https://plato.stanford.edu/archives/fall2018/entries/reasoning-moral/

Rubens, A. J., Oleckno, W. A., & Cisla, J. R. (1998). Knowledge, attitudes, and behaviors of college students regarding organ/tissue donation and implications for increasing organ/tissue donors. *College Student Journal, 32*, 167–178.

Schwartz, S. H. (1977). Normative influences on altruism. In L. Berkowitz (Ed.), *Advances in experimental social psychological* (pp. 221–279). Academic.

Schwartz, S. H., & Howard, J. A. (1981). A normative decision – Making model of altruism. In J. P. Rushton & R. M. Sorrentino (Eds.), *Altruism and helping behavior* (pp. 189–211). Lawrence Erlbaum.

Stryker, S. (1968). Identity, salience, and role performance: The relevance of symbolic interaction theory for family research. *Journal of Marriage and the Family, 30*, 558–564.

Stryker, S. (1997). "In the beginning there is society": Lessons from a sociological social psychology. In C. McGarty & S. A. Haslam (Eds.), *The message of social psychology* (pp. 315–327). Blackwell.

Zentrales Knochenmarkspender-Register Deutschland, ZKRD. (2018). *Jahresbericht 2018. Annual report 2018. zkrd-online.* Download 11 Feb 2020 https://www.zkrd.de/

Zentrales Knochenmarkspender-Register Deutschland, ZKRD. (2021). *ZKRD.* Download 13 May 2021 https://www.zkrd.de/

Zeyer, A. (2012). A win-win situation for health and science education: Seeing through the lens of a new framework model of health literacy. In A. Zeyer & R. Kyburz-Graber (Eds.), *Science| environment| health. Towards a renewed pedagogy for science education* (pp. 147–173). Springer.

Chapter 14
Mapping Adolescents' Nutritional Knowledge

Martha Georgiou and Matina Moshogianni

14.1 Introduction

Food choices affect the health of every human being and if not balanced, could lead to various diseases or malfunctions of the organism (Botero & Wolfsdorf, 2005). Nowadays, we are able to access a variety of information about healthy eating in every stage of life. Regarding childhood, it is argued that a healthy diet during these years will contribute to good health and physiological development, thereby leading to better school performance (Worobey & Worobey, 1999). According to Neufeld et al. (2023, 21) "a healthy diet is health-promoting and disease-preventing. It provides adequacy, without excess, of nutrients and health-promoting substances from nutritious foods and avoids the consumption of health-harming substances." Moreover, according to Pérez-Rodrigo and Aranceta (2001), the adoption of healthy eating habits by adolescents is accompanied by healthy life practices, including physical activity, avoidance of smoking, and stress management, which could reduce the effects of chronic diseases later in adulthood. However, a variety of factors, including cultural and economic ones, play a role in the composition of a healthy diet because the respective combination of foods that make up the diet is context-specific (Neufeld et al., 2023).

However, do adolescents consume healthy food? When talking about adolescents' nutrition, it is important to distinguish between what they should eat and what they generally prefer to eat. Obviously, the latter arises from both personal preferences and the strong social norms that are created by peers, which are often adopted both voluntarily and involuntarily due to the vulnerability of adolescents (Pirouznia, 2001). For example, very often adolescents' outings include junk food (i.e. fast food) and snacks that are particularly high in calorific value, but low in

M. Georgiou (✉) · M. Moshogianni
Department of Biology, National and Kapodistrian University of Athens, Athens, Greece
e-mail: martgeor@biol.uoa.gr; matinamosh@biol.uoa.gr

K. Korfiatis et al. (eds.), *Shaping the Future of Biological Education Research*,
Contributions from Biology Education Research,
https://doi.org/10.1007/978-3-031-44792-1_14

nutritional value (e.g., chips, soft drinks, doughnuts, ice cream, pizzas, and others) (Word Health Organization, 2018). It is evident that such a diet contributes to obesity, and may cumulatively cause other types of diseases. On the other hand, there are other adolescents who also acquire health-damaging eating habits: Very strict diets are generally popular among the ranks of adolescents since it is believed that diets can ensure a very lean and, by social standards, beautiful body. Therefore, they are eager to reduce their daily food intake, sometimes resulting in cases of malnutrition which can lead to very serious conditions (e.g., anorexia nervosa) (WHO, 2018).

These practices lead adolescents far away from following a balanced diet that contributes to good health and the proper functioning of the body. In fact, they have an increased need for minerals such as calcium and iron during this stage of life due to the development of the musculoskeletal system and the onset of menstruation in girls. Vitamin intake should also be at high levels (e.g., vitamins D and K). Regarding energy requirements, it is not possible to generalize because there are significant variations from person to person, depending on their particular characteristics and choices (i.e., growth rate, physical activity, and other factors). (Brown, 2016).

14.1.1 International Research of Adolescents' Nutrition Knowledge

Several studies (Barzegari et al., 2011) have shown that, nutrition knowledge and attitudes towards nutrition are rarely not directly related and, more specifically, there is generally a positive correlation between nutrition knowledge and adolescents' eating habits. In other words, if adolescents had adequate knowledge about nutrition, they could probably make better choices about the foods they consume. This lack of basic knowledge about nutrition leads to many misconceptions, which are not limited to issues relating to food composition, but also extend to eating habits (e.g., recommended portions from each food group) (Sakamaki et al., 2005). In a recent survey of adolescents in Turkey, Saribay and Kirbas (2019) found that more than 65% of the participants had a very low level of knowledge about nutrition concepts. Similar results were reported in Pirouznia's (2001) research conducted in the US: Once again, it was shown that adolescents' nutrition knowledge was poor. Finally, Kostanjevec et al. (2012) identified that Slovenian adolescent students also had limited knowledge of nutrition concepts. These results were in accordance with those of Barzegari et al. (2011) in Iran, and the recent findings of Mizia et al. (2021) in Poland.

Moreover, nutrition knowledge has been found to vary according to the field of study chosen by secondary school students: In countries where nutrition and physiology programs are implemented, mainly through the subject of physical education, students participating in those programs had an improved perspective of nutrition. Nevertheless, even their performance level is not satisfactory, since similar performance is recorded for students with different educational orientations (Barzegari et al., 2011).

14.1.2 Nutrition Knowledge of Greek Adolescents

In Greece, research on adolescent nutrition (Varelas, 2006; Tsamita et al., 2007) focuses mainly on habits rather than nutrition knowledge, and even that is quite limited. Hassapidou and Bairaktari (2001) found that Greek adolescents do not choose a well-balanced Mediterranean diet, resulting in a reduced intake of essential nutrients from the foods they prefer to eat, thus preventing proper physiological development. At the same time, there is an increase in the consumption of unhealthy foods which are rich in fat and sodium (Hassapidou & Bairaktari, 2001) in their daily lives, and a reduction of the consumption of fruits and vegetables (Bebetsos et al., 2015). It is no coincidence that Greece ranks first in Europe in childhood (7–9 years old) obesity and third in adolescent (15 years old) obesity (WHO, 2018). Finally, the curriculum dealing with the concepts of nutrition involves only four teaching hours in K7, and those occur in the home economics class, rather than in the science class. If we combine all of the above, it becomes clear why teaching students about food and dietary choices beginning in the early school years is really important: This kind of knowledge could contribute to healthier habits during and after adolescence.

In order to enhance Greek adolescent students' knowledge about nutrition, we decided to create educational opportunities outside of and/or complementary to the formal education framework: We planned to introduce educational games, treasure hunts, and other activities as a means of evaluating students' nutrition knowledge after these tools had been implemented in the classroom. However, in order for the tools to effectively achieve our goal, it was first necessary to identify students' existing knowledge, especially since this had not been systematically assessed in Greece. Thus, the purpose of this study was to capture a first picture of Greek adolescent students' knowledge about nutrition in order to better guide our next step, i.e., the creation of appropriate educational tools. Details of this first phase are presented below.

14.2 Research Design and Method

14.2.1 Participants

In this study, 269 students aged 15–16 (50.9% girls, 49.1% boys) from different public schools across Athens participated. Schools were randomly selected, and the students were all of middle socioeconomic status; according to their teachers, they were average achievers. We decided to work with students of this age since it marks the end of compulsory education in Greece. Moreover, we restricted the sample size because our research is not a pure statistical study, but rather an initial collection of trends in order to accumulate some useful information on where to focus the educational tools we plan to create in the next phase.

14.2.2   Research Tool

To conduct our survey, we used the General and Sport Nutrition Knowledge questionnaire (GeSNK) (Calella et al., 2017), a questionnaire of 29 close-ended questions created for secondary students. The questionnaire is divided into two parts: The first refers to general knowledge about nutrition, and the second explores attitudes of students involved in different sports. For the purpose of our study, we used only the first part of this questionnaire as the two parts are separate and, more importantly, our intention was not to evaluate the differences in nutrition knowledge between students who are athletes and those who are not. It should be noted that while the first part of the questionnaire had 29 major questions, some of them contained sub-questions, which resulted in a total of 63 questions. For each question, the participants could choose the answer they considered correct from the given options. Two additional questions were also included in the research tool: the gender of the participants, and their source of information on nutrition issues (indicated by choosing from among different options, and/or by adding a source of their own). The questionnaire was translated into Greek by one researcher, and then translated back into English by the second researcher to ensure the correctness of the translation; it was then converted into electronic form to ensure the anonymity of the students, and for easier and safer distribution in compliance with COVID-19 health measures then in effect.

14.2.3   Data Collection and Analysis

The research was conducted in January 2022. The questionnaires were completed by the students within 45 min in each school's computer room in the presence of a teacher, who did not participate in the process except to supervise its smooth execution. The students' answers were automatically saved and uploaded after students' submissions, and the data were retrieved and analyzed by the researchers. More specifically, the analysis was initially done for all the students as a group, and then each student's responses were analyzed individually. Each correct answer was awarded one point. Therefore, as explained above, the maximum possible score for each student was 63 points, and for the total number of students it was 16,947 points (269 $(\times)$ 63). Answers were also analyzed on a question-by-question basis to show the possible lack or sufficiency of the students' knowledge regarding the various issues included in the questionnaire. Finally, we examined whether there were differences in participants' scores based on gender. This type of analysis aims to provide a more complete picture of Greek adolescent students' general nutrition knowledge, which, as has been discussed above, is something that does not exist at present. The results will point to areas where more weight should be given when designing similar teaching interventions, educational programs, and teaching and learning materials.

14.3 Results

The results of the aggregate scores of the participating student population for all questions are presented first, along with the percentages of students who scored correct answers, followed by the students' results on some of the individual questions, i.e., mainly those questions that showed maximum or minimum score values, or those that had a broader categorization. Cronbach's alpha coefficient for the GeSNK General Nutrition section, which we used in the present study, was 0.857.

14.3.1 Findings for the Aggregate of the Participants

Students correctly answered slightly more than half of the questions. The overall percentage of correct answers for the aggregate of the participants was found to be 57.3%, i.e., the total score of all the students combined was found to be 9704 points out of a maximum possible score of 16,947.

Table 14.1 presents the aggregate results in percentages of correct answers in the total population of participants, as well as the distribution of scores between genders. It is important to mention that the maximum score was 85.7% and was achieved only once, while the second maximum score of 84.1% was achieved three times. The minimum score of 0% was achieved once as well, with the second minimum score of 6.3% achieved four times.

Table 14.1 Aggregate results of correct answers to the GeSNK questionnaire

% correct answers	N (entire population)	% (entire population)	N Boys	N Girls
0–40	32	11.9	17	15
40–50	51	17.1	28	23
50–60	64	23.8	31	33
60–70	73	27.1	38	35
70–80	38	14.1	15	23
>80	11	4.1	2	9

14.3.2   Analysis of Answers to Individual Questions

In the first eight questions (including the sub-questions), students were asked about the nutrient contents of different foods. The answers to each question were analyzed individually, and the results are shown in Table 14.2.

Regarding the carbohydrate content of six different foods, a total of 65.3% correct answers were found, although 197 students considered tomatoes to be rich in carbohydrates, while 198 gave the same answer for cereals; the majority (205 students) correctly identified bread as a high carbohydrate food. In the protein category, 64% of the answers were correct, but only 128 students correctly identified beans, while cod was considered by 93 students as containing low/no protein. Low scores were observed for fiber, i.e., only 30.5% of the answers were correct, and for calcium and potassium content (47.21% and 45.91% correct answers, respectively).

On the other hand, the finding of 63.8% correct answers about iron was encouraging. However, a deeper look showed that many students considered apples (67 students) and honey (72 students) to be sources of iron. It was also encouraging that 61.86% identified foods containing high amounts of salt (i.e., canned foods such as peas, tuna, and others), although 166 participants did not consider bread to contain added salt. Finally, the category of fats had the highest percentage of correct answers: 76.27%. Surprisingly, some 37 students did not consider salami as a fatty food; 32 students did not recognize butter as being fatty; while the same was true of 59 students' idea about mayonnaise. However, according to 102 students, jam was noted as a food rich in fat.

Questions 9–23 were true-false questions without a specific categorization, which resulted in both encouraging and disappointing findings, e.g.: 53.53% answered that "egg white has a high cholesterol content," and 68.4% thought that "a high energy food is necessarily a fatty food." Furthermore, 46.5% noted that "dairy products are a good source of iron," while only 44.2% believed that "a variety of foods did not contain added salt." Furthermore, about half (47.3%) did not know that "omega-3 and omega-6 are fatty acids," and more than one-third (33.5%) did not know that "dried fruit is a good source of essential fatty acids." However, a large percentage (72.9%) responded that "carrots are a good source of vitamin A," that

Table 14.2 Percentage of correct answers per nutritional component of different foods

Nutritional ingredient of various foods	% correct answers	Notable qualitative deviations from the correct answer (N participants: answer)
Carbohydrates	65.3	Tomato (197: rich in carbohydrates)
Proteins	64.0	Beans (141: low/no proteins)
Fats	76.27	Jam (102: rich in fat)
Fibers	30.5	Honey (170: rich in fibers)
NaCl (salt)	61.86	Bread (166: low/no salt)
Ca (calcium)	47.21	Turkey breast (118: rich in calcium)
Fe (iron)	63.8	Apple, honey (67/72 respectively: rich in Fe)
K (potassium)	45.91	Pasta (110: low/no potassium)

"our bodies produce vitamin D through our skin directly from sunlight while we are outdoors" (72.1%), and that "canned products contain more salt than dried products" (71%).

The last six questions (24–29) dealt with nutrition and health issues. For the question as to whether "in obesity, diet plays an important role, while physical exercise does not," 77.7% answered correctly, while for the question as to whether an "unbalanced diet is the only risk factor for the development of cardiovascular diseases," the correct answers were found to be lower (59.48%). Finally, 58% noted that "in order to achieve a healthy weight loss, carbohydrates should not be removed from the diet."

Regarding their sources of nutrition information, the Internet ranks first (62.1%), followed by parents (61.3%), by coaches (39.1%), and then by teachers and school nutrition programs, both with exactly the same percentage (33.5%). Friends, television and extracurricular nutrition programs complete the main sources of information, while many people noted that they have no knowledge of healthy eating. Also, several students noted other sources, such as a nutritionist, a pediatrician, and other specialists.

14.4 Discussion

Through this research involving 269 Greek adolescent students, we found that their level of nutrition knowledge is only moderate. In fact, the aggregate number of correct answers to our research tool was only 57.3%, indicating that the number of students lacking nutrition knowledge is significant. Our findings are in line with those of other international studies, which also identify deficits in adolescent students' knowledge of nutrition issues. As mentioned above, studies in European countries, as well as in Asian and American countries (Barzegari et al., 2011; Mizia et al., 2021; Pirouznia, 2001; Sakamaki et al., 2005; Vaitkeviciute et al., 2015), have found that adolescent students are unaware of basic nutrition concepts.

In the present research, this is reflected through a limited number of correct answers on our research tool, leading us to identify what the distribution of the students' levels of successful (correct) answers was (Table 14.1). The analysis shows that 52.8% of the students managed to correctly answer up to 38 questions out of the 63 in the questionnaire (i.e., 60% of possible correct answers). This means that a large number of the students have a very basic knowledge of nutrition issues. However, if we look at the percentage of students who achieved higher scores (i.e., >70%), then the percentage of students with a very basic knowledge of nutrition issues drops to 18.2%. In other words, only 49 out of the 269 students managed to correctly answer more than 45 of the 63 questions, while only 11 managed to correctly answer more than 50 questions. While we did not expect students to achieve a score of 100%, certainly a performance close to 50% by almost half of the participants cannot be considered satisfactory, as it indicates that there is a significant lack of nutrition knowledge.

In addition to getting a first overview of the level of the students' nutrition knowledge, it was important for our purposes to identify the aggregate success rate of the participating students. As can be seen in Table 14.1, the majority of students (i.e., 27.1 + 23.8 = 50.9%) are ranked in the 60–70% and 50–60% success ranges. This could be considered a positive sign in the sense that the majority of the students' scores are not concentrated in lower ranges. However, if we observe the rates above 70% and below 50%, we find that for <50% we have a population of 29%, while for >70% we have 18.2%, i.e., students who have insufficient nutrition knowledge outnumber those who have a higher level of knowledge. It is therefore clear that within the participating student population there are several students of different knowledge levels. This certainly needs to be taken into account when designing an educational proposal on nutrition.

In order to further identify weaknesses and/or strengths in the students' nutrition knowledge, an analysis of each question individually followed. Again, the results varied considerably, but it now became more apparent what the students know as absolutely basic knowledge, what possibly constitutes for them more specialized information about something they already know, or even what information they have previously not come into contact with. For example, the general and widely known categories of nutritional content were found to be more familiar to the students, and therefore we observed relatively satisfactory rates of successful responses. In particular, carbohydrates, proteins, and fats were obviously within the students' range of knowledge, so success rates relating to questions about these elements were roughly in the range of 64–76%. However, there were striking qualitative deviations from the correct answers, as has been discussed in detail in Sect. 14.3.2 above. Therefore, if we look at the general picture of what the students know about nutrition, we may be partially satisfied, but an in-depth look reveals the misconceptions (or perhaps the ignorance) that exist among the students.

Furthermore, the variety of answers to the true-false questions ultimately shows once again that the adolescent students have fragmented knowledge about nutrition: They do know some things about vitamins (e.g., where we get vitamin A from, or how vitamin D is produced), but they assume that any food with a high calorific value is necessarily high in fat. Specifically, a number of the students (>2/3 of the participants) consider that "dried fruit is a good source of essential fatty acids". If we combine this answer with the one about jam previously discussed, we would say it is likely that the students think that any food with a high calorific value includes fat. It is therefore possible that sugar and fat are not sufficiently distinguished by many of the students, who associate them with their high calorific yield and not with their other, substantive characteristics. If this is indeed the case, it could have dangerous implications for their health.

Regarding the students' answers in relation to nutrition-health issues, as represented by the last six questions of the research tool, we found that while many (77%) acknowledge that both poor nutrition and lack of exercise are responsible for obesity, fewer (59.5%) believe that the same is true for cardiovascular diseases. Therefore, according to the beliefs of the students, exercise is a habit to avoid obesity rather than a habit that promotes health. Let us not forget, as previously mentioned, it has already been found (WHO, 2018) that adolescent students (especially girls) resort to extreme diets, investing heavily in their physical appearance, which requires a slim body according to social norms; in addition to diet, exercise is considered as a way to achieve this. Cardiovascular diseases and their prevention are not among the students' interests, so that their knowledge about these issues is more limited. Finally, it is worrying that 42% of the students believe that carbohydrates should be eliminated from their diet in order to lose weight. Obviously, this skewed view not only indicates a lack of knowledge, but also raises the alarm about possible health disorders that may arise in adolescents if they follow such practices.

It is also worth commenting on the fact that no particular gender differences were observed between the number of students who performed similarly, with the exception of the high percentages achieved by more girls than boys (in the sense of absolute numbers rather than strict statistical tests, which are beyond the scope of this research). This slight predominance of girls is also confirmed in Pirouznia's (2001) survey. However, what is noteworthy is the sources of nutrition information reported by the students: While parents come second after the Internet as a source of information for the students, school and teachers come quite a bit later, although they should be a primary source. The students' thirst for learning about nutrition is reflected by the Internet being first in the list of information providers. The gap that exists between the students' level of nutrition knowledge and the school programs focusing on this is, therefore, evident. However, there are also several individual responses: "I looked it up myself," "from my pediatrician," "from a nutritionist," "I looked it up while in quarantine," and others.

All this suggests that while the students obviously would like and need to have more knowledge about nutrition, their sources of information can be fortuitous and possibly not always reliable. For example, although one would expect that the vast majority of the students would know that omega-3 and omega-6 are fatty acids, mainly because there is a daily bombardment of television advertisements featuring products enriched with these ingredients, in the end only about half (47.2%) were aware of this, even though television is among their sources of information.

It is noteworthy that even students who achieved high scores for their responses to the research tool noted, in addition to indicating their nutrition information sources, that "I have no knowledge of healthy eating." This shows a lack of confidence about what they know, and could even harbor a warning that they might have chosen answers randomly, which may need further investigation. The need not just for knowledge acquisition, but also for socio-scientific reasoning skills to be cultivated around similar issues is also reflected (Georgiou et al., 2020a, b; Maniatakou et al., 2020). At the same time, we must also consider the limited number of participants in the study, the results of which may indicate some trends in the Greek adolescent students' nutrition knowledge, but is not generalizable to the total population of Greek adolescent students.

As has already been mentioned, the present research is a first step in the subsequent construction of appropriate teaching tools in order for students to come into contact through education with concepts relating to nutrition more effectively. However, we believe that our findings are not limited to Greece: The lack of knowledge about nutrition is an issue that reflects not only on the cognitive level of students, but also on health issues they may face as human beings due to eating habits. It is, therefore, important to conduct similar surveys and compare the results to identify possible patterns in the deficiencies that students may have. In this way, educational tools that transcend the borders of a single country and that can be shared and implemented in larger populations could be built/designed, with the first aim of improving personal nutrition knowledge, and the ultimate goal being the permanent adoption of desirable attitudes and habits.

In our study of Greek students, only one part of the questionnaire (i.e., general nutrition) was tested, and an in-depth analysis of the data was carried out, constituting the Greek context of students' nutrition knowledge. It should be noted that the GeSNK questionnaire has already been tested for its validity and reliability by researchers in Spain (Manzano-Felipe et al., 2022), Turkey (Gokensel Okta & Yildiz, 2021) and Romania (Putnoky et al., 2020), who created the Spanish, Turkish and Romanian versions of the GeSNK, respectively. In Spain it was found that 59.1% of the respondents have a low level of knowledge, while similar findings were found in Turkey and Romania. In line with previous surveys (with other tools) described in Sect. 14.1.1 of this chapter, a lack of knowledge in this area has been confirmed. Additionally, the need to create appropriate educational tools is further intensified and their effectiveness can be tested even through comparative studies and collaboration of the research community.

14.5   Conclusion

The present research was conducted to obtain a first picture of the level of nutrition knowledge of Greek adolescent students in order to prepare appropriate learning tools for nutrition concepts (e.g., educational games, treasure hunts, and other activities). The students' responses to nutrition-related questions on the GeSNK research tool allowed us to identify areas in which the students had a lack of nutrition knowledge, leading them to have misconceptions about food values and healthy eating habits. In fact, we found that it should not be assumed that any basic nutrition knowledge can be excluded on the grounds that it is already known to students, since our analysis identified students of different knowledge levels. The study also revealed that the students do not receive adequate nutrition information from their school program, but turn to other sources for information, some of which may not be valid or reliable. These findings will guide us in planning and implementing appropriate learning and teaching tools which will meet students' needs.

References

Barzegari, A., Ebrahimi, M., Azizi, M., & Ranjbar, K. (2011). A study of nutrition knowledge, attitudes and food habits of college students. *World Applied Sciences Journal, 15*(7), 1012–1017.

Bebetsos, E., Zorzou, A., Bebetsos, G., Kosta, G., & Karamousalidis, G. (2015). Children's self-efficacy and attitudes towards healthy eating. An application of the Theory of Planned Behavior. *International Journal of Sports and Physical Education, 1*, 1–8.

Botero, D., & Wolfsdorf, J. I. (2005). Diabetes mellitus in children and adolescents. *Archives of Medical Research, 36*(3), 281–290.

Brown, J. E. (2016). *Nutrition through the life cycle*. Cengage Learning.

Calella, P., Iacullo, V. M., & Valerio, G. (2017). Validation of a general and sport nutrition knowledge questionnaire in adolescents and young adults: GeSNK. *Nutrients, 9*(5), 439.

Georgiou, M., Mavrikaki, E., & Constantinou, C. P. (2020a). Is teaching biology through socio-scientific issues enough for the development of argumentation skills? In B. Puig, P. Blanco Anaya, M. J. Gil Quílez, & M. Grace (Eds.), *Biology education research: Contemporary topics and directions* (pp. 177–186). Servicio de Publicaciones Universidad de Zaragoza. https://doi.org/10.26754/uz.978-84-16723-97-3

Georgiou, M., Mavrikaki, E., Halkia, K., & Papassideri, I. (2020b). Investigating the impact of the duration of engagement in socioscientific issues in developing Greek students' argumentation and informal reasoning skills. *American Journal of Educational Research, 8*(1), 16–23. https://doi.org/10.12691/education-8-1-3

Gokensel Okta, P., & Yildiz, E. (2021). The validity and reliability study of the Turkish version of the general and sport nutrition knowledge questionnaire (GeSNK). *Progress in Nutrition, 23*(1), e2021027.

Hassapidou, M. N., & Bairaktari, M. (2001). Dietary intake of pre-adolescent children in Greece. *Nutrition & Food Science, 31*(3), 136–140.

Kostanjevec, S., Jerman, J., & Koch, V. (2012). The influence of nutrition education on the food consumption and nutrition attitude of schoolchildren in Slovenia. *US China Education Review, 11*, 953–964.

Maniatakou, A., Papassideri, I., & Georgiou, M. (2020). Role-play activities as a framework for developing argumentation skills on biological issues in secondary education. *American Journal of Educational Research, 8*(1), 7–15. https://doi.org/10.12691/education-8-1-2

Manzano-Felipe, M. Á., Cruz-Cobo, C., Bernal-Jiménez, M. Á., & Santi-Cano, M. J. (2022). Validation of the General and Sport Nutrition Knowledge Questionnaire (GeSNK) in Spanish adolescents. *Nutrients, 14*(24), 5324.

Mizia, S., Felińczak, A., Włodarek, D., & Syrkiewicz-Świtała, M. (2021). Evaluation of eating habits and their impact on health among adolescents and young adults: A cross-sectional study. *International Journal of Environmental Research and Public Health, 18*(8), 3996.

Neufeld, L. M., Hendriks, S., & Hugas, M. (2023). Healthy diet: A definition for the United Nations food systems summit 2021. In *Science and innovations for food systems transformation* (pp. 21–30). Springer.

Pérez-Rodrigo, C., & Aranceta, J. (2001). School-based nutrition education: Lessons learned and new perspectives. *Public Health Nutrition, 4*(1a), 131–139.

Pirouznia, M. (2001). The association between nutrition knowledge and eating behavior in male and female adolescents in the US. *International Journal of Food Sciences and Nutrition, 52*(2), 127–132.

Putnoky, S., Banu, A. M., Moleriu, L. C., Putnoky, S., Șerban, D. M., Niculescu, M. D., & Șerban, C. L. (2020). Reliability and validity of a general nutrition knowledge questionnaire for adults in a Romanian population. *European Journal of Clinical Nutrition, 74*(11), 1576–1584.

Sakamaki, R., Toyama, K., Amamoto, R., Liu, C. J., & Shinfuku, N. (2005). Nutritional knowledge, food habits and health attitude of Chinese university students–a cross sectional study. *Nutrition Journal, 4*(1), 1–5.

Saribay, A. K., & Kirbas, S. (2019). Determination of nutrition knowledge of adolescents engaged in sports. *Universal Journal of Educational Research, 7*(1), 40–47.

Tsamita, I., Kontogianni, P., & Karteroliotis, K. (2007). Evaluation of students' eating habits in a city of the Greek province. *Searches in Physical Education & Sport, 5*(1), 105–115 (in Greek).

Vaitkeviciute, R., Ball, L. E., & Harris, N. (2015). The relationship between food literacy and dietary intake in adolescents: A systematic review. *Public Health Nutrition, 18*(4), 649–658.

Varelas, A. (2006). *Nutritional education: A longitudinal study of the eating habits of Greek students and investigation of the possible effects of the first decade of health education programmes in Greece.* PhD thesis, Aristotle University of Thessaloniki. Department of Primary Education (in Greek).

World Health Organization. (2018) *Spotlight on adolescent health and well-being.* Findings from the 2017/2018 Health Behaviour in School-aged Children (HBSC) survey in Europe and Canada. International report. Volume 2. Key data. Geneva; 2020. Available from https://apps.who.int/iris/handle/10665/332104

Worobey, J., & Worobey, H. S. (1999). The impact of a two-year school breakfast program for preschool-aged children on their nutrient intake and pre-academic performance. *Child Study Journal, 29*(2), 113–113.

Part III
Outdoor and Environmental Education

Chapter 15
Implementing Climate Change Education: The Role of Inter-sectorial Collaborations

Nofar Naugauker, Orit Ben-Zvi-Assaraf, Daphne Goldman, and Efrat Eilam

15.1 Introduction

Climate Change (CC) is an existential threat confronting all global societies, natural ecosystems, and the fabric of life on Earth (Intergovernmental Panel on Climate Change [IPCC], 2020). Anthropogenic-led global CC is at a critical point, where natural forces in the climate system react to cause further warming. Weather and climate extremes have detrimental socio-economic and ecological impacts. The rising temperatures in the Eastern Mediterranean and Israel in the past few decades have already brought about significant damage through forest fires, ecological impairment, decimation of water resources and heat-related morbidity and mortality, especially among vulnerable populations (Yosef et al., 2020). It is widely agreed that pupils in today's world need to be equipped with the appropriate knowledge, skills, values and attitudes, which will enable them to cope with the challenges CC poses in informed and responsible ways. Accordingly, education systems are expected to address these needs, by developing and implementing effective CC educational programs (UNESCO, 2017). The enormity of the threat and the crucial role of education in addressing it have been endorsed by numerous international and

N. Naugauker (✉) · O. Ben-Zvi-Assaraf (✉)
Department of Science and Technology Education, Ben-Gurion University of the Negev, Beersheba, Israel
e-mail: ntorit@bgu.ac.il

D. Goldman
Department of Environmental Science and Agriculture, Beit Berl College, Kfar Saba, Israel
e-mail: dafnag@beitberl.ac.il

E. Eilam
College of Arts and Education, Victoria University, Footscray, VIC, Australia
e-mail: efrat.eilam@vu.edu.au

© The Author(s) 2024
K. Korfiatis et al. (eds.), *Shaping the Future of Biological Education Research*,
Contributions from Biology Education Research,
https://doi.org/10.1007/978-3-031-44792-1_15

regional bodies, who have been calling governments to boost Climate Change Education (CCE) within their curricula (UNESCO, 2021a, b).

Studies examining CC representation in the curriculum confirm that in most nations' curricula CC is represented in Science and Geography (UNESCO, 2021a, b). However, a study by Dawson et al. (2022) that compared CC curricula of seven different countries revealed diverse approaches to curricular inclusion. These include the embedding of CC (i) under the concepts of "sustainability" or "environmental literacy"; (ii) within one or few disciplinary subjects; or (iii) across many disciplinary subjects. Each approach has its challenges and limitations. Acknowledging CC as a distinct discipline has also emerged as an option (Eilam, 2022). In Israel CCE is included in both Science and Technology (as a topic of "environmental literacy") and in Geography (as a topic of "sustainable development"), with more significant inclusion in Geography (Dawson et al., 2022). In the USA, the National Center for Science Education and the Texas Freedom Network Education Fund (2020), conducted a review of the quality of CC inclusion in the science curricula of 50 states. The review found that only 27 states earned a score of B+. The Next Generation Science Standards (NGSS), itself, earned only a B+ for their representation of CC (NGSS Lead States, 2013). However, the State of New Jersey is an exception. In June 2020, the State Board of Education adopted the "2020 New Jersey Student Learning Standards", making New Jersey the first state in the USA to incorporate K–12 CCE across content areas (New Jersey Climate Change Education Hub, n.d.).

Various studies have directed attention to the important role of biology education in addressing key CC issues (Rushton & Walshe, 2022). CC impacts all levels of biological systems, from individual organisms to biomes, reducing the efficiency of ecosystem services (Bellard et al., 2012; Nunez et al., 2019; Zhou et al., 2020). CC has already begun to reduce the resilience of ecosystems, a necessary condition for species survival (Nunez et al., 2019). CC has also been shown to decrease the genetic diversity of populations due to directional selection and rapid migration, which, in turn, could affect ecosystem functioning and resilience (Bellard et al., 2012). When considering biology education's contribution to CCE, Rushton and Walshe (2022) suggested that beyond providing general CC knowledge and directing students to green jobs, biology education is particularly well positioned to address issues concerning biodiversity conservation and enhance students' connection to nature, an important affective aspect of CCE.

From a policy level perspective, the development of CC curricula is recognized as a multi-stakeholder effort. International and regional bodies such as UNESCO and the European Commission highlight the important role that inter-sectorial collaborations play in promoting CCE in various subjects, including Science and Biology (European Commission, 2022; UNESCO 2021b). The European Commission (2022) applied a survey for examining the extent of collaborations between education and training institutes and the wider communities, in relation to environmental sustainability programs, including CCE. The findings suggest that this form of cross-sectorial collaborations is widespread, with 52% of respondents reporting on collaborations. The report emphasizes the importance of inter-sectorial collaborations, noting that: "in the EU, there has been a growing emphasis on the benefits and potential of [...] cross-sectoral partnerships between the public and private sector and/or civil society. A dedicated strand under the Erasmus+ programme fosters cooperation among organisations and institutions and many projects have had a strong focus on cross-sector collaboration as a way to strengthen learning for environmental sustainability" (European Commission, 2022, p. 49).

In the UK, a Policy Exchange publication highlights the need for governments to take a mediating role in supporting cross-sectorial collaborations (Blake, 2014). UNESCO (2021a) further suggests that it is good practice when ministries collaborate in advancing CCE, thus extending collaborations from the inter-sectorial to the intra-sectorial. Interestingly, data collected in the EU countries suggest that the nature of collaboration is impacted by the number of collaborators. When cross-sectorial partnerships involve multiple stakeholders, the collaboration tends to become more embedded in the system and there is more resource-sharing. When the cross-sectorial collaboration involves only two partners, the focus tends to be project-based (Mulvik et al., 2022). This observation resonates with Kwauk's (2020) observations in the USA, suggesting that while dual-sectors grassroot collaborations between NGOs and the education sector are helpful in many ways, they are often isolated and overly dependent on individual pioneers or local advocacy. In Israel, it was found that inter-sectorial collaborations play a key role in adding value to policy implementation (Gali & Schechter, 2021). However, there is a lack of information regarding the value of such collaborations in relation to CCE.

Research thus far has given little attention to the perceptions of education policy makers and educators involved in developing and implementing CCE. More clarity is needed regarding the ways in which these key CCE actors respond to the international and grassroot calls for enhancing the representation of CC in national curricula and the ways in which they perceive and navigate inter-sectorial collaborations. Addressing this gap, our study utilizes a case study consisting of Israeli CCE policy makers and educators in order to understand some of the political "push and pulls", professional considerations and key collaborations involved in including CC in the national curriculum. Overall, this study aims to give CCE makers a voice.

Research Objectives
The objectives of the study are to (i) analyze the perceptions of education policy-makers and education professionals involved in implementing CCE; (ii) identify existing opportunities and challenges to effective implementation; and (iii) characterize the inter-sectorial collaborations involved in CCE implementation.

15.2 Methods

This case study focuses on examining the perceptions of policymakers in the Israel Ministry of Education (MoE), coordinators of middle school teachers' professional development, and secondary school teachers who are currently active in CCE. This study employs a qualitative-phenomenographic research approach to qualitatively map and explain the participants' perceptions and interpretations of their lived experiences (Marton, 2005).

Semi-structured interviews were employed for the purpose of eliciting the participants' perspectives regarding a range of issues concerning CC curriculum development and implementation, including their perceptions regarding how CC should be represented in the curriculum and how it is presently represented; enablers and inhibitors for implementing CCE, networking and collaborations in CCE. A total of 17 participants gave their consent and were interviewed for the study. The participants represented three populations involved in CCE in the Israeli educational system:

Policymakers in the MoE (P). These included Chief Directors of subjects relevant to CCE who are responsible for overseeing all aspects related to the pedagogical content knowledge of their subjects, and Inspectors who are responsible for overseeing the compatibility between the formal curriculum and its teaching in schools.

Teacher professional development coordinators (D). This group included heads of teacher professional development (PD) courses, which provide in-service training for teachers. In Israel, in-service training is provided for teachers via several channels: designated regional PD centers managed directly by the Ministry of Education (termed 'Pisga'), National Teacher Centers for Science & Technology ('Malam'), and Master of Education programs which operate as a form of in-service PD training for teachers within a master's degree.

Secondary school teachers (T) in year levels 7–12. Five of the teachers also hold various leadership positions in the MoE, such as mentoring roles. Their disciplinary specialization ranges across science, geography and environmental science. The teachers who participated in this study teach CC in their schools.

Some of the participants belong to more than one group. In these cases, the participants are coded by two letters, signifying their dual roles. For example, a policymaker who also performs an additional role as a professional development coordinator is coded PD. Table 15.1 summarizes the number of participants per group.

Table 15.1 Participants by code role and number

Code	Role	Number of participants
P	Policymakers in the MoE	5
D	Teacher professional development coordinators	3
T	Secondary school teachers	4
PD	Policymakers in the MoE + teacher professional development coordinators	1
PT	Policymakers in the MoE + secondary school teachers	2
DT	Teacher professional development coordinators + secondary school teachers	2
Total		**17**

The data were analyzed thematically and inductively, creating themes and categories. Initial coding by the first author and subsequent coding by the other authors, with discussions and negotiation through an iterative process of revisiting the coded texts, led to forming seven main themes, of which four themes are discussed in this paper. These reflect the dominant aspects that emerged from the data and constitute the participants' perceptions regarding CC education.

It is important to note that the data for this study were collected in late 2021. A short time after the completion of the data collection, the Israel MoE and the Israel Ministry of Environmental Protection jointly announced the mandating of CCE across the K-12 curriculum. This announcement was followed by CC curriculum reform. The findings of this study reflect the perceptions of the participants prior to the curriculum reform.

15.3   Findings

The findings are organized by the four emerging themes reflecting the participants' perceptions concerning: The importance of CC and the role of CCE; the scope of CC curriculum; challenges for effective implementation; and inter-sectorial collaborations in CCE. Appendix shows examples of citations of these themes. A detailed explanation of each theme, which includes citations, is provided in the following.

15.3.1   The Importance of CC and the Roles of CCE

The participants expressed broad consensus regarding the urgency of CC as a threat and the importance of CCE. For example, a participant claimed that:

> We all experience it, and it's going to have consequences on our lives here in Israel. And from an ecological perspective, too, which is less interesting as far as the average person is concerned. But it's there in the social, personal, and economic realms, too; our quality of life and well-being, our personal welfare, and health (D2).

This highlights a profound understanding of the urgency of CC, its current and future presence in our everyday life, and the pervasiveness of CC impacts across all human and non-human systems. Accordingly, there was wide agreement concerning the need to educate about CC. A policy maker outlined her vision for what may be regarded as a CC-literate school graduate:

> The ultimate goal is that every child has heard about the climate crisis. They understand the basis, the causes, the models, the scientific consensus, understand the nature of science, understand solutions, technological and non-technological, are able to offer their own solutions, understand policy issues, and challenges […] and the impacts (P1).

One participant identified the various purposes that CCE serves, stating: "There are lots of reasons to educate these students both as the future generation who'll be doing research on these issues, also as citizens, and also as those who will be involved in the decision-making" (DT4). According to this perception, CCE served the purposes of educating for general CC literacy for everyday citizenry; preparing future researchers in CC; and cultivating the necessary mindset of future decision-makers. This perception resonates well with Roberts's (2007) scientific literacy types, similarly, conveying the idea that CC literacy is more than one conceptual construct.

15.3.2 The Scope of CC Curriculum

In relation to the question of what needs to be included in CCE, there seemed to be wide agreement that the science foundation is important in CCE, however, not as a stand-alone, but alongside other aspects relevant to CC. For example, a policy maker stated that "Science must be there, social studies must be there, civics must be there, and economy. But all the time new things join in: geography, borders, migration, …" (P1). Most participants agreed that CC is multidisciplinary and that CCE needs to be taught across the curriculum. For example, a policy maker stated that:

> I […] reiterate this to teachers and students, so they understand that CC is not a topic on its own. It is all around us. In physics, chemistry, biology, earth sciences […] I mean in every domain […] you can find how to connect the topic to CC (P4).

Similarly, a teacher noted that:

> I would say to the students, 'Notice which of the subject areas deals with global warming. Look at the subjects and make a list.' And then I'll say, 'So, what's the subject? Which experts discuss it?' And then, all of a sudden, they realize it's everywhere (T6).

Many of the participants highlighted the importance of integrating knowledge, skills, attitudes, values and student activism as well as student agency. For example, a policy maker who is also a teacher specified:

> These are the three focal points that every student in Israel needs to know—the scientific knowledge, the impact on the environment and society, and at a more advanced stage, about the economy, too, and the solutions to the climate crisis, and what we, as students, can do in this regard (PT7).

The importance of values in CCE were also highlighted: "It's not only because of the content but because of all the moral values that go with it" (DT5); "There are inherent values. Values toward the environment which we must commit to. Values of cultivation, reduced consumption, efficiency…commitment" (DP6). Finally, much emphasis was placed on behavioral acquisition and activism. For example, a teacher explained that CC "includes aspects of taking responsibility, action taking, and change. Even in personal life…. Education for sustainability and social justice, education for environmental activism…" (T1). Overall, the participants cast a broad scope for CCE.

15.3.3   Challenges for Effective CCE Implementation

The findings revealed various challenges to CC inclusion in the curriculum. The first relates to systemic change, where CC curriculum reform challenges the MoE at the system level. The second challenge relates to CCE epistemic ambiguity, and additionally, a challenge concerning the curriculum users' experience.

From a policymaking perspective, the main challenge was described as mobilizing the system. A policymaker described MoE as a massive system that is hard to change from its present track; thus it tends toward conservatism. Curriculum reform in CCE requires more flexibility than the system naturally tends to offer. A policy maker equated MoE to a big ship: "Turning a big ship from its course takes time. It's not a sailboat that can zip through the wind. But things are happening […] There's a lot of determination, and you can also see things in writing that are under way" (P2).

Another potential barrier is the lack of clarity and agreement regarding CC epistemology and ways for its inclusion in the curriculum. For example, a policymaker explained the confusion that exists even at the terminology level:

> It used to be called the Greenhouse Effect, then Global Warming [...]. Two years ago [it was] Climate Change. Last year, and this year, the expression has taken on a more urgent tone [climate crisis]. So, it's not fair to ask us how come you're not teaching it… and talk about a concept that's essentially new and constantly changing (P1).

Concerning CC inclusion in the curriculum, the data revealed some tension among policymakers over ownership, where the Biology Curriculum specialist suggested that Biology is the curricular frame for CCE, stating that "climate change is Biology, only more general- global biology" (P2). Similarly, the Geography Curriculum specialist suggested that Geography needs to host CC, stating that "Geography is the core of climate change; climate is actually a geographical subject in its essence" (PT8).

The lack of conceptual agreements regarding the epistemology of CC poses a challenge for integrating the subject within the curricula. The fact that educational professionals responsible for the biology curriculum identify biology as an umbrella field for CCE has implications concerning the didactics of teaching CC within biology education. Additionally, when taught within the framework of Geography, this may lead to the omission of aspects of CC concerning biological systems.

Importantly, various participants across the three groups suggested that CC is somewhat sidelined in the curriculum and does not receive sufficient attention and resources, suggesting that CC "gets mentioned, but it doesn't get any major attention in teachers' lessons. They can skim through it. They don't have to delve into it" (DT5). The sidelining of CC poses a challenge from a curriculum user perspective, as it forms a barrier for educators who wish to devote more time in the curriculum to addressing CC, yet they are not supported to do so by the curriculum, and, consequently, by the system.

15.3.4   Inter-sectorial Collaborations in CCE

The analysis revealed that the MoE is involved in inter-sectorial collaborations across three sectors, including: other governmental bodies, such as the Ministry of Environmental Protection (MoEP); Non-Governmental Organizations (NGOs); and the scientific and academic community. The relationships between these groups and the MoE are described below.

Other governmental bodies. Governmental bodies such as the Israel Meteorological Service, the MoEP, and the Electricity Company, collaborate with the MoE in diverse ways, including funding projects, developing educational resources and supporting their implementation. For example, a participant stated: "The MoEP is a great partner. They fund the MoE educational programs with 10 million NIS, supporting all the educational programs and a list of annual programs" (PD6). Analysis also revealed that while these collaborations are appreciated by the MoE, various tensions arise around policymaking ownership. For example, a participant described how a MoEP representative may "direct us: 'Next year I want you to work on one, two, and three.' How is this possible?" (P1).

Non-Government Organizations. A range of NGOs are active in developing resources and implementing CCE. Some of their activities are funded by the MoE. They also play an important role in applying pressure on the Ministry to implement CCE and not procrastinate in this endeavor. A policymaker, who is also a teacher, explained "the topic of climate change is … a 'hot' topic that the Green organizations and social groups are extremely active in implementing. Sooner or later CCE will become imbued deep within the education system" (PT8). NGOs have the advantage of not being constrained by the formal curriculum. While their contribution is appreciated, here too criticism arose regarding educational risks associated with NGOs imposing their agenda, and challenges arose regarding quality assurance, for example: "There are materials, but I will not say that they are sufficient, I will not say that experts have read them all and checked their scientific or pedagogical correctness" (D3).

The scientific and academic community. Strong and trusting relationships exist between the MoE and academic institutions and scientists. Policymakers trust the scientists' views and seek their consultancy, for example: "It is very important for me to be in touch with all the Green academicians, because they are going to help us identify the 'moving target' [metaphor for the issue of climate change]" (P1). Academicians not only have influence on shaping the curriculum, but they are also actively involved in the classroom implementation of CCE. Overall, the findings suggest a network of inter-sectorial collaborations. These collaborations are by no means simple. Policymakers describe their complexity as mutually beneficial on one hand, yet the blurring of boundaries also involves the blurring of responsibilities and authority division between the sectors.

15.4 Discussion and Conclusions

This study provides insights into the perceptions of CC educators who are actively involved in policymaking and implementation in Israel. The findings reveal that at the time of the interviews, there was a broad consensus regarding the importance and the need for CC curriculum reform. These sentiments emphasize the importance of CCE in educating students to become change agents as part of the means for a deep social transformation toward sustainability, reflecting an environmental citizenship approach (Sarid & Goldman, 2021), in order to withstand CC calamities.

At the time of the data collection, CC was implemented in the Israeli curriculum mainly in secondary Science and Geography subjects. However, the study reveals a gap between the actual approach to implementation and what educators perceive as the best practice which is the cross curricular approach. Previous studies revealed similar gaps. The evidence suggests that while the cross-curriculum approach is highly advocated (e.g., European Commission, 2022; Mulvik et al., 2022; UNESCO, 2021a, b), contrary to the prominent expectation, only scarcely do countries implement this approach in their curricula, and it rarely trickles down into implementation (European Commission, 2022; UNESCO, 2021b). In England, for example, CC in the national curriculum is confined to secondary Science and Geography subjects. However, like Israel, a study among teachers clearly showed the teachers' preference toward the cross-curriculum implementation approach (Howard-Jones et al., 2021).

When considering the challenges for CCE curriculum development and implementation, epistemological vagueness seemed to act as a barrier. This epistemological vagueness seems to permeate CC discourse at all levels. Vagueness concerning CC epistemology was also found in the State of Victoria Australia upper-secondary curriculum documents (Eilam et al., 2020).

Finally, this study contributes to shedding light on the under-researched aspect of inter-sectorial collaborations in CCE. While previous literature has stressed the importance of such collaborations (Mulvik et al., 2022), thus far studies identifying and characterizing specific collaborations between Ministries of Education and other sectors in CCE are scarce. The present study identified three different sectors with which the MoE collaborates, and characterizes the nature of these collaborations. The findings suggest complex relationships. On one hand they support the findings of Gali and Schechter (2021), by which Ministries in Israel acknowledge the contributions of inter-sectorial collaborations. On the other hand, these relationships often come at a cost and are often difficult to navigate. The tension concerning authority or who decides what goes into the curriculum can also take the form of a struggle over who holds the authoritative knowledge. In the present study, policymakers expressed concerns that the quality of materials delivered in schools by the NGOs may not align with the Ministry's standards and may suffer from inaccuracies. Here, the policymakers perceive themselves as bearing the responsibility for and authority over the quality of education delivered in their schools. Thus, it was revealed that the aspect of authority is not only a question of power of decision making, but also a question of responsibility and authority regarding knowledge and best educational practice. Currently, worldwide, NGOs play a major role in CCE resources development and implementation (Mulvik et al., 2022). However, thus far little is known regarding the knowledge-authority relationships between the formal and informal education systems. This knowledge gap suggests that further research is required to develop deeper understanding concerning the knowledge-authority in the relationships between the formal and informal education sectors.

To conclude, by giving a voice to CCE policymakers and educators, this study provides a meaningful contribution to understanding some of the challenges involved in CCE policy development and implementation, and points out areas requiring further investigation. Particularly, it was found that more research is required to understand best practices for the inclusion of CC in the curriculum, and identify the scope of CCE and effective ways for implementation. In light of the current curriculum reform concerning CCE that occurred subsequent to this study, it is important to further explore how this reform influenced inter-sectorial collaborations around CCE, as well as aspects concerning its implementation as perceived by the actors involved in its updated implementation.

Appendix

Additional examples of citations for each theme

The importance of CC and the role of CCE	The scope of CC curriculum	Challenges for CCE effective implementation	Inter-sectorial collaborations in CCE
They witness destructive phenomena in some form or other almost every day, both near and far. There's evidence almost daily. Climate change happens almost every day in every corner of the world, and the phenomena are getting to be more extreme (PT8)	We have five core [geography] topics that include: ecosystems and human-environmental interactions, water resources, air resources, waste from the resources, and noise and radiation. The climate crisis is a part of almost all of the core topics, except noise and radiation. That's why it can be incorporated into any topic (PT7)	I don't think they've defined any yet. And a committee has been set up on the subject, and it's really a fledgling committee. But it's expected to swing into high gear in geography and in the sciences, and also to set goals […]. As far as I know, it means that no objectives have been defined and set as yet by the Ministry of Education. It's in the process [of happening] (PT8)	There are collaborations to boost this issue and also requests from them to collaborate; to work with students and with teachers. I mean that for their part, they often initiate in-service PD courses to consolidate the existing information and to get the teachers to take action (P4)

(continued)

The importance of CC and the role of CCE	The scope of CC curriculum	Challenges for CCE effective implementation	Inter-sectorial collaborations in CCE
I think it's mostly about how it's connected. It's something that's happening around us, and it can't be separated from our daily lives. And it's our role as science teachers to connect it to what's encountered on a daily basis, and to put it into a scientific context (T3)	I would like it to have more of a presence in the [science and technology] curriculum [...]. I agree that it hasn't been proclaimed in any way. Teachers who are more involved can add their own input to the subject matter and link it to the climate crisis (P4)	I think this will entail a bigger job, because this subject doesn't appear again in the curriculum in a structured form, with exercises and references, like there are in all the other subjects (T1)	Currently, outside of the ministry, there are a lot of well-intentioned agencies that sometimes pull a little in other directions, but by and large, it's possible to collaborate with them (P2)

References

Bellard, C., Bertelsmeier, C., Leadley, P., Thuiller, W., & Courchamp, F. (2012). Impacts of climate change on the future of biodiversity. *Ecology Letters, 15*(4), 365–377. https://doi.org/10.1111/j.1461-0248.2011.01736

Blake, J. (2014). *Completing the revolution. Delivering on the promise of the 2014 National Curriculum.* Policy Exchange. https://policyexchange.org.uk/wp-content/uploads/2018/03/Completing-the-Revolution.pdf

Dawson, V., Eilam, E., Tolppanen, S., Assaraf, O. B. Z., Gokpinar, T., Goldman, D., et al. (2022). A cross-country comparison of climate change in middle school science and geography curricula. *International Journal of Science Education, 9*, 1379–1398. https://doi.org/10.1080/09500693.2022.2078011

Eilam, E. (2022). Climate change education: The problem with walking away from disciplines. *Studies in Science Education, 57*(3), 231–264. https://doi.org/10.1080/03057267.2021.2011589

Eilam, E., Prasad, V., & Widdop Quinton, H. (2020). Climate change education: Mapping the nature of climate change, the content knowledge and examination of enactment in upper secondary Victorian curriculum. *Sustainability, 12*(2), 591. https://doi.org/10.3390/su12020591

European Commission. (2022). *Learning for the green transition and sustainable development: Staff working document.* Accompanying the document Proposal for a Council Recommendation on learning for environmental sustainability. https://op.europa.eu/en/publication-detail/-/publication/db585fc7-ed6e-11ec-a534-01aa75ed71a1/language-en/format-PDF/

Gali, Y., & Schechter, C. (2021). NGO involvement in education policy implementation: Exploring policymakers' voices. *Journal of Educational Administration and History, 53*(3–4), 271–293. https://doi.org/10.1080/00220620.2021.1957792

Howard-Jones, P., Sands, D., Dillon, J., & Fenton-Jones, F. (2021). The views of teachers in England on an action-oriented climate change curriculum. *Environmental Education Research, 27*(11), 1660–1680. https://doi.org/10.1080/13504622.2021.1937576

Intergovernmental Panel on Climate Change (IPCC). (2020, September 22). *IPCC Chair Hoesung Lee Keynote Speech City Week London 2020.* https://www.ipcc.ch/site/assets/uploads/2020/0 9/2020SpeechHL_22092020.pdf

Kwauk, C. (2020). *Roadblocks to quality education in a time of climate change.* BRIEF Center for Universal Education at the Brookings Institution. https://www.brookings.edu/wp-content/ uploads/2020/02/Roadblocks-to-quality-education-in-a-time-of-climate-change-FINAL.pdf

Marton, F. (2005). Phenomenography: A research approach to investigating different understandings of reality. In R. R. Sherman & W. B. Rodman (Eds.), *Qualitative research in education: Focus and methods* (pp. 140–160). Routledge.

Mulvik, I., Pribuišis, K., Siarova, H., Vežikauskaitė, J., Sabaliauskas, E., Tasiopoulou, E., Gras-Velazquez, A., Bajorinaitė, M., Billon, N., Fronza, V., Disterheft, A., & Finlayson, A. (2022). *Education for environmental sustainability: Policies and approaches in European Union Member States: Final report.* European Commission, Directorate-General for Education, Youth, Sport and Culture. https://doi.org/10.2766/391

National Center for Science Education and the Texas Freedom Network Education Fund. (2020). *Making the grade? How state public school science standards address climate change.* https:// ncse.ngo/files/MakingTheGrade_Final_10.8.2020.pdf

New Jersey Climate Change Education Hub. (n.d.). https://njclimateeducation.org/

Next Generations Science Standards (NGSS) Lead States. (2013). *Next generation science standards: For states, by states.* National Academies Press.

Nunez, S., Arets, E., Alkemade, R., Verwer, C., & Leemans, R. (2019). Assessing the impacts of climate change on biodiversity: Is below 2° C enough? *Climatic Change, 154*(3), 351–365. https://doi.org/10.1007/s10584-019-02420-x

Roberts, D. A. (2007). Scientific literacy/science literacy. In S. K. Abell & N. G. Lederman (Eds.), *Handbook of research on science education* (pp. 729–780). Lawrence Erbaum.

Rushton, E. A., & Walshe, N. (2022). Climate change, sustainability and the environment: The continued importance of biological education. *Journal of Biological Education, 56*(3), 243–244. https://doi.org/10.1080/00219266.2022.2116843

Sarid, A., & Goldman, D. (2021). A value-based framework connecting environmental citizenship and change agents for sustainability – Implications for education for environmental citizenship. *Sustainability, 13*, 4338. https://doi.org/10.3390/su13084338

United Nations Educational, Scientific and Cultural Organization (UNESCO). (2017). *Educating for sustainable development goals – Learning objectives.* https://unesdoc.unesco.org/ ark:/48223/pf0000247444

United Nations Educational, Scientific and Cultural Organization (UNESCO). (2021a). Learn for our planet. *A global review of how environmental issues are integrated in education.* https:// unesdoc.unesco.org/ark:/48223/pf0000377362

United Nations Educational, Scientific and Cultural Organization (UNESCO). (2021b). Getting every school climate-ready. *How countries are integrating climate change issues in education.* https://unesdoc.unesco.org/ark:/48223/pf0000379591

Yosef, Y., Baharad, A., Uzan, L., Furshpan, A., & Levi, Y. (2020). *Israel temperature projections by 2100.* Research report no. 4000-0802-2020-0000044, Israel meteorological service (in Hebrew).

Zhou, Z., Wang, C., & Luo, Y. (2020). Meta-analysis of the impacts of global change factors on soil microbial diversity and functionality. *Nature Communications, 11*(1), 1–10. https://doi. org/10.1038/s41467-020-16881-7

Chapter 16
From Curriculum to Enacted Teaching of Photosynthesis, the Carbon Cycle and Sustainability in an Upper Primary School Class

Lina Varg

16.1 Introduction

A recent interview study indicated a lack of clarity in science teachers' articulation of their work to offer opportunities for upper primary students (grades 4–6, age 10–12) to practice reasoning (Varg et al., 2022). In addition, grade 6 science teachers were found to view practical work as the most important aspect of science education, while simultaneously implying that their teaching practices mainly consisted of whole-class discussions (Lidar et al., 2019). These findings expose a need to further explore how upper primary school teachers' views of important elements in science education influence their classroom teaching practices. If, for example, practical work or student reasoning are considered fundamental, how does this show in the classroom? A previous Australian case study of a secondary school science teacher enacting reformed curriculum to teach sustainability showed that rather than teaching according to his own convictions, his teaching was strongly influenced by the pressures caused by time constraints and external assessment (Tomas et al., 2022). These findings from secondary school science raise questions about how intentions or convictions and instruction relate to one another in upper primary school science. Teachers often spend considerable time constructing lesson plans to guide their teaching (Ziebell & Clarke, 2018). However, a recent U.S. case study suggests that many teachers devote excessive amounts of time to planning, while expert teachers tend to rely on different strategies, rather than strictly adhering to elaborate plans, to guide their teaching (Hatch & Clark, 2021). Examples of strategies found in their study were the use of open-ended questions and encouraging students to elaborate on their answers. Whether teaching practices are enacted as planned, thereby providing opportunities for students to develop the intended

L. Varg (✉)
Umeå University, Umeå, Sweden
e-mail: lina.varg@umu.se

K. Korfiatis et al. (eds.), *Shaping the Future of Biological Education Research*,
Contributions from Biology Education Research,
https://doi.org/10.1007/978-3-031-44792-1_16

knowledge, is an interesting question. It is relevant to study teachers' work and ability to select and use teaching strategies which enable students to develop according to intentions. The present case study of how one primary science teacher enacts her intention for students to practice reasoning around sustainability issues is a contribution to research on the congruence between lesson plans and enacted teaching. Guiding the study were the following research questions:

- How congruent are a primary school science teacher's intentions and the implemented teaching practices?
- What factors impact the congruence between intentions and teaching practices?

16.1.1 Background

A recent study suggested that teachers use different strategies when planning their teaching, such as consulting colleagues, strictly adhering to curriculum, or following the textbook (Hatch & Clark, 2021). Regardless of the chosen path and resources used to plan, teachers' ability to teach in ways that provide adequate opportunities for students to develop the intended knowledge might differ. Teachers governed by national syllabus are navigating a zone, or "space of tensionality" (Lewthwaite et al., 2014), between the intended and enacted curriculum. There is a widespread belief that congruence between intended and enacted curriculum is crucial for reaching educational goals (Pepin et al., 2013; Ziebell & Clarke, 2018). However, the factors impacting this congruence likely vary among different teachers and contexts. Therefore, the need to study possible factors was emphasized as an important step to enable a reduction or elimination of their impact (Tobin et al., 1998). Findings from a case study of a secondary school science teacher show that as he worked to plan and implement a new curriculum on sustainable development, he experienced a narrowing of the space of tensionality, which manifested in feelings of reduced autonomy (Tomas et al., 2022). The teacher further identified the two main factors impacting the congruence as time constraints and the need to cover curricular content to prepare students for an external assessment (Tomas et al., 2022).

Alignment studies researching the congruence between intended, enacted and assessed curricula are quite common, while studies looking closer at the planning processes and influencing factors are rarer (Hatch & Clark, 2021; Ziebell & Clarke, 2018). There are several models for looking at alignment. Porter (2004) proposed four levels of curricula that could be compared: intended, enacted, assessed and learned. However, studying different levels of curricula are bound to produce different results concerning degrees of alignment. For example, using Webb's (1997) model, which assumes that if standards and assessment align, the instruction must be aligned with the curricula, means restricting the view to include only the intended and assessed curricula, while excluding the enacted and learned curricula. Such a view possibly overlooks key details in the teacher's process to reconceptualize intended curricula into teaching practices. Nevertheless, most studies focusing on the relationships between curricula standards and assessment point to a poor alignment (Ziebell & Clarke, 2018). Ziebell and Clarke's (2018) comparative case study included a closer look at the underlying reconceptualization processes. They used categories to explore the types of performances that were explicit throughout curricula, instruction, and assessment to identify promoted performance types. A deeper understanding of teachers' transformation of curricula into efficient teaching is important to identify where there is potential for implementing development efforts. Research on what impacts the congruence between lesson plans, as a teacher's interpretation of the intended curricula, and the opportunities offered through teaching practices has received less attention (Tobin et al., 1998). Reaching an understanding of the planning and how the plans are enacted through teaching requires an insight into the perspective of the teacher responsible for the enacted curricula, rather than an exclusive reliance on assessment data. This study presents an attempt to get a broad sense by following one upper primary science teacher as she moves from the national science syllabus, via her selection and interpretation of it in the form of teaching unit and lesson planning, and finally in her implementation of certain teaching practices in her grade 6 classroom.

16.2 Research Design and Method

To deeply explore and understand how one science teacher reconceptualized and enacted the intended curriculum, an intrinsic case study was conducted (Stake, 1995). This is the case of one science teacher who interprets curricula, plans lessons and teaches in an upper primary school class. As a single case, the aim is not to produce generalizable results. Rather, it provides an example of how various factors impact the congruence between this upper primary science teacher's intentions and teaching practices. The results, in full or in part, could be used and transferred to inform or enrich research and practice. Data were gathered from teacher interviews, documents, and classroom observations. The following sections contain descriptions of the participant, the data gathering process and the analytical approach.

16.2.1 Participants and Setting

The search for a participating teacher for this study was initiated by an e-mail sent to a group of 14 upper primary school science teachers who had previously participated in an interview study (Varg et al., 2022). Anna, which is used as a pseudonym for the teacher in this paper, was planning to teach a teaching unit of suitable length and timing. Therefore, she was asked and accepted to participate. Although this is a convenience sample, Anna did not stand out as significantly different in her approach to teaching science compared to other teachers who participated in the aforementioned interview study (Varg et al., 2022). She had worked at the present small-town school since graduating as a certified grade 4–6 science teacher 3 years earlier. The observations were conducted in a grade 6 class, whose 22 students Anna had taught science since the fourth grade. She described the class as well-functioning and noted that although there were many students with special needs, she and her colleague had worked hard to support the students' improved work effort over the past two and a half years. A letter, containing information about the purpose and design of the study, as well as their rights as participants (Swedish Research Council, 2017), was provided to all participants. The students' parents and Anna also signed a consent form.

The observed teaching unit, called 'Substances around us', was an integrated science topic revolving around for example the carbon cycle, combustion and photosynthesis, human exploitation of natural resources, and human impact on climate change. The lesson content was varied and an overview of one example lesson is provided in Table 16.3. In terms of coverage of the national science syllabus, the subject matter was comprehensive, providing opportunities for different teaching practices. The inclusion of topics, ranging from a submicroscopic to a macroscopic perspective, meant that students had to grasp challenging content (Sirhan, 2007) and this rendered the teaching unit suitable for the study purposes.

16.2.2 Gathering Data

This paper focuses on one aim of the Swedish national science syllabus namely that students practice and develop the ability to search for and evaluate information, communicate, and take a stand on environmental issues. The choice to look closer at this aim was validated by Anna's indication that student reasoning was prioritized in the current teaching unit. Data was gathered from multiple sources during 7 weeks. Semi-structured teacher interviews (Kvale, 1997), documents (national syllabus, Anna's planning documents), and lesson observations enabled the analysis of Anna's transformation of content through the different stages, from curriculum through planning and finally as enacted teaching practices (shown in Fig. 16.1).

Anna initiated the planning process by copying pertinent excerpts from the syllabus text and pasting them in a document she called a *local pedagogic plan* (LPP). LPP offers a planning structure used to transform vague syllabus text into more explicit and tangible teaching methods. Anna's LPP included general objectives, subject content, and competences for students to develop throughout the teaching unit, as these were formulated in the science syllabus (The Swedish National Agency for Education, 2018). Anna organized the plan into separate lessons and

Fig. 16.1 Stages and respective data gathered at each stage

included her interpreted and clarified objectives, teaching activities, assessment methods and knowledge requirements. An example of one transformed lesson plan is provided in Table 16.3.

Two teacher interviews were conducted. The purpose of interview 1, conducted before the observation period, was to gain insights into Anna's plans and intentions for the teaching unit. The interview questions focused on the teaching unit, Anna's teaching objectives, and her routines for planning science teaching. The second interview took place 3/4 into the teaching unit. It was designed to provide insights into Anna's thoughts on science teaching and views on student learning development as the teaching unit progressed. The questions explored whether the observed teaching matched Anna's typical teaching, her perception of the congruence between her intentions and instruction, her rationale for determining lesson objectives, and methods for student assessment. The interviews lasted 28 and 42 min, respectively. They were audio-recorded and transcribed verbatim, resulting in 15 pages of transcripts.

During the seven-week teaching unit, one of two lessons per week was observed to enable a comparison between Anna's intentions, as expressed in planning documents and interviews, and her actual teaching. The lessons were audio-recorded and the verbatim transcripts were combined with written observational notes resulting in approximately 60 pages of narrative records. The observations were conducted to provide data which would enable an exploration of how Anna's initial intentions were transformed into classroom teaching.

16.2.3 Analysis

Data analysis consisted of two parts. The first part was a content analysis (Krippendorff, 2019) comparing all the data to determine what types of learning categories were promoted at the different stages of the transformation. Five selected learning categories that students could be expected to develop in science were used (Table 16.1). The first and last categories originate from Webb's (1997) *depth of knowledge levels* and the middle three are *performance type categories* developed by Ziebell et al. (2017). From the textual data (curriculum, planning documents, interview transcripts, and lesson narratives), units which appeared to promote one of the five learning categories were extracted. For example, a lesson objective formulated in planning documents as "To be able to talk about the carbon cycle, and human impact on it" was categorized as *Recall and Reproduction (RR)*, while "I really want to focus on reasoning, talking and discussing questions..." (pre-interview) was categorized as *Reasoning*. Since Anna used text copied from the syllabus and placed this in a column next to her own interpretations of syllabus text, the parts of the syllabus intended to be taught and learned during each lesson were easily identified in her plan. The abstracted text units were summarized and rough proportions were estimated to provide an overview of the extent to which each learning category was promoted at each stage. This resulted in a figure (Fig. 16.2),

Table 16.1 Learning categories

Learning categories	Description
[a]Recall and reproduction (RR)	Students reproduce previously taught content.
Performing	Students reproduce previously taught methods or procedures.
Communicating	Students describe, discuss, and represent concepts, use models and diagram.
Reasoning	Students draw conclusions, test hypotheses, make judgements and generalizations.
[a]Extended thinking	Students use higher order thinking processes such as synthesis, reflection, assessment and adjustment of plans over time to solve real-world problems with unpredictable outcomes.

Adapted from Webb[a] (1997) and Ziebell et al. (2017)

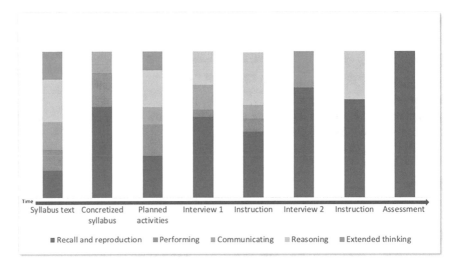

Fig. 16.2 Occurrence of learning categories in different stages over time (visualization)

which should be considered a visualization of the analysis results, rather than a statistical diagram.

The second part of the analysis consisted of an inductive thematic analysis (Guest et al., 2012) of the interview transcripts from the two teacher interviews. These were read several times in an attempt to capture Anna's views of and experiences from teaching science, resulting in the key themes: *Intentions for teaching the topic*, *Planning processes*, and *Factors impacting level of congruence* with accompanying codes (Table 16.2). Excerpts relating to these themes were organized and compared. Some example quotes are presented in Table 16.2, as well as Table 16.3. An important aspect to address here is the concept of "intentions". The intended curriculum usually refers to the "curriculum-as-written" and teachers do not always agree with curricular intent (Lewthwaite et al., 2014). However, Anna expressed a clear

Table 16.2 Themes and codes from interview data

Themes	Codes	Example quotes
Intentions for teaching the topic	Recall and reproduction (RR)	"What is good [about filling in worksheets together in whole-class settings] is that I feel like the whole class is with me. When you [fill in worksheets together, the students] really focus on the white board. Drawing, writing, following along. It feels like a guarantee that everybody finishes it." (RR, Interview 2) "[The students] will grow up and live with [ideas of sustainable development] and therefore it's important to be able to take a stand, communicate, information. It's important. They need to learn and take a stand." (Reasoning, Interview 2)
	Performing	
	Communicating	
	Reasoning	
	Extended thinking	
Planning processes	Use of science syllabi	"I usually create an LPP and I select [from syllabi] the skills that the students will get to practice, and the content, I mean what is in line with the chapter that we will work on." (Use of science syllabi, Interview 1)
	Assessment	
Factors impacting level of congruence	Stress	"I'm 'locked in' during class and that makes it hard, I mean if I said [in the beginning of class] that we will get to the questions, I really want to get to the questions now, let's go now." (Stress, Interview 2)
	Notion of repetitive nature of science education	

Table 16.3 Empirical example: Reasoning

Lesson objective (syllabus)	Clarified objective (Anna)	Planned activity	Interview 1	Enacted activity	Interview 2
"… use knowledge of biology to examine information, communicate, and take a stand **on questions (e.g. natural resource use and ecological sustainability)."**	Photosynthesis, combustion, and some other basic reactions.	Conversation and discussion about human impact on the climate.	"I want to focus on students **discussing and reasoning a lot**. Do more conversation exercises than we've done previously."	[a]See excerpt below from discussion on use of energy.	"About human impact […] it turned out very brief […] I'm thinking **we could have discussed it a lot deeper**. Like what, how can we change our use of these fossil fuels for example."
	Fossil fuels and renewable fuels. Their significance for energy use and impact on climate.				

(continued)

Table 16.3 (continued)

[a]The following episode comes from the example lesson. During one of the reasoning sessions that took place between the readings, one student suggested using electricity instead of fossil fuels to reduce our impact on climate change, and the suggestion resulted in the following interaction (*A* Anna, *S* Students)

A: "What is so good about electricity then? How can we make electricity? In what 2. ways can we get electricity? How is [it] produced?"

S1: "Hydropower."

A: "Hydropower, exactly."

S2: "Wind power."

A: "Wind power. S3?"

S3: "The sun."

A: "And solar energy, that's right […] Absolutely, that was a good suggestion. […] Can you think of anything else? What other ways can we think of so that we don't have to emit as much carbon dioxide?"

S4: "Reduce the demand for energy."

A: "Yes, and how can we do that? Do you have any suggestion for how we could 13. reduce it?"

S4: "Perhaps, like, I don't know. Use cell phones and such things less."

A: "So you can reduce your use of energy, exactly, and electricity and such. Absolutely. Now let's look at this picture…"

agreement with the syllabus objectives. Therefore intentions, in this case study, refer to syllabus intentions, which seemed to be fully adopted by Anna.

16.3 Results

Results are presented in two parts, one stemming from each type of analysis. The content analysis of all data sources highlights the degree of promotion of the five learning categories (Table 16.1) in different stages of the transformation process from intentions to teaching. An empirical example of how *Reasoning* was approached in planning and teaching is included to enrich these results. In addition, results from the inductive thematic analysis of interview data are used to present Anna's perception of how her intentions and teaching practices match and what factors impact the level of congruence.

16.3.1 Promoted Performances – Content Analysis

All five learning categories were represented in Anna's planning documents. Figure 16.2 shows a rough estimate of the proportion of the five categories as they appeared over time, from selecting syllabus text through planning, teaching and finally assessment. The syllabus text that Anna drew from was slightly dominated by *Reasoning*, while *RR* was more emphasized after the transformation of syllabus text into clarified objectives. All learning categories resurfaced among the planned activities. In interview 1, Anna mentioned *RR* more frequently than the other categories, however, she emphasized *Reasoning* as the main intention for the teaching unit. During instruction, *RR* was most prominent in the classroom, while opportunities for students to practice *Communicating* and *Reasoning* were less common. The two categories *Performing* and *Extended thinking* remained invisible in the classroom throughout the observation period. Final assessment of students' knowledge consisted of a written test asking them to define terms and concepts from a list, which exclusively engaged students in *RR*.

The empirical example presented next serves to deepen the understanding of how *Reasoning* was represented in the planning documents, Anna's expressed intentions (interview), and the classroom teaching. The example reports on a lesson that had slightly varied objectives, depending on where these were found (see Table 16.3 for a summary), which was not uncommon. The syllabus objective was that students should learn to "use knowledge of biology to examine information, communicate, and take a stand on questions (e.g. natural resource use)". Anna shifted focus when transforming the syllabus objective into a clarified objective, which instead listed factual content such as photosynthesis, combustion, fossil fuels and their significance for energy use and climate impact. During the enacted lesson, Anna explicitly told the class, and wrote on the whiteboard, that the lesson objective was that the students would become "familiar with the concept of the carbon cycle and know a little about the processes of the carbon cycle" (Anna, lesson 3 transcript). The activity involved students taking turns to read aloud from the textbook about the carbon cycle. This was followed by short sessions where students summarized and drew conclusions from the readings. The learning category *Reasoning* was strongly emphasized in the selected syllabus text, but less emphasized after the transformation into the clarified syllabus and presentation of objective to the class. In interview 1, reasoning was highlighted as a main intention for the teaching unit. However, as the lesson transcript below shows, questions with the potential to engage students in *Reasoning* were often reformulated into questions of *RR* character during interactions. The interactions seen in the transcript were not unique among the observed lessons, but is rather representative of an overall pattern. In interview 2, Anna addressed this lack of student engagement in *Reasoning*, as described in the section on inductive thematic analysis below.

Anna initiated the interactive sequence with an open-ended question with the potential to promote *Reasoning* (row 1), but quickly changed it into questions which induced *RR*-answers (rows 1–6). In rows 12–13, another open-ended question was asked and one student provided an answer of possible *Reasoning* character. Rather than allowing students to question or build on this and practice *Reasoning* about why and how to reduce energy demand, Anna wrapped things up with a short recap (rows 15–16) and moved on. In short, there were opportunities to engage students in *Reasoning*, but for some reason Anna adjusted the questions and rephrased them as *RR*-questions instead. In interview 2 she reflected upon this and acknowledged that "we could have discussed it a lot deeper" (Table 16.3).

16.3.2   Impacting Factors – Inductive Thematic Analysis

Anna implied in the first interview that "there are always changes" in school and this seven-week teaching unit was no exception. Anna's planned activities included two conversation exercises specifically intended for students to practice and develop their *Reasoning*. However, none of these exercises were conducted due to changes in the schedule implemented to accommodate a mandatory national Swedish test and a field day. Changes like these are inevitable in schools, and when faced with the task of prioritizing which activities to reduce, Anna chose to omit *Reasoning* rather than *RR*.

One factor that had an impact on the congruence between intentions and teaching, and which surfaced in the interviews, was stress. Anna identified the vast core content and time constraints as the main sources of stress. For example, when asked why they rushed through complex questions and/or answers during class discussion, Anna replied: "You have to move on, move on, move on. Like, 'well good that I got an answer' and then you move on" in order to cover the content that needs to fit into the teaching unit. Another factor causing incongruence which was identified in the interviews was a reassuring sense of repetitiveness. Anna expressed a notion of science as a subject that contains a lot of repetition for students during their compulsory schooling. This underpinned a sense of calm that rested on the assurance that students would practice and develop their *Reasoning* skills in secondary school, if not in upper primary school. For example, she said that "[s]ome things come back all the time, for example sustainable development", implying that if the students don't grasp the concepts the first time around, there will be more opportunities as they progress through school. On the other hand, Anna indicated that she felt uncertain about what students were expected to know before entering secondary school science. She talked about preparing students by encouraging them to independently search for answers to questions in texts, an activity that primarily requires skills in *RR*.

While the observation results support the interview findings regarding the time limitation of science lessons, they show that most of the lesson time is spent on teaching practices aimed at students learning to recall and reproduce scientific terms and concepts. An example comes from the transcript above, where open-ended questions were replaced with recall questions, and student reasoning was acknowledged, but not further elaborated. Anna suggested that students need "some knowledge about how things work too, to be able to take a stand. They need some background, I mean some knowledge". Another example was prioritizing teaching terms and concepts while omitting *Reasoning* exercises to make room for extracurricular activities. In summary, while the basis for Anna's LPP was copied from the science syllabus and thereby showed great congruence, the emphasized learning category *Reasoning* was excluded both from the clarified syllabus and the objectives presented to the class.

16.4 Discussion

In this study, it was found that Anna's main intention to offer opportunities for students to engage in *Reasoning* about subject matter such as photosynthesis, the carbon cycle, and human impact on climate were not efficiently transformed into enacted teaching practices. This resembles the finding from the study of a secondary school science teacher who, despite having "positive dispositions towards and knowledge of ESD", was not able to "make ESD happen" (Tomas et al., 2022, p. 11). There were opportunities for *Reasoning* through Anna's open-ended questions in class. However, the rephrasing of these into *RR*-eliciting questions suggests that factual content was more prioritized and this notion is further supported by the fact that when lesson time was devoted to extracurricular activities, the lessons revolving primarily around *Reasoning* were omitted, while those centered on *RR* remained. This may relate to the results of a previous interview study, which indicated that the most influential teacher role in primary school science is that of "The Encyclopaedia", whose objective is to share established scientific facts, theories, and concepts (Varg et al., 2022). It may also be a sign of the reduced teacher autonomy experienced by the teacher in Tomas et al. (2022) who struggled to cover all factual content of the teaching unit, ultimately at the expense of ESD. The space of tensionality (Lewthwaite et al., 2014) appears to allow mainly concrete factual content to seep through into the clarified syllabus and the enacted teaching. Regardless of whether the activities focused on scientific models such as the carbon cycle or more complex issues such as ESD, Anna hustled to cover factual content to prepare her students for secondary school science. Such a stance is a natural consequence of the "standardized accountable environment" that is encouraged within the current educational discourses (Ryan & Bourke, 2013, p. 412). In this case study, the first transformation occurred right at the beginning of the planning process, when syllabus text referring to *Reasoning* was filtered out while *RR* dominated Anna's clarified syllabus. The learning category *Reasoning* seemed more difficult to transform into clear objectives than *RR*. Understandably, one possible reason for Anna's preference for *RR* is that she was a new teacher and possibly relied on the textbook, which essentially offered content suitable for memorization.

Anna identified two main reasons for the lack of congruence between the intended and enacted curriculum. In addition to the stress caused by the extensive subject matter (as discussed above), Anna expressed a sense of relief in knowing that different learning categories are constantly reappearing throughout compulsory school science education. She expected that if students did not develop proper reasoning skills in upper primary school, they would be able to do so in secondary school. At the same time, she expressed an uncertainty about what was expected of the students when they entered secondary school science. This is similar to what the teacher in Tomas et al. (2022) experienced when considering what the mandatory tests would examine. The observations made it possible to draw some alternate, or complementary, conclusions about the lack of congruence. Although time was a limiting factor in this case, where science lessons were relatively few and short, most of the instruction and all assessment was focused on students being trained to recall and reproduce science facts. Was there a lack of access to suitable teaching strategies to encourage classroom talk of *Reasoning* character or was this emphasis on *RR* a sign of strong academic traditions defining upper primary science (Lidar et al., 2019). Teaching strategies have been found among expert teachers, who resort to these instead of careful lesson planning (Hatch & Clark, 2021). Although Anna asked open-ended questions, her habit of asking several questions in a row resulted in students answering the last ones which tended to be phrased as *RR*-questions. Her planning documents were ambitious and elaborate, but ultimately, and as indicated by Anna (Table 16.2), the lesson plan may have presented an obstacle rather than a tool in her attempts to realize the intention for students to practice and develop *Reasoning* abilities as she hurried to cover factual content instead.

This case study, although small and including only one teacher in one classroom, makes an important contribution to inform researchers, teachers and teacher educators about potential pitfalls to consider during the transformation of intentions into teaching. This may apply particularly to less established content, such as sustainable development and higher order learning categories, like *Reasoning*. This is because their positions within science educational culture are not as pronounced as traditional subject content, which is well promoted in both curriculum and textbooks, and which tends to elicit *RR* (van Eijck & Roth, 2013). Teacher education and professional development efforts could benefit from using these study results to support pre- and in-service teachers, not only in their development of teaching strategies that promote students' reasoning, but also in their efforts to navigate, interpret and transform science syllabi within the space of tensionality. This could help to strengthen teachers' reflexivity and increase their agency, thereby giving them the autonomy needed to handle constant changes in the context and/or culture within which they work.

References

Guest, G., MacQueen, K. M., & Namey, E. E. (2012). *Applied thematic analysis*. SAGE.
Hatch, L., & Clark, S. K. (2021). A study of the instructional decisions and lesson planning strategies of highly effective rural elementary school teachers. *Teaching and Teacher Education, 108*, 103505.
Krippendorff, K. (2019). *Content analysis: An introduction to its methodology* (4th ed.). SAGE.
Kvale, S. (1997). *Den kvalitativa forskningsintervjun*. Studentlitteratur.
Lewthwaite, B., Doyle, T., & Owen, T. (2014). 'Did something happen to you over the summer?': Tensions in intentions for chemistry education. *Chemistry Education Research and Practice, 15*(2), 142–155.
Lidar, M., Engström, S., Lundqvist, E., & Almqvist, J. (2019). Undervisningstraditioner i naturvetenskaplig undervisning i relation till svenska utbildningsreformer i skolår 6. *Nordina: Nordic studies in science education, 15*(2), 174–192.
Pepin, B., Gueudet, G., & Trouche, L. (2013). Investigating textbooks as crucial interfaces between culture, policy and teacher curricular practice: Two contrasted case studies in France and Norway. *ZDM, 45*(5), 685–698.

Porter, A. C. (2004). *Curriculum assessment*. Vanderbilt University.

Ryan, M., & Bourke, T. (2013). The teacher as reflexive professional: Making visible the excluded discourse in teacher standards. *Discourse: Studies in the Cultural Politics of Education, 34*(3), 411–423.

Sirhan, G. (2007). Learning difficulties in chemistry: An overview. *Journal of Turkish Science Education, 4*(2), 2–20.

Stake, R. E. (1995). *The art of case study research*. SAGE.

Swedish Research Council. (2017). *Good research practice* (VR1710). Swedish Research Council.

The Swedish National Agency for Education. (2018). *Curriculum for the compulsory school, pre-school class and school-age educare 2018*. The Swedish National Agency for Education.

Tobin, K., McRobbie, C., & Anderson, D. (1998). Dialectical constraints to the discursive practices of a high school physics community. *Journal of Research in Science Teaching, 34*(5), 491–507.

Tomas, L., Mills, R., & Gibson, F. (2022). 'It's kind of like a cut and paste of the syllabus': A teacher's experience of enacting the Queensland Earth and Environmental Science syllabus, and implications for Education for Sustainable Development. *Australian educational researcher., 49*(2), 445–461.

van Eijck, M., & Roth, W. (2013). *Imagination of science in education: From epics to novelization*. Springer.

Varg, L., Näs, H., & Ottander, C. (2022). Science teaching in upper primary school through the eyes of the practitioners. *Nordina: Nordic studies in science education., 18*(1), 128–142.

Webb, N. (1997). *Criteria for alignment of expectations and assessments in mathematics and science education*. National Institute for Science Education Publications.

Ziebell, N., & Clarke, D. (2018). Curriculum alignment: Performance types in the intended, enacted, and assessed curriculum in primary mathematics and science classrooms. *Studia Paedagogica (Brno), 23*(2), 175–203.

Ziebell, N., Ong, A., & Clarke, D. (2017). Aligning curriculum, instruction and assessment. In T. Bentley & G. Savage (Eds.), *Educating Australia: Challenges for the decade ahead* (pp. 257–276). MUP Publishing.

Chapter 17
Primary Students' Visions Regarding Environmental Factors Influencing Biodiversity in Specific Environments

Chadia Rammou, Arnau Amat, Isabel Jiménez-Bargalló, and Jordi Martí

17.1 Introduction

In this chapter we present and discuss which factors primary students take into account when they think of the biodiversity of particular environments. This study is part of a larger research project with the main goal of understanding how primary students conceptualize biodiversity, the main factors of biodiversity loss and which collective and individual actions could promote an improvement of biodiversity. Specifically, our research questions are: *Which factors do primary school children take into account when we discuss the biodiversity of a particular environment?* and *What kind of factors do they consider as biodiversity enhancers or limiters?*

Furthermore, this study is framed in an educational innovation project called Patis Biodivers, with two major goals. On the one hand, it endeavours to promote processes of authentic inquiry practices among kindergarten, primary and secondary students, such as: collecting empirical data of their schoolyard, drawing conclusions, and promoting actions to improve the diversity of the species. On the other hand, it aims to use this empirical data from schoolyards to do scientific research. Last year, in a pilot project, the materials of the project were created collaboratively from an interdisciplinary team of researchers from the authors' university, with the support of more than 40 teachers, from 13 different primary and high schools.

C. Rammou (✉) · A. Amat (✉) · I. Jiménez-Bargalló (✉) · J. Martí (✉)
Universitat de Vic – Universitat Central de Catalunya, Barcelona, Spain
e-mail: chadia.rammou@uvic.cat; arnau.amat@uvic.cat; isabel.jimenez@uvic.cat; jordi.marti@uvic.cat

© The Author(s) 2024
K. Korfiatis et al. (eds.), *Shaping the Future of Biological Education Research*,
Contributions from Biology Education Research,
https://doi.org/10.1007/978-3-031-44792-1_17

17.2   Rationale Behind the Research

After the United Nations Decade on Biodiversity (2011–2020), the Convention on Biological Diversity published a first draft framing the Post-2020 Global Biodiversity Strategy. The document, which is aligned with 2030 Sustainable Development Goals, sets out a plan to implement numerous actions to transform society's relationship with biodiversity (Convention on Biological Diversity, 2021).

In this respect, learning about biodiversity is seen as a key element of this strategy, which is based on four main arguments: (a) emotional, creating personal meanings through discovery and experiencing biodiversity; (b) ecological, understanding the global interdependencies between the different elements of the ecosystem; (c) ethical, dealing with values and taking a moral stance on environmental issues; (d) political, making choices and developing action competence (Van Weelie & Wals, 2002).

Thus, learning about biodiversity is not only related to learning facts from different sources of information, but also it should be regarded as experiential learning, involving participants in a community, hands-on activities, and contact with nature. For this reason, Patis Biodivers focuses on how to help teachers to sustain authentic inquiry to learn about biodiversity. According to Afkin and Black (2003), authentic scientific inquiry involves "doing science" resembling the actual practice of scientific communities. Roth and Lee (2004) also pointed out that authentic inquiry in school settings ought to be motivated by the same goals within the scientific or local community where the inquiry is taking place.

Understanding the environmental factors influencing biodiversity is crucial to promoting informed actions for its preservation. In this respect, in a previous study where primary students were asked to consider niche variables to estimate the likelihood that wolf populations become established in the Pyrenees, the results showed that 10-to 11-year-old students only took into account few variables (Jiménez et al., 2023). Before any kind of instruction, primary students considered shelter and water as the most important abiotic factors, while availability of primary consumers was seen as the key biotic element. Anthropic factors were also considered by students, thus contemplating how human beings could disturb a wolf population settlement. The study reports sophistication of students' ideas after three weeks of engagement in an authentic inquiry project, therefore demonstrating the potential of this kind of instruction.

As reported by Kilinc et al. (2013), the students pointed to some human factors as one of the main factors in the loss of biodiversity, whether through hunting or through the imbalance of nature. However, vegetation is mentioned as an enhancer of biodiversity. In their study, 245 students were interviewed with the aim of identifying students' conceptions of biodiversity loss.

Moreover, regarding how young children make sense of the living beings, it is important to highlight personification as a common thinking strategy. According to Inagaki and Hatano (2002), when children do not have enough knowledge of a particular animate object, they tend to construct an explanation using the person analogy in a constrained way. As children are familiar with humans they use this knowledge as the source to predict the reactions of less familiar animate objects.

17.3 Methodology

This study is situated within the frame of an interpretative paradigm, since its aim is to show how primary students consider factors to explain biodiversity in specific environments. In this way, the main objective is explored through the use of different questions that start from the subjective experiences of the students.

In this study, data was obtained from 49 primary school students (8–12 years old), from three urban schools which participated in the Patis Biodivers project. For students participating in the study it was the first time that they were conducting inquiries and learning about biodiversity. Each school decided to carry out a biodiversity inquiry of one group of living beings according to their curriculum requirements. Two of these schools worked on birds, while the other one worked on invertebrates. Just before starting the inquiry-based learning project with the children, 13 focus groups, with the participation of four or five children for each group, were conducted to collect data regarding the aim of the research. One of the main aims was to identify the key factors that children mention constraining or promoting the biodiversity of specific environments. The same focus group also aimed to identify children's perceptions of the definition of biodiversity and what actions they choose to deal with biodiversity loss.

During the focus groups discussions, students were provided with some depictions of different environments. After looking at them, they were required to discuss which one of them sustained more biodiversity. Specifically, they were asked: *Which of these environments do you think is the most biodiverse? Which of these environments do you think is the least biodiverse? And why do you choose these ones?* The depictions of the environments presented are described below:

– Desert: *an arid region with a large amount of sand and dunes, where precipitation is not common.*
– Jungle: *an area covered with dense vegetation.*
– Forest: *an area where trees predominate.*
– Polar areas: *an area which is mostly covered by frozen landscape.*
– Savannah: *an ecosystem characterized by trees being widely separated.*
– Grassland: *an area dominated by vegetation. Most of the land is covered by grass. Sedge and rush can also be found.*
– Marine ecosystems: *an area covered by water.*

Even though the Patis Biodivers project is based on studying the living beings of their schoolyard, children's explanations of why they considered a specific environment more or less biodiverse elicited their ideas about factors responsible for the gain or loss of biodiversity.

In addition, children's discussions were recorded and transcribed. A qualitative content analysis approach, using an inductive category development strategy, was conducted (Mayring, 2000). Firstly, the transcription was divided into significant discourse units, considering only those quotes that provide meaningful information. After that, each significant unit was labeled using an emergent code family system (as shown in Table 17.1) to make sense of the data collected.

Table 17.1 Codes from the view of primary students about the enhancers and limiters they identify in a certain environment

Systemic network				
Enhancers	Abiotic factors	Sunlight	Refers to the light that comes from the sun.	2/94
		Humidity	When the ground is damp, and the air feels heavy.	4/94
		Temperature	Refers to the environmental temperature.	2/94
		Rivers	The water flows in a long line over the land.	4/94
		Shadow	Refers to those places where the light of the sun does not arrive.	5/94
	Biotic factors	Requirements linked to welfare	Include calmness and comfort of the different groups of living beings.	16/94
		Vital requirements	In terms of essential needs, such as water or oxygen to survive.	8/94
		Trees for nesting	Allude to the tree as a place where birds can build their nests.	5/94
		Presence of vegetation	Relating to the presence of a large quantity of plants, grass, trees, and shrubs in a particular area.	25/94
		Trees as an attraction of other living beings	Allude to trees as a natural structure that attracts invertebrates, birds, etc.	1/94
		Trees as a shelter	Allude to trees as a place where some living beings use as a refuge, as protection from danger.	1/94
		Vegetation as producers	Regarding the vegetation which can be eaten.	11/94
		Specific adaptation of some living beings	These include various types of animals and plants that have developed certain abilities and skills to adapt to a particular place.	1/94
		Absence of carnivores	It is the non-presence of carnivores.	1/94
	Anthropic factors	Absence of humans	It is the non-presence of humans.	7/94

Limiters			
Abiotic factors	High temperature	A high measurement of the temperature.	8/57
	Low temperature	A low measurement of the temperature.	9/57
	Sand storms	A large amount of sand in the air.	2/57
	Drought	A lack of water.	2/57
	Absence of earth surface	A lack of earth surface to support life-beings	2/57
	Absence of sunlight	A lack of sunlight.	2/57
Biotic factors	Absence of vegetation	The non-presence of vegetation.	7/57
	Absence of vital requeriments	The non-presence of water or oxygen.	10/34
	Absence of trees for nesting	The non-presence of trees.	1/57
	Absence of welfare requeriments	The lack of calmness, liberty and comfort.	1/57
	Presence of carnivores	The presence of some carnivores.	3/57
	Absence of vegetation as producers	The lack of vegetation to eat.	5/57
Anthropic factors	Mainly caused for humans	The problem is directly caused by some negative actions of people.	2/57
	Aggravitation of natural phenomena due to human factor	Relating to some indirect actions of people that affect the positive development of living beings.	2/57
	Presence of humans	The presence of people.	1/57

Table 17.2 System network of jungle environment with enhancers and limiter limiting factors of biodiversity

Jungle	Enhancers (67/70)	Abiotic factors (14/70)	Sunlight	2/70
			Humidity	4/70
			Temperature	2/70
			Shadow	3/70
			Rivers	3/70
		Biotic factors (46/70)	Requirements linked to welfare	14/70
			Vital requirements	8/70
			Trees for nesting	4/70
			Presence of vegetation	11/70
			Trees as an attraction for other living beings	1/70
			Vegetation as producers	8/70
		Anthropic factors (7/70)	Absence of humans	7/70
	Limiters Limiting factors (3/70)	Biotic factors (1/70)	Presence of carnivores	2/70
		Anthropogenic factors (2/70)	Mainly caused by humans	1/70

Having defined the code families, each environment was analysed using a system network (Bliss et al., 1983) as presented below (Table 17.2). The system network was organized in two dimensions: On the one hand, the enhancers represent key factors that improve the biodiversity of particular environments; on the other hand, the limiters as key factors are associated with the lack of biodiversity. For each of these two dimensions, there were also three categories: abiotic, biotic, and anthropic. The first category refers to the elements that are part of the biotope of the environment, such as environmental conditions. The second category refers to living organisms, such as trees as nesting sites. Finally, the anthropic category refers to all elements that are concerned with humans.

17.4   Results

As shown in Fig. 17.1, children mainly chose the forest and the jungle as the most biodiverse environments, identifying vegetation as a key factor to enhance biodiversity. Most of the enhancer factors are identified in these two areas, considering them as suitable places where animals can find the necessary requirements, from environmental to vital conditions (e.g., *"I think the jungle is one of the places I chose because there are many animals and no people. Also, it is humid there and, in some areas, there is even sunlight. There are a lot of flowers and plants. They feel better in the absence of people, in peace, and with the food they need"*). On the other hand, polar areas and deserts are considered the least biodiverse environments because of the lack of vegetation and the environmental conditions (e.g., *"Because there aren't any plants"*). Having said that, it is important to remark that only one student mentioned savannah as an environment with biodiversity, without specifying any factor (e.g., *"Because there are a lot of animals"*). It is also relevant to state that children considered marine environments as non-biodiverse environments, since they did not take into account the presence of vegetation in their discussions. Just one student referred to the presence of vegetation underwater (e.g., *"I choose marine environment because there are a lot of fish and there is vegetation underwater"*).

In the group discussion, factors promoting biodiversity were more salient than factors with negative effects on biodiversity. In total, 151 significant discourse units were identified: while 94 of them were related to enhancing factors, only 57 were related to negative factors. Most of these units were described in the jungle environment (see Fig. 17.1).

Regarding the enhancers, as it can be seen in Fig. 17.2, biotic factors topped all other categories. Within this category, a total number of 69 significant discourse units were identified. It is noteworthy that children mostly chose these factors as enhancers for animal biodiversity, not as enhancers for biodiversity in general. Some of the most important factors to explain the biodiversity in particular environments are related to welfare and vital requirements (e.g., *"I have chosen the forest because there are many birds there, because it is calm and quiet"*), but also the absence of carnivores in certain places.

However, vegetation is considered the most influential factor to promote the biodiversity of the specific area, which was mentioned in 31 significant discourse units, (e.g., *"Because it's calm and there is a lot of vegetation"*). Consequently, the absence of these elements in a specific environment is defined as an unsuitable place for animals or plants to survive. It should also be noted that the children acknowledged diverse roles of vegetation as enhancers of biodiversity. The children mentioned trees as an important element from different perspectives: on the one hand,

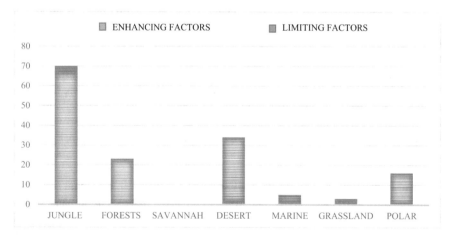

Fig. 17.1 Number of significant discourse units mentioned by students as enhancing or limiting biodiversity factors in relation to a particular environment

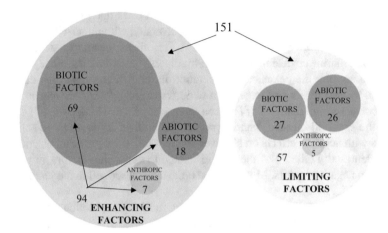

Fig. 17.2 Representation of the predominance of each dimension and categories. The larger the circle, the more significant discourse units are mentioned by the children

trees were seen as structural elements of the habitat that provide shelter for animals in general or a place to nest for birds (e.g., *"I think that there is a great diversity of species in the forest and jungle. The birds can build their nests in the trees)*; on the other hand, vegetation was considered as a dynamic element that provides food for animals.

Regarding abiotic factors, in the group discussions, primary students took into account environmental factors that promoted biodiversity, such as humidity, shadow, sunlight, environmental temperature and the presence of rivers. Among them, humidity was the most important factor to promote biodiversity in the focus group (e.g., *"The jungle, because there is moisture and there are many insects, a great diversity of species, water..."*).

As far as limiting factors are concerned, abiotic factors are the most important ones that account for the lack of biodiversity in some environments. Children understood extreme temperatures as a key factor that constrains the biodiversity of deserts and polar environments. They acknowledged that animals cannot survive in such extreme conditions (e.g., *"I think in the desert the temperatures are very high and it is one of the places where it is very hot. An animal cannot survive there"* or *"Because it is very cold there"*). In addition, the lack of vegetation was also mentioned as a limiting biotic factor, together with the absence of trees. Mostly, they brought it up when discussing desert, polar and marine environments (e.g., *"I believe that the polar environment is not very rich in species due to low temperatures and lack of vegetation."*).

Apart from abiotic and biotic factors, the children also took into account anthropic factors as limiters of biodiversity. In particular, the mere presence of humans, which affects animals' tranquillity, or the anthropic impact, which unbalances ecosystems, were important factors to justify the limitation of biodiversity in certain places. In fact, this last factor was relevant to explain why students considered the marine environment as one of the least biodiverse. Students claimed that humans pollute the ocean, thus causing a decline in biodiversity (e.g., *"I think the problem is that there is a lot of plastic in the marine environment, so there are fewer animals"*).

Results also reveal that students sometimes used personification to explain the biodiversity in a particular environment. They considered those factors by taking human needs as a reference point (e.g., *"Because it is a quiet place"*). Mostly, they identified human needs with animal and plant needs.

17.5 Conclusions

In this paper, the focus of attention was on what kind of factors primary students take into account in terms of enhancing or limiting the biodiversity of a specific environment. The results show that students identified above all, the presence of vegetation as an important element that allows the specific area to be diverse, mainly in terms of animal diversity. It is important to note that children tended to exclude humans as a living organis. This narrow view is reflected in their discourses, where, in most of the cases, they identified the enhancing or limiting factors in terms of animal biodiversity, rather than using a broader term, such as biodiversity in general. In consonance with other studies (Morón et al., 2021), these findings suggest the need to look for spaces where children can acquire the necessary tools to avoid this blind view, and consider biodiversity in a more comprehensive sense. Consequently, a more comprehensive and holistic education about biodiversity and ecosystem functioning should be taught.

To a lesser extent, the need for vegetation as some source of food was also included. Moreover, variables from abiotic factors, such as a suitable weather, were also considered to promote biodiversity, whereas too low/high temperatures were seen as a limiting factor. Similar results were found in a previous study on the wolf population requirements (Jiménez et al., 2023), in which the children also considered the weather as a limiting factor for the wolf.

In addition, the lack of space in natural habitats was also cited as an indirect limiting factor that made students concerned about the loss of biodiversity. This concern could be seen when they identified the presence of trees as an important element, especially when the students described the trees for nesting. A similar outcome was also observed in other studies (Kilinc et al., 2013), since the lack of habitat is one of the factors that children identify as an important element to consider.

Furthermore, welfare factors were the elements that children considered the most. Students usually mentioned them in the group discussions to argue that animals cannot survive without these factors. Most of the time, they put themselves in the situation and spoke from their own perspective. In accordance with Inagaki and Hatano (2002), these ideas reflect the way of reasoning of children using a person analogy. Since the children did not have a good knowledge of how an ecosystem works, they tended to map the factors taking into account their own experiences as humans.

Besides, the answers referred to limiting and enhancers factors were really simple. According to the study carried out by Jiménezt et al. (2023), they also found that all the factors considered by children as necessity for the wolf to be established were straightforward because in the course of their explanations they mostly developed a linear cause-effect relationship. Students mentioned a certain factor, negative or positive, and connected it with its respective effect in the environment, without including other variables.

It is also important to note that the children took into account animals, mostly among the other living beings of the ecosystem, to describe how species-rich a particular environment was. In this regard, the elements that they did not mention would suggest that students focused only on particular groups of living beings of the environment, ignoring a broad range of possible ecological relationships. These results are consistent with other studies which have shown that young children count animals first, and then plants and other living beings, when describing the elements of a particular ecosystem (Kilinc et al., 2013).

Regarding the human factors, participants were really concerned and referred to humans only as one of the main limiting factors to enhance biodiversity, by only describing factors such as a subsequent environmental pollution, deforestation and other negative impacts. Moreover, since they never included themselves as an active agent in the discussions, we can conclude that conservation actions were not taken into account. In this respect, other studies (Kilinc et al., 2013) pointed out that conservation education needs to be taught in schools to change this view and give children the tools they need to act.

In light of the results obtained and discussed with the related literature in this study, it seems clear that there is a need to promote science teaching practices to assist children's development of a more comprehensive meaning of biodiversity. The anthropocentric and simplistic view observed in this and similar studies should be taken into account to review current teaching practices. Students' early experiences are of great importance for the development of their ability to discern biodiversity. Therefore, understanding misunderstandings and incompletions that occur when children try to acknowledge such a complex concept should help us bring a wider concept of biodiversity into formal education in order to empower students to decide on current sociocientific issues.

Acknowledgements This research was supported by Fundación Española para la Ciencia y la Tecnología - Ministerio de Ciencia e Innovación.

References

Afkin, J. M., & Black, P. (2003). *Inside science education reform: A history of curricular and policy change*. Teachers College Press.
Bliss, J., Monk, M., & Ogborn, J. (1983). *Qualitative data analysis for educational research: A guide to uses of systemic networks*. Croom Helm.

Convention on Biological Diversity. (2021). *First draft of the post-2020 global biodiversity framework*. Convention on Biological Diversity. https://www.cbd.int/doc/c/abb5/591f/2e46096d3f0330b08ce87a45/wg2020-03-03-en.pdf

Inagaki, K., & Hatano, G. (2002). *Young children's naive thinking about the biological world*. Psychology Press.

Jiménez, I., Amat, A., & Codony, L. (2023). The wolf is back! Revisiting the concept of the ecological niche to predict the viability of the wolf population with elementary students. *Journal of Biological Education, 57*(4), 791–810. https://doi.org/10.1080/00219266.2021.1979626

Kilinc, A., Yeşiltaş, N. K., Kartal, T., Demiral, Ü., & Eroğlu, B. (2013). School students' conceptions about biodiversity loss: Definitions, reasons, results and solutions. *Research in Science Education, 43*(6), 2277–2307. https://www.infona.pl/resource/bwmeta1.element.springer-3eac6cde-0aa7-3b1e-8c18-1d6c4a332748

Mayring, P. (2000). Qualitative content analysis. *Forum Qualitative Social Research, 1*(2), Art 20. http://nbn-resolving.de/urn:nbn:de:0114-fqs0002204

Morón, H., Hamed, S., & Morón, M. (2021). How do children perceive the biodiversity of their nearby environment: An analysis of drawings. *Sustainability, 13*, 1–19. https://doi.org/10.3390/su13063036

Roth, W. M., & Lee, S. (2004). Science education as/for participation in the community. *Science Education, 88*(2), 263–291.

Van Weelie, D., & Wals, A. (2002). Making biodiversity meaningful through environmental education. *International Journal of Science Education, 24*(11), 1143–1156. https://doi.org/10.1080/09500690210134839

Chapter 18
Learning Biology in the Early Years Through Nature Play in the Forest: An Exploratory Study from Slovenia

Marjanca Kos, Sue Dale Tunnicliffe, Luka Praprotnik, and Gregor Torkar

18.1 Introduction

In early years, children mostly learn through play. Outdoor environments, especially natural environments with their characteristics of diversity, variability, unpredictability, openness and richness in sensory stimuli are known to be especially stimulating for children's play and learning. The potential offered by the outdoor environment is supposed to be most effectively used for education through unstructured, child-initiated and child-led play (Maynard & Waters, 2007; Wilson, 2012).

In recent research, the term 'nature play' has been used to define freely chosen, unstructured, and open-ended playful interactions with and in nature (Erickson & Ernst, 2011; Ernst et al., 2021). The purpose of this study is to contribute empirical data about the importance of nature play in early childhood, especially for learning biology.

18.1.1 The Benefits of Children's Engagement with Nature

Childhood experiences in the world of nature are crucial for children's life. Several studies have identified nature as a significant space, which supports physical and mental health, as well as emotional well-being, and fosters children's holistic

M. Kos (✉) · L. Praprotnik · G. Torkar
Faculty of Education, University of Ljubljana, Ljubljana, Slovenia
e-mail: Marjanca.Kos@pef.uni-lj.si; Luka.Praprotnik@pef.uni-lj.si;
Gregor.Torkar@pef.uni-lj.si

S. D. Tunnicliffe
UCL Institute of Education, London, UK
e-mail: S.Tunnicliffe@ucl.ac.uk

© The Author(s) 2024
K. Korfiatis et al. (eds.), *Shaping the Future of Biological Education Research*,
Contributions from Biology Education Research,
https://doi.org/10.1007/978-3-031-44792-1_18

development in all the developmental domains (Adams & Savahl, 2017; Gill, 2014; Sahlberg & Doyle, 2019; Wilson, 2012). Children's engagement with nature leads to better learning because learning in nature improves learners' attention, levels of stress and self-discipline, and it provides a more supportive, quieter and safer context of learning (Kuo et al., 2019). Play and learning in natural environments also improve the learning of disadvantaged children with learning disabilities and with underachieving academic scores (Maynard et al., 2013).

Many studies have consistently shown that children are more physically active when outdoors (Dankiw et al., 2020; Gill, 2014; Torkar & Rejc, 2017). Walking over rough terrain, climbing trees and running around in the natural environment positively affects children's motor development: children's gross and fine motor skills and stamina; children's coordination and balance; improves health-related fitness (Santana et al., 2017). Additionally, it reduces obesity risks (Dankiw et al., 2020; Herman et al., 2009).

Nature has prosocial effects as it fosters warmer, more cooperative relations (Dankiw et al., 2020; Scott et al., 2018). Natural environments bring children more freedom and sense of autonomy (Adams & Savahl, 2017; Dankiw et al., 2020; Kuo et al., 2019). Outdoor play and learning have a positive effect on children's self-esteem, self-confidence and self-awareness, as well as on how they make choices and take risks (Gill, 2014; O'Brien & Murray, 2007). It helps children acquire skills that are so important later in adulthood, such as perseverance, self-efficacy, resilience, teamwork, leadership and communication (Kuo et al., 2019). Nature reduces their behavioural problems (Fiskum & Jacobsen, 2012); it showed to be effective in the reduction of disruptive episodes and dropouts among 'at-risk' children (Ruiz-Gallardo et al., 2013). Children who ordinarily struggle when indoors emerge as leaders and a low-performing child gets more of a chance to build an image of a strong, competent person (Maynard et al., 2013; Kuo et al., 2019).

In the affective field of development, research has shown an increase in mood and better emotional regulation: a decrease in depression and aggression (Brussoni et al., 2017; Gill, 2014). Children generally love playing in natural environments (Wilson, 2012). In the nature, children have been observed to laugh and smile more than those in highly structured play environments (Singer, 1994).

Outdoor experiences foster environmental ethics. There is good evidence of a link between time spent in nature in early childhood and adult environmental attitudes (Barrable & Booth, 2020; Gill, 2014; Torkar, 2014). Frequent outdoor experiences and contact with nature strengthen children's empathic relationship with nature, as well as promote their intrinsic care, emotional connection and appreciation of nature. A lack of regular positive experiences in nature is associated with the development of discomfort, fear and the dislike of the natural environment, as well as a failure to develop a personal connection to the world of nature (Adams & Savahl, 2017; Ernst et al., 2021; Gill, 2014; Jørgensen, 2016). Connectedness with nature should be advanced in early childhood, i.e. under the age of 11, as environmental education programmes have been found to be more sustainable amongst this age group (Lieflander et al., 2013). Nature experiences also improve environmental knowledge (Gill, 2014; Kuo et al., 2019).

18.1.2   The Role of Play in Early Childhood

Most of what children need to learn in their early childhood could be discovered through play. Play is a fundamental avenue for early childhood learning and is deeply rooted in early childhood education as the primary way to meet children's developmental needs (Sahlberg & Doyle, 2019; Wilson, 2012). Research supports the idea that play is so important to our development and survival that the impulse to play has become a biological drive and that it is internally generated. Children are naturally drawn to the low-risk scenarios of play in order to learn, grow, adapt and thrive (Brown, 2009). Neuroscience studies in recent years have also linked childhood play to brain development and proper functioning (Yogman et al., 2018). Play helps shape the brain. Animal play researchers who have studied the effects of play on brain development in depth suggest that the brain actually develops a sense of itself during play through stimulation and testing. This helps to explain why play is most prevalent during childhood, which is also the most important period of brain development (Brown, 2009; Loebach & Cox, 2020; Yogman et al., 2018).

18.1.3   Learning Biology in Early Years

Young children actively construct a coherent worldview and knowledge based upon their personal experiences. Therefore, active engagement is crucial for children's learning. Essential qualities of early biology education are to be hands-on and interest based. Adults should create a context in which pre-schoolers can have worthwhile, meaningful, cooperative and fun learning experiences; play is one of such contexts (Curriculum for preschools, Ministry of Education and Sport, 1999). Loebach and Cox (2020) observed children's behaviour in different outdoor spaces (not exclusively natural environments) and developed typology for capturing children's play behaviour (Loebach & Cox, 2020). Among nine different types of outdoor play behaviour (physical, exploratory, imaginative play, play with rules, expressive play, digital play, restorative play and non-play), they described a new play type – *bio play* – which included playful interactions between children and living things: 'Although these experiences might also be recorded in another category, such as *exploratory play*, the significance of these natural experiences is profound enough to warrant capturing these interactions through a distinct play type' (Loebach & Cox, 2020, p. 5611). This new type, named 'bio play' was divided into three play subtypes: 'Plants', 'Wildlife' and 'Care'.

18.1.4 Nature: A Stimulating Place for Children's Play and Learning

Quality play requires access to a rich learning environment. Natural environments provide a variety of spaces and ground cover, loose parts that children can manipulate, the possibility of 'random' events and can be characterised by natural elements, such as plants, animals, rocks, mud, sand, gardens, forests and ponds, or water. Thus, nature provides the diversity, variability, openness, rich and varied sensory stimuli needed to engage and challenge young children, which enhances the opportunity for learning and development through play (Dankiw et al., 2020; Ernst et al., 2021; Klofutar et al., 2022; Wilson, 2012).

Spending time in nature promotes developmental forms of play. Many researchers have demonstrated significant increases in constructive, dramatic, imaginative and symbolic forms of play (Kuo et al., 2019; Wilson, 2012). Wojciehowski and Ernst's (2018) study measured the creative outcomes of play in the natural environment. Results showed significant increases in originality, play fluency and imagination compared to controls who played in traditional play spaces.

The natural environment is a very appropriate learning environment for a wide range of activities: from highly structured, adult-led activities to unstructured, free, child- initiated and child-led play. Finally, the term 'nature play' has been used and is defined as freely chosen, unstructured and open-ended playful interactions with and in nature (Erickson & Ernst, 2011).

The natural environment with its features is a place where high-quality interactions between a child and an adult can develop (Maynard et al., 2013). The adult is encouraged to respond to children's initiatives and base activities on their interests and prior knowledge (Waters & Maynard, 2010). The educator should always be ready for sensitive interventions, while allowing children to play freely without interference (Waller, 2007). The educator enters nature play as a facilitator of learning and a more experienced partner. At the right moment, he or she engages with it, elevating it to a higher level and enabling more effective learning (Maynard & Waters, 2007; Maynard et al., 2013).

The importance of play in early childhood learning has been well established and the natural environment has been recognized as one of the most stimulating environments for it. Free (unstructured, child-initiated and child-led) play is the category of play in which the potential offered by the natural environment is supposed to be most effectively used. In the field of early science education (and especially in the field of early biology education), little research has been done on the importance of free, unstructured play in the natural environment (nature play) for gaining initial science experiences, acquiring science literacy, developing science skills, and environmental awareness (Beery & Jørgensen, 2018; Jørgensen, 2016; Tunnicliffe, 2020). The main purpose of our research was to contribute empirical data on what children can learn about biology through nature play in the forest.

18.2   The Exploratory Case Study

The aim of this exploratory case study was to find out how nature play in the forest provides preschool children with play episodes where they experienced biological phenomena and living organisms. The following research questions were formed:

1. How frequently did bio play episodes occur during nature play?
2. What living beings did the children interact with during the bio play episodes?
3. Which types of bio play episodes were detected?

One of the main reasons for choosing this research methodology was to provide empirical evidence on daily educational practice and, in particular, educational context. Case study methodologists stress that teachers always teach in particular places, specific groups of students and under conditions that significantly shape and temper teaching and learning practices (Freebody, 2003). Early exploratory case studies are set to explore any phenomenon in the data which serves as a point of interest to the researcher, where variables are still unknown and the phenomenon not at all understood (Meredith, 1998).

A non-random sample of 21 four-to six-year-old children (11 girls, 10 boys) from a public preschool in a suburb region of Ljubljana, Slovenia, participated in the study. The European commission's ethical rules in social science research were considered. Informed consent was obtained from the parents and teachers to conduct the study.

The preschool is situated less than a kilometre from a semi-natural deciduous forest. The forest was bright, mostly flat, with a little woody undergrowth and a few fallen trees. The group of children was accompanied by their two preschool teachers. The two researchers, who made the observations, were also present. The children were told to play in the forest. As a guideline for defining the role of the educator in the children's nature play, we used the principle of 'least intrusive involvement' (Kostelnik et al., 2007), which means providing only the level of support actually needed to extend children's engagement in an activity and acting as a facilitator, not an instructor of the learning process. Children's nature play was observed over a period of four consecutive days, with each session lasting approximately 1 h and a half at the same time of the day (morning). The weather was warm, sunny or partly cloudy throughout.

The study was documented through video records gathered with small video cameras attached to children's heads to automatically record sound and images from their perspective. Three randomly selected children in the group were equipped each time with small video cameras (Fig. 18.1) attached to their heads to automatically record sound and images from their perspective. Altogether 12 children were equipped over the 4 days.

Evidence from video and photographic images and unstructured narrative observations are presented together in a back-and-forth fashion. The video recordings were then transcribed, independently reviewed and analysed by two researchers using 'a read re-read' process to identify and consequently analyse the data.

Fig. 18.1 Children with small video cameras in the forest

Table 18.1 Description of analysed video recordings

Day	1	2	3	4
Children's code (letter) & gender: 1-boy, 2-girl)	A-1, B-2, C-1	D-1, E-1, F-2	G-2, H-2, I-2	J-1, K-1, L-2
Minutes of play	88:00	98:00	105:00	98:00
Minutes of transcripts	264:00	294:00	315:00	294:00
Minutes of bio play episodes	3:20	86:21	71:49	13:41
Number of bio play episodes	5	14	12	9
Minute of overlaps of 2 or 3 cameras	0:00	24:18	45:01	0:00

Altogether, 12 video recordings, for a total of 1167 min of transcripts were analysed. Of these, recordings identified as play episodes (N = 39) where the children experienced living organisms and biological phenomena (bio play episodes) were analysed and overlaps of 2 or 3 cameras in these episodes were recorded (Table 18.1). A tool for observing play outdoors (TOPO) was used to analyse nature play episodes (Loebach & Cox, 2020). Only bio play episodes were further analysed. For example, if children treated the wooden stick only as an object and a play tool, and they did not perceive it as part of a plant (i.e. a living being), we did not recognise such an episode as bio play. There was a perfect agreement on the number of identified bio play episodes between the two independent coders (Cohen's kappa = 1.0). We focussed on all three subtypes of bio play (plants, wildlife, care). The emotionality and care of children for living beings was identified through their verbal and nonverbal expressions showed in the play episodes. We also looked at the intersection of bio play with other types of outdoor play described by Loebach and Cox (2020). During the analysis of episodes, we allowed the possibility of creating new codes that were not previously foreseen. Children's codes are represented with capital letters and gender (1-boy, 2-girl).

18.3 Results

All bio play episodes, their frequency, and total time are shown in Table 18.2. In addition, the number of play episodes in which the teacher was asked by the children to enter play as a facilitator and experienced partner was recorded. Altogether, 14.9% of the total time (excluding overlaps of 2 or 3 cameras) were recordings identified as play episodes in which the children experienced biological phenomena involving naturally occurring biofacts and living organisms: episodes are named after the main object(s) observed by the children (e.g. snail, bird egg, moss).

The episodes described and their frequency are largely driven by the characteristics of the learning environment (forest), but also provide information about the objects the children were attracted to in the forest. They paid the most attention to the butterfly, the slug, the brown frog, the bird egg and the salamander. The number of episodes and the time devoted to animals is longer compared to plants and fungi. For some animals, such as the earthworm and snail, one might expect more and longer episodes, but this was not the case. We assume that they already had more experience with them, which was confirmed by their preschool teacher. It is noticeable that many of the episodes in which the teachers were involved were longer. They helped to facilitate the observation and conversation of the children in the group (Fig. 18.2).

Table 18.2 Bio play episodes, their frequency, their total time, the number of play episodes in which the teacher was asked to join, types and subtypes of TOPO

Name of the bio play episode	Total number of bio play episodes	Total time of all bio play episodes	Number of bio play episodes where teachers entered as a facilitator	Bio play subtypes	Intersections with other play types
Slug	5	57:33	3	Wildlife, care	Exploratory-sensory, exploratory-active, expressive-conversation
Butterfly	4	1:10:35		Wildlife, care	Exploratory-sensory, exploratory-active, exploratory-constructive, expressive-conversation
Salamander	3	13:30	3	Wildlife, care	Exploratory-sensory, exploratory-active
Bird egg	2	11:30	2	Wildlife	Exploratory-sensory

(continued)

Table 18.2 (continued)

Name of the bio play episode	Total number of bio play episodes	Total time of all bio play episodes	Number of bio play episodes where teachers entered as a facilitator	Bio play subtypes	Intersections with other play types
Spider	2	4:50		Wildlife, care	Exploratory-sensory
Ant	2	4:07		Wildlife	Exploratory-sensory
Spruce tree	2	3:55	1	Plants	Exploratory-sensory, exploratory-active, imaginative-symbolic
Butterbur	2	2:31	1	Plants	Exploratory-sensory, exploratory-active
Fern	2	1:55		Plants	Exploratory-sensory
Moss	2	1:07		Plants	Exploratory-sensory
Black elder	2	1:03		Plants	Exploratory-sensory
Brown frog	1	10:52	1	Wildlife	Exploratory-sensory, exploratory-active
Flowers	1	6:00		Plants	Exploratory-sensory, exploratory-active
Bird feather	1	5:07		Wildlife	Exploratory-sensory
Oak galls	1	2:13		Plants	Exploratory-sensory
Earthworm	1	0:29		Wildlife	Exploratory-sensory, exploratory-active
Blueberry	1	0:28		Plants	Exploratory-sensory
Centipede	1	0:15		Wildlife	Exploratory-sensory
Snail	1	0:10		Wildlife	Exploratory-sensory
A wood-decay fungus	1	0:18		Wildlife	Exploratory-sensory
Blackberries	1	0:13		Plants	Exploratory-sensory
Bird dropping	1	0:05		Wildlife	Exploratory-sensory

Various types and subtypes of bio play episodes, where the children experienced living organisms and biological phenomena, were observed (Table 18.2). Concrete examples are given in the following sections to illustrate what children learned and experienced. Most commonly, bio play intersects with exploratory-sensory play (e.g. slug, butterfly), exploratory-active play (e.g. slug, salamander, butterfly), exploratory-constructive play (e.g. butterfly) and expressive conversation play, which was most evident in the episode where children interacted with a butterfly and a slug. Bio-care play was most evident from children interaction with an injured butterfly. Codes of children with head cameras (see Table 18.1) are used in the descriptions of examples and other children (without cameras) are coded with the letter CH and a number (1-boy, 2-girl), e.g. CH-1.

While playing and exploring in the woods, small ground animals caught the children's attention, though sometimes only briefly. Other times, the observation lasted

Fig. 18.2 Observing a salamander

longer and developed into an activity. They invited and shared their observations with their peers. This is evident in the following conversation in which the children closely observed a snail, its tentacles, and its pneumostoma. After the initial sensory experiences, the teacher was asked to answer some questions and provide information to support the children's observations. This example beautifully demonstrates that the teacher can support learning when the children are sensually engaged in experiential learning.

Slug

1. H-2: Slug, a black one, a tiny slug. ((She wants to relocate it with a butterbur leaf, but she accidentally touches the animal. Then she grabs the slug and the snail begins to crawl on her arm.)) ((smiles)) That tickles me! ((She carries it to a group of children playing near the teacher.))
2. CH-2: Yuck! ((Some children initially show a negative feeling when they look at the slug and back away. But then they approach and begin to observe the animal.))
3. H-2: It is licking me.↑ Will it fall out of my hand?
4. CH-2: This will not happen.
5. H-2: Yes, because it is stuck to my hand with slime. ((She holds her hand upright.)) It is still crawling up the vertical surface. ((She then walks with the slug in her hand, overcoming obstacles on the ground. She explains to the other children:)) The slug is with me because it loves me. ((She runs her fingers over the front of its body.)) When you do this to it then it hides the tentacles.
6. Teacher: Observe carefully how many tentacles the slug has?
7. H-2: One, two, three… and four.

8. Teacher: Eyes are at the end of the tentacles.
9. CH-2: ((She stands next to Maja, observes the slug and suddenly says in amazement.)) <u>Hey, it has a hole here</u>!
10. H-2: <u>Oh</u>, the hole has opened and closed again.
11. Teacher: Why does slug have this hole?
12. H-2: There are little slugs in it.
13. Teacher: The slug breathes through this hole.

Shortly after, the group found a salamander and began observing the animal, passing it from hand to hand and sharing their observations. The children's handling of the snail and salamander shows their respect for living animals and their interest in animal morphology and behaviour. The teacher provides instructions for handling the animal and takes care of animal's wellbeing.

Salamander
((Children found a salamander. They are excited at first; then they calm down and begin to look at the animal.))

1. K-1: ((He takes the animal into his hands, looks at it and smiles.)) Do you dare to touch it and hold it? ↑
2. Teacher: ((She joins the group.)) Whoever is holding the salamander should crouch down so that the animal does not accidentally fall down. Be gentle when handling the salamander. And do not touch their eyes and mouths when handling the salamander.
((K-1 hands the salamander to other children... The salamander walks and moves from hand to hand. Children are loud))
3. CH-1: <u>You are scaring it</u>. ((He hands the salamander back to K-1.))
4. CH-2: See, with K-1 it has calmed down completely.

In the children's direct experiences with animals described above, their biological ideas and emotions can also be recorded. The most vivid example is the conversation of a group of children who found an injured butterfly. They discussed the reasons why the butterfly cannot fly. They tried to take care of the injured animal by providing it with suitable shelter and food. The paragraph illustrates what children collectively knew about the animal's welfare and ecology, which they took into account in the activities. An example of an ecological idea is their discussion about the predator-prey relationship between spider and butterfly.

Butterfly
((F-2 walks through the forest and comes to CH-2 who tells her she found a butterfly on the ground. F-2 squats down and carefully places the butterfly on her palm.))

1. F-2: Do you see how beautiful it is.
((They meet another C-2 who screams at the sight of the butterfly.))
2. F-1: What are you screaming about? ↑ The butterfly is not hurting anyone.
3. CH-1: <u>Yes</u>.
((The butterfly falls from F-2's hands on the ground.))

4. CH-2: The butterfly cannot fly.
5. CH-1: It fell into the water, got wet, and cannot fly.
6. F-1: Let us put it on the moss. Do not scream, because you can see it's scared.
7. CH-1: The butterfly has lost the dust on its wings and it should be placed on a flower blossom. ((He finds a large flower blossom, for which F-2 praises him.))
8. CH-1: Let's put the butterfly in a place where it is warm and that the butterfly would like.
9. F-1: <u>Yes</u>. I will go this way; it will not be slippery here. ((Holding the butterfly in her hands.))
((CH-1 shows F-2 the den the children built the day before.))
10. F-2: Yes, this is where we want to put the butterfly.
11. CH-1: Oh yes, here, on the moss.
12. F-1: It will be called Čara. ((F-1 carefully places the butterfly on the moss)).
13. CH-2: We should build a small shelter for the butterfly. ↑
 ((The girls build a house out of sticks, but F-2 thinks it is too small and begins to build a new house with the boys.))

Spider

1. CH-1: Here is a spider's web. A spider can eat the butterfly here.
((They finished the house and they want to put the butterfly inside.))
2. F-1: Watch out, there's a spider on the house. The spider can eat the butterfly.
((CH-2 gently removes the spider.))
3. F-2: The house is still too small. ((CH-1 tears down the house and he starts building a new one.))

The group dynamic of a child being in the company of friends and peers can be in some cases crucial in overcoming prejudice. This can be observed in the above-mentioned narratives about the encounters with the slug (sequence 2) and the butterfly (sequences 1–2, 6). Such peer (social) learning is certainly one of the first steps to overcoming prejudices. During the nature play, the children also noticed other arthropods (ant, centipede) and expressed various emotions – from fear to excitement. This usually happened incidentally during the play. They observed something and shortly shared their experiences with each other. The children also found the shells of birds' eggs and wanted to know the bird species from the teacher.

During the nature play, the children came across different plants and fungi which they mostly perceived as a playing material (exploratory constructive play), not as living beings at first sight. Sometimes they named them or briefly talked about their usefulness. Among the plants observed, there were some that were edible or medicinal (e.g. blueberry, black elder, blackberry), which the children also pointed out in their conversations. They built dens, animal houses, bridges and other structures out of branches and shrubs. They added leaves, cones, flowers and fruits to the construction, covered the roofs with bark, made ornaments by gluing together flowers and fern leaves with spruce resin, and covered the surfaces with moss, etc. This aspect of playing with natural materials is also very interesting for research, but we do not discuss it in detail in the present paper.

Mushroom

1. C-1: I still need to cover it ((the house)) with leaves.
((CH-1 brings a wood mushroom to Miha.))
2. C-1: We have two of those mushrooms at home.

Medical Plants
((B-2 had a very small wound and showed it to the educator. The educator suggested covering the wound with a plantain leaf.))

1. CH-1: This is that grass. ((He brings the plant.))
2. Teacher: Tie the leaf around the wound.
3. CH-1 ((to B-2)): You have to leave this wound in peace and let me bandage it. Then the wound will heal and you can throw the bandage away. ((He bandages the wound.))
4. B-2: So, the wound is healed.
5. CH-1 ((After a while he runs to B-2)): I know another medicinal herb. Should I use it to bandage your foot?

18.4   Discussion and Conclusion

Freely chosen, unstructured, and open-ended playful interactions with and in nature define nature play (Erickson & Ernst, 2011; Ernst et al., 2021) and make learning outcomes highly open-ended and unpredictable. The opportunity for nature play is a valuable activity for young children, as it provides opportunities to gain first-hand experiences in biology.

The list of bio play episodes in Table 18.2, which represent 15% of the total playing time, shows that children can experience species diversity in the forest. Children focused on the animals, plants and fungi which they observed on the forest floor. There is no focus on the biodiversity in the treetops during their play. Similarly, the results of research with Norwegian kindergarten children (Beery & Jørgensen, 2018) support the idea that childhood interaction with variation and diversity with living and nonliving items from nature allows children important learning opportunities, including deeper understanding of biodiversity.

A comment is also necessary regarding the frequency and time devoted to different organisms: It is noticeable that children focus more on animals than on plants and fungi. Many adults have 'plant blindness' – the inability to see or notice the plants in one's own environment, leading to the inability to recognise the importance of plants in both the biosphere and human affairs, and the inability to appreciate the aesthetic and unique biological features of plants (Wandersee & Schussler, 2001). This phenomenon was also observed in our research. Not only do they not see or notice the plants, in line with the research of Gatt et al. (2007), it is evident that some children even struggle with the concept of plants being alive. Perhaps that is why in their play they looked at plants more as materials for building, decorating, fencing, etc., and not as living beings.

The differences in children's experiences of plants, animals, and fungi are also evident in the different types and subtypes of the play episodes. Findings of the study show that bio play commonly intersects with exploratory-sensory and exploratory-active types of play. This is in line with the findings of Loebach and Cox (2020). Exploratory play helps develop children's observational skills, to make a transition from seeing to observing and developing scientific process skills and science concepts (Klofutar et al., 2022; Tomkins & Tunnicliffe, 2007). An example of this was children observing the tentacles of a slug. Furthermore, it is important to highlight the role of teachers in improving the observational skills of children by providing extended time for careful observations and directing them with questions and encouragement (Eberbach & Crowley, 2009; Tomkins & Tunnicliffe, 2007).

Experiences with observed animals are perceived more emotionally and lead to deeper learning about them. With plants and fungi, on the other hand, children are content with knowing their names and uses. Strgar (2007) and Pany et al. (2018) also point out the importance of emphasising the usefulness of plants for effective teaching and learning.

In the recordings identified as play episodes in which children experienced living organisms and biological phenomena, teachers played an important role as facilitators and experienced play partners. The play episodes in which teachers were involved were often longer. Teachers were very sensitive to the children and allowed them to play freely without interfering (Waller, 2007) and facilitating a more effective learning experience (Maynard et al., 2013).

Two limitations of the study should be highlighted. The first limitation was the lack of standardised analysis of the data collected and the selective access to the children in the sample due to the use of only three head cameras per group; the second was that the small sample and specific learning environment further limited the generalisability of the findings.

In conclusion, despite the limitations of the exploratory study methodology, this study has demonstrated the importance of nature play for the spontaneous learning of biology in the early years. It has improved the understanding of the phenomenon under study, which is not yet clearly defined. Our findings may contribute to more systematic and analytical studies needed to further investigate the role of nature play in biology learning.

References

Adams, S., & Savahl, S. (2017). Nature as children's space: A systematic review. *Journal of Environmental Education, 48*(5), 291–321.

Barrable, A., & Booth, D. (2020). Increasing nature connection in children: A mini review of interventions. *Frontiers in Psychology, 11*, 492.

Beery, T., & Jørgensen, K. A. (2018). Children in nature: Sensory engagement and the experience of biodiversity. *Environmental Education Research, 24*(1), 13–25.

Brown, S. (2009). *Play: How it shapes the brain, opens the imagination, and invigorates the soul.* Penguin Group.

Brussoni, M., Ishikawa, T., Brunelle, S., & Herrington, S. (2017). Landscapes for play: Effects of an intervention to promote nature-based risky play in early childhood centres. *Journal of Environmental Psychology, 54*, 139–150.

Curriculum for preschools. (1999). Ministry of Education and Sport.

Dankiw, K. A., Tsiros, M. D., Baldock, K. L., & Kumar, S. (2020). The impacts of unstructured nature play on health in early childhood development: A systematic review. *PloS one, 15*(2), e0229006.

Eberbach, C., & Crowley, K. (2009). From everyday to scientific observation: How children learn to observe the biologist's world. *Review of Educational Research, 79*(1), 39–68.

Erickson, D., & Ernst, J. (2011). The real benefits of nature play every day. *Exchange, 33*, 97–99.

Ernst, J., McAllister, K., Siklander, P., & Storli, R. (2021). Contributions to sustainability through young children's nature play: A systematic review. *Sustainability, 13*(13), 7443. https://doi.org/10.3390/su13137443

Fiskum, T. A., & Jacobsen, K. (2012). Individual differences and possible effects from outdoor education: Long time and short time benefits. *World Journal of Education, 2*(4), 20–33.

Freebody, P. (2003). *Qualitative research in education*. Sage.

Gatt, S., Tunnicliffe, S. D., Kurtsten, B., & Lautier, K. (2007). Young Maltese children's ideas about plants. *Journal of Biological Education, 41*(3), 117–121.

Gill, T. (2014). The benefits of children's engagement with nature: A systematic literature review. *Children, Youth & Environments, 24*(2), 10–34.

Herman, K. M., Craig, C. L., Gauvin, L., & Katzmarzyk, P. T. (2009). Tracking of obesity and physical activity from childhood to adulthood: The physical activity longitudinal study. *International Journal of Pediatric Obesity, 4*(4), 281–288.

Jørgensen, K. A. (2016). Bringing the jellyfish home: Environmental consciousness and 'sense of wonder' in young children's encounters with natural landscapes and places. *Environmental Education Research, 22*(8), 1139–1157.

Klofutar, Š., Jerman, J., & Torkar, G. (2022). Direct versus vicarious experiences for developing children's skills of observation in early science education. *International Journal of Early Years Education, 30*(4), 863–880.

Kostelnik, M. J., Soderman, A. K., & Whiren, A. P. (2007). *Developmentally appropriate curriculum*. Pearson/Prentice Hall.

Kuo, M., Barnes, M., & Jordan, C. (2019). Do experiences with nature promote learning? Converging evidence of a cause-and-effect relationship. *Frontiers in Psychology, 10*, 305.

Liefländer, A. K., Fröhlich, G., Bogner, F. X., & Schultz, P. W. (2013). Promoting connectedness with nature through environmental education. *Environmental Education Research, 19*(3), 370–384.

Loebach, J., & Cox, A. (2020). Tool for observing play outdoors (TOPO): A new typology for capturing children's play behaviors in outdoor environments. *International Journal of Environmental Research and Public Health, 17*(15), 1–34.

Maynard, T., & Waters, J. (2007). Learning in the outdoor environment: A missed opportunity? *Early Years, 27*(3), 255–265.

Maynard, T., Waters, J., & Clement, J. (2013). Child-initiated learning, the outdoor environment and the 'underachieving' child. *Early Years: An International Journal of Research and Development, 33*(3), 212–225.

Meredith, J. (1998). Building operations management theory through case and field research. *Journal of Operations Management, 16*(4), 441–454.

O'Brien, L., & Murray, R. (2007). Forest school and its impacts on young children: Case studies in Britain. *Urban Forestry & Urban Greening, 6*(4), 249–265.

Pany, P., Lörnitzo, A., Auleitner, L., Heidinger, C., Lampert, P., & Kiehn, M. (2018). Using students' interest in useful plants to encourage plant vision in the classroom. *Plants, People, Planet, 1*, 261–270.

Ruiz-Gallardo, J.-R., Verde, A., & Valdés, A. (2013). Garden-based learning: An experience with "at risk" secondary education students. *The Journal of Environmental Education, 44*(4), 252–270.

Sahlberg, P., & Doyle, W. (2019). *Let the children play: How more play will save our schools and help children thrive*. Oxford University Press.

Santana, C. C. A., Azevedo, L. B., Cattuzzo, M. T., Hill, J. O., Andrade, L. P., & Prado, W. L. (2017). Physical fitness and academic performance in youth: A systematic review. *Scandinavian Journal of Medicine and Science in Sports, 6*, 579–589.

Scott, J. T., Kilmer, R. P., Wang, C., Cook, J. R., & Haber, M. G. (2018). Natural environments near schools: Potential benefits for socio-emotional and behavioral development in early childhood. *American Journal of Community Psychology, 62*(3–4), 419–432.

Singer, J. L. (1994). Imaginative play and adaptive development. In J. H. Goldstein (Ed.), *Toys, play, and child development* (pp. 6–26). Cambridge University Press.

Strgar, J. (2007). Increasing the interest of students in plants. *Journal of Biological Education, 42*(1), 19–23.

Tomkins, S., & Tunnicliffe, S. D. (2007). Nature tables: Stimulating children's interest in natural objects. *Journal of Biological Education, 41*(4), 150–155.

Torkar, G. (2014). Learning experiences that produce environmentally active and informed minds. *NJAS – Wageningen Journal of Life Sciences, 69*, 49–55.

Torkar, G., & Rejc, A. (2017). Children's play and physical activity in traditional and forest (natural) playgrounds. *International Journal of Educational Methodology, 3*(1), 25–30.

Tunnicliffe, S. D. (2020). The progression of children learning about 'nature', our living world. In M. F. Costa & B. V. Dorrio (Eds.), *Discovering and understanding the wonders of nature* (pp. 216–220). Hands-on Science Network.

Waller, T. (2007). The Trampoline Tree and the Swamp Monster with 18 heads': Outdoor play in the Foundation Stage and Foundation Phase. *Education 3–13, 35*(4), 393–407.

Wandersee, J. H., & Schussler, E. E. (2001). Toward a theory of plant blindness. *Plant Science Bulletin, 17*, 2–9.

Waters, J., & Maynard, T. (2010). What's so interesting outside? A study of child-initiated interaction with teachers in the natural outdoor environment. *European Early Childhood Education Research Journal, 18*(4), 473–483.

Wilson, R. A. (2012). *Nature and young children: Encouraging creative play and learning in natural environments*. Routledge.

Wojciehowski, M., & Ernst, J. (2018). Creative by nature: Investigating the impact of nature preschools on young children's creative thinking. *International Journal of Early Childhood Environmental Education, 6*(1), 3–20.

Yogman, M., Garner, A., Hutchinson, J., Hirs-Pasek, K., Golinkoff, R. M., Baum, R., Gambon, T., Lavin, A., Mattson, G., & Wissow, L. (2018). The power of play: A pediatric role in enhancing development in young children. *Pediatrics, 142*(3), 1–16.

Chapter 19
Personal Relevance in Secondary School Students' Nature Experiences

Marcus Hammann

19.1 Introduction

Direct and personal interaction with nature is diminishing, in an ongoing alienation termed the "extinction of experience" (Pyle, 2003). Its causes are loss of opportunity (e.g., space and time for exploring nature) and loss of orientation (e.g., positive feelings and attitudes towards nature) (Soga & Gaston, 2016). Loss of orientation is particularly relevant for biology education because educational programs can develop and foster secondary school students' positive feelings and attitudes towards nature (Baird et al., 2022). Concerning positive attitudes towards nature, research has focused mainly on nature connectedness, which is a multi-dimensional psychological trait that refers to a person's belief about the extent to which they are part of nature, their emotional relationship with nature, and their experience with it (Richardson et al., 2019). For school-aged children, nature connectedness declines after early childhood before recovering in adulthood (Richardson et al., 2019). This is problematic because for school-aged children nature connectedness predicts interest in participating in nature-based activities (Cheng & Monroe, 2012), and high levels of nature connectedness are associated with pro-environmental behavior (Hughes et al., 2018).

The present study applies a recent relevance framework for science education (Priniski et al., 2018) to the field of research in secondary school students' nature experiences. The authors of the framework argue that personal relevance is a key driver of motivation. The motivation to seek contact with nature is a neglected – but potentially important – aspect of school-aged children's extinction of experience. More specifically, the extent to which the perceived personal relevance of a nature experience contributes to the motivation to engage in a nature experience of the

M. Hammann (✉)
Muenster University, Muenster, Germany
e-mail: hammann.m@uni-muenster.de

same kind is poorly understood. For example, Tal and Morag (2009) report on secondary school students' lack of motivation to engage in outdoor education, but researchers have so far neglected to research the role of motivation in this context in more detail. Therefore, this study examines secondary school students' out-of-school, free-choice nature experiences from the perspective of perceived personal relevance. More specifically, the focus of this study lies on variables related to students' perception of the personal relevance of their own nature experience. For educators, knowledge of these variables is useful because they can use them as pathways to increase the personal relevance of nature experiences in outdoor teaching and increase the students' motivation to make more frequent contact with nature (for a similar pathway approach, see Lumber et al., 2017).

19.2 Theoretical Background

The authors of a recent framework for relevance research define personal relevance as an individual's perception of the degree to which a stimulus (e.g., having a nature experience) has some relation to the individual personally (Priniski et al., 2018). More specifically, personal relevance is an important variable in three major motivation theories, the four-phase-model of interest development, expectancy-value theory, and self-determination theory. In the four-phase-model of interest (Hidi & Renninger, 2006), for example, personal relevance may trigger situational interest and support the development of individual interest. Furthermore, the authors of the framework predict that the more personally relevant a person perceives a stimulus to be, the more motivated she or he is to reengage with it (Priniski et al., 2018). Analogously, this study examines the hypothesis that secondary school students who perceive their nature experiences as personally relevant are more likely to re-engage in similar experiences than students who perceive their nature experiences as less personally relevant. This hypothesis is relevant because research has shown that the frequency of nature experiences is correlated with nature connectedness, as well as pro-environmental attitudes and behavior (Oh et al., 2021).

Furthermore, the authors of the framework conceptualize personal relevance as a continuum, ranging from personal association to identification (see Fig. 19.1): At the low end of the continuum, *personal association* describes the perception that the stimulus is connected to some other object or memory that is personally relevant.

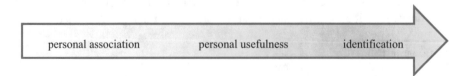

personal association   personal usefulness   identification

Fig. 19.1 Personal relevance as a continuum ranging from personal association to identification. (According to Priniski et al., 2018)

Thus, the perception of relevance is indirect because the stimulus is not perceived relevant in and of itself, but rather through association to something else. For example, an individual might find a nature experience relevant because it involves something else – bringing guinea pigs outside to play with them, for example – which is fun for the student and makes going out into nature indirectly personally relevant. *Personal usefulness* is in the middle of the continuum. It describes the perception that the stimulus can be used to fulfill an important personal goal. For example, a person finds going out into nature personally relevant because the natural setting is beautiful and fulfills the desire to be in an unspoiled place with no cars and no loud noises. At the high end, *identification* describes the perception that the stimulus is highly relevant because it is integrated into one's identity. For example, an individual who identifies as a nature lover (someone who loves birds, plants, etc.) finds nature experiences highly relevant because they are an opportunity to confirm one's identity.

This application of the framework for relevance research is compatible with broader frameworks which define ecological identity as "one part of the way in which people form their self-concept: a sense of connection to some part of the non-human-natural environment" (Clayton, 2003, p. 46). More specifically, the perceived personal relevance of the nature experience investigated in this study can be considered to be one aspect of the ecological self-concept, which is defined as a collection of beliefs one holds about oneself and nature.

19.3   Research Rationale and Purpose

This study explores variables related to secondary school students' perceived personal relevance of their nature experiences and the associations between perceived personal relevance and the frequency of such nature experiences. Methodologically, this study differs from prior studies on students' nature experiences which used preformulated items to investigate the different dimensions of high school students' anticipated nature experiences (Bögeholz, 1999) or retrospective designs, with adults commenting on their childhood nature experiences (Chawla, 1999). In this study, in contrast, we asked secondary school students to portray a moment or situation in nature that really occurred, rate its perceived personal relevance and indicate how frequently such moments or situations in nature are.

This study has the following research questions:

1. How personally relevant do the secondary school students of this sample perceive their nature experience and what variables are associated with the perceived personal relevance?
2. To what extent are the perceived personal relevance of the nature experience and its frequency associated?

19.4 Research Design and Method

In order to describe the methodology of the study, we first report on the participants and setting (see Sect. 19.4.1), before explaining the "draw and write" research method (Angeli et al., 2015) which we used – in combination with questionnaire items – to elicit secondary school students' nature experiences, their perceived personal relevance of the nature experience, and the frequency of such nature experiences (see Sect. 19.4.2).

19.4.1 Participants and Setting

The sample of this study consisted of a total of 70 secondary school students (41 female; 29 male, aged 10–18 years) from three classes of one secondary school in a small town (30,000–40,000 inhabitants) located in an area in Lower Saxony, Germany. More specifically, 23 students (11 female; 12 male) were in grade 5 (aged 10–11 years), 26 students (18 female; 8 male) in grade 8 (aged 13–15 years) and 21 students (12 female; 9 male) in grade 11 (aged 16–18 years).

The school was located in a district close to the border with the Netherlands which was less densely populated (107 inhabitants per square kilometer) than the average in Lower Saxony (168 inhabitants per square kilometer). We chose this secondary school randomly to test the "draw and write" method for eliciting students' nature experiences because we plan to use the test instrument with a larger sample of secondary schools. The school is located in a district dominated by agriculture (65% of the district), which is above the average for Lower Saxony (61%). Woodlands account for 17% of the district, which is below the average for Lower Saxony (21%). Furthermore, the district has a relatively strong industrial workforce (92 industrial jobs per 1000 inhabitants), which ranks above the average in Lower Saxony (75 industrial jobs per 1000 inhabitants) (Landesamt für Statistik Niedersachsen, 2007).

19.4.2 Assessment Instrument and Data Collection

We used a standardized questionnaire for data collection. The assignment was: "Draw an image and write a text about a moment or a situation in nature, which you remember particularly well." Because the title and the assignment used the term "nature", we informed the students that the term included all kinds of nature, for example places inside and outside the city, untouched nature (for example wilderness) and human-made nature (for example botanical gardens). Furthermore, we asked the students to rate the following questions: How often are such moments/ situations in nature for you? (very often, often, rare, very rare); How personally relevant are such moments/situations in nature for you? (very relevant, relevant, not so relevant, not relevant at all); Did the moment/situation occur in the context of biology instruction? (yes, no); Should biology instruction provide more frequent opportunities for such moments/situations? (yes, no). Finally, we asked the students to give a reason for the last question (open response).

The author of this study obtained permission from the head of the high school to collect data. For minor students, parents were asked to give written consent. Additionally, students were informed about the aims of the study and their right to decline to participate. The questionnaire was administered during regular biology classes, and the students were provided the full class period of 45 minutes to respond. Data collection took place in the summer of 2019.

19.4.3 Data Coding and Analysis

Data coding focused on inner and outer aspects of the secondary school students' nature experiences (see Sect. 19.4.3.1). For data analysis, we related these variables to the students' perceived personal relevance of the nature experience and investigated the association between the perceived personal relevance of the nature experience and its frequency (see Sect. 19.4.3.2).

19.4.3.1 Nature Experience

In order to characterize the students' nature experiences, data coding focused on outer aspects (setting, type of interaction, entities encountered and social context), as well as inner aspects (emotions, aesthetic judgments) of the nature experience. This distinction was guided by Matthew Zylstra, who described meaningful nature experiences as "a continuous shifting interaction between the inner and outer" dimensions (Zylstra, 2014, 241). More specifically, we analyzed the following aspects:

- the *setting* of the nature experience (Where did the nature experience happen? Was it a natural setting or a human-made setting? Was the setting close to school or a vacation setting?),
- the student's *interaction* with nature (What did the student do? Which senses were involved?),
- the *entities* in nature the student encountered (Did the nature experience involve plants, animals, the landscape?),
- the *social context* (Which people were involved in the nature experience?),
- the *emotions* and *aesthetic judgments* elicited by the nature experience (Did the students describe positive or negative emotions? Did the students comment on the beauty of nature?).

Concerning the middle and high school students' interactions with nature, our analyses were guided deductively by the following five types of interactions (Bögeholz, 1999, 2006; Lude, 2001):

- the aesthetic dimension: enjoying the beauty of plants animals, landscapes,
- the exploratory dimension: observing animals, looking closely at plants and exploring natural places,
- the instrumental dimension: cultivating and using parts of plants, as well as breeding or caring for domestic animals,
- the recreational dimension: playing and doing leisure activities in nature,
- the social dimension: having a strong emotional relation to a pet.

Concerning the inner aspects (emotions, aesthetic judgments of the nature experience) of the secondary school students' nature experiences, our categories of analysis were formed deductively by the nature connectedness inventory (Lumber et al., 2017), which distinguishes between the positive emotions of happiness (e.g., "Being outdoors makes me happy") and amazement (e.g., "I find being in nature really amazing"), as well as aesthetic evaluations of beauty (e.g., "I always find beauty in nature"). Inductively, we formed the categories "cool", "fun", "relaxing", and "interesting", which we categorized as positive evaluations.

To characterize the diversity of nature experiences, two independent coders, the author of this paper and a person not involved in the research, independently aligned the 70 nature experiences with the five dimensions of nature experience. The Cohen's kappa coefficient was 0.84, which we interpreted as almost perfect inter-rater reliability according to Landis and Koch (1977).

19.4.3.2   Statistical Analyses

In order to investigate which variables are associated with the perceived personal relevance of the nature experience (RQ 1 and RQ 2), we used Mann-Whitney-Wilcoxon and Kruskal-Wallis H Tests because the data for the variables "perceived personal relevance of the nature experience" and "frequency of the nature experiences" was not normally distributed. We report on effect size r, which is small for $0.10 - < 0.30$, medium for $0.30 - < 0.50$ and large for ≥ 0.50. In non-parametric tests, the observed data is converted into ranks. Nevertheless, when we report on the significance level of the Mann-Whitney-Wilcoxon and Kruskal-Wallis H Test, we also report on means and standard deviation to show differences between the groups of students we compared which are invisible when mean ranks are reported. To report means and standard deviations, we coded the categories "very often" and "very relevant" as 4 and "very rare" and "not relevant at all" as 1. Accordingly, high mean values signal high perception of personal relevance and high frequency of the nature experience. In the context of RQ 2, we used Spearman's Rho to measure the strength of association between "perceived personal relevance of the nature experience" and "frequency of the nature experiences." We considered correlation coefficients between 0.50 and 0.70 as moderate, and between 0.3 and 0.5 as low. We used T-Test statistics only once for comparing the mean age of students in the recreational and exploratory groups of nature experiences. All statistical analyses were performed using SPSS statistical software version 28.

19.5   Results

To report the results of this study, we provide an overview of students' nature experiences (see Sect. 19.5.1). Then we present findings to answer the two research questions, first on inner and outer aspects of the nature experience and their association with perceived personal relevance (see Sect. 19.5.2), and second on the extent to which the perceived personal relevance of the nature experience and its frequency are associated (see Sect. 19.5.3).

19.5.1   Overview: Secondary School Students' Nature Experiences

The secondary school students described a broad range of differing personal experiences, which occurred in a variety of natural and human-made *settings*, and consisted of clearly recognizable *interactions* with nature. The majority of the nature experiences were exploratory (N = 34; 47%) and recreational (N = 29; 40%). Among the recreational experiences, there was a fairly large group of nature experiences focusing on building huts or tepees from branches in the woods. Fewer nature experiences were instrumental (n = 2; 3%), social (n = 2; 3%), and purely aesthetic (N = 5; 7%). Generally, visual impressions dominated the descriptions, for example looking at a sunset, although there were also a few descriptions mentioning other sensations, like feeling the sunshine, listening to birds or music, and tasting self-grown vegetables. Furthermore, the descriptions differed in terms of the *entities* involved in the nature experience: plants (n = 14; 19%), animals (n = 20; 28%), both plants and animals (n = 7; 10%), and other aspects of nature, like the landscape or the sunset, but neither plants nor animals (n = 31, 43%). Thus 57% of the secondary school students described nature experiences involving plants and/or animals, whereas 43% of the secondary school students described nature experiences without plants and/or animals.

Furthermore, secondary school students' nature experiences had the following social contexts: family (n = 22; 31%), one (best) friend (n = 14; 19%), unspecified group of people (n = 14; 19%), kindergarten group (n = 5; 7%), and school class (n = 2; 3%). Fifteen secondary school students (21%) reported that they were alone during their nature experience. For other students, spending time with friends, or being with one's best friend, was an essential aspect of the nature experience. The majority of the secondary school students (n = 49; 71%) evaluated their nature experiences explicitly positively. Three students (4%) commented negatively on their nature experience, and 17 students (25%) described their nature experience in neutral terms, without using any positive or negative evaluations. The students whose texts conveyed positive evaluations mainly used evaluative adjectives, like "fascinating", "interesting", "nice", and "fun", but also explicit descriptions of feelings and reflections of the experience. The most frequently used evaluative adjectives were "beautiful" (n = 22), "cool" (n = 18), "fun" (n = 13), "relaxing" (n = 10), and "interesting" (n = 6). Furthermore, the students stated that their nature experiences provided opportunities for learning (n = 1) and inspiration (n = 1), made them feel connected to nature (n = 2), happy (n = 1), and good being in one's favorite place (n = 1), as well as having a new experience (n = 1). Very few students (n = 3; 4%) evaluated their nature experiences negatively. For example, one student described a day hike in the mountains (including the trees, the trail covered with ground bark and the beautiful sky), but wrote that he did not like the nature experience.

Often, the secondary school students' nature experience elicited *emotions*, for example, "happiness" and "awe", which the students explicitly mentioned. Furthermore, the descriptions contained *aesthetic judgments*, mainly descriptions of natural beauty. Except for one student who provided a description of the pattern of moss on one side of the tree, we did not identify any specialized biological knowledge. Instead, the students' descriptions of nature experiences clearly conveyed a personal feeling of lived experience, and not of school biology.

19.5.2   Findings for RQ 1: How Personally Relevant Do the Secondary School Students of this Sample Perceive Their Nature Experience and What Variables Are Associated with the Perceived Personal Relevance?

In our sample, secondary school students rated the perceived personal relevance of their own nature experiences (N = 69; M 3.10; SD .91) well above the theoretical mean of the scale (M 2.5). We observed considerable variation in the ratings as revealed by the standard deviation.

19.5.2.1   Age and Gender

We observed statistically significant age differences (p = .001) between students in grade 5 (M 3.65 SD .48), grade 8 (M 2.84 SD .89), and grade 11 (M 2.81 SD 1.03). The significant differences were between students in grade 5 and 8 (p = .003), and between students in grade 5 and 11 (p = .008), that is, students in grade 5 assessed higher personal relevance compared to students in grade 8 and in grade 11. Both effect sizes are medium (d = .46 and d = .45). Looking at the complete sample, gender differences were not statistically significant (p = .058), although females (N = 40; M 3.30 SD .75) rated their nature experiences as more personally relevant than males (N = 29; M 2.83 SD 1.03). In grade 5, however, females (N = 11; M 3.91 SD .30) perceived their nature experiences significantly (p = .027) more personally meaningful than males (N = 12; M 3.42 SD .51). The effect size is large (r = .50). In grade 8, differences between females (N = 11; M 3.06 SD .65) and males (N = 8; M 2.38 SD 1.18) were not significant (p = .107). Likewise, in grade 11 differences between females (N = 12; M 3.08 SD .90) and males (N = 9; M 2.44 SD 1.13) were not significant (p = .189) either.

19.5.2.2 Setting of the Nature Experience

79% of the nature experiences occurred in natural settings and 21% in human-made settings. Furthermore, we distinguished between nature experiences occurring in settings close to school/home (n = 46, 64%), settings within the range of a day trip from home (n = 8; 11%), and settings far away from home, which the students visited on a longer vacation (n = 18; 25%). The latter were often exotic settings, like the Namib, Antarctica and the Grand Canyon.

Secondary school students perceived their nature experiences as more personally relevant (p = .025) when it occurred in human-made settings (N = 16; M 3.50 SD .81) than when it occurred in natural settings (N = 53; M 2.98 SD .90). The effect size is small (r = .27). We observed no significant differences in the students' perceived personal relevance of their nature experiences concerning the distance of the setting of the nature experience from home (close to home, within a daytrip from home, vacation) (p = .213).

19.5.2.3 Interactions with Nature

The main types of interactions were the exploratory (N = 32) and the recreational dimension (N = 28). We observed no statistically significant differences (p = .120) regarding the students' ratings of personal relevance when we compared exploratory (N = 32; M 3.03 SD .86), recreational (N = 28; M 3.00 SD 1.01) and "other" nature experiences (N = 9; M 3.67 SD .50). The category "other" nature experiences combined students from the aesthetic, instrumental and social dimensions.

19.5.2.4 Entities Encountered in Nature

We observed no significant differences (p = .722) in the students' perceived personal relevance of their nature experiences when we compared secondary school students whose descriptions mentioned plants and/or animals (N = 39; M 3.12 SD .92) and secondary school students whose descriptions did not mention plants and/or animals (N = 29; M 3.07 SD .90).

19.5.2.5   Social Context of the Nature Experience

We observed no statistically significant differences (p = .09) between secondary school students whose nature experience occurred with a friend (N = 14; M 3.43 SD .64), with family members (N = 20; M 3.35 SD .67), alone (N = 15; M 3.07 SD .88), in an unspecified group (N = 14; M 2.86 SD 1.09), and a school class/kindergarten group (N = 6; M 2.17 SD 1.16).

19.5.2.6   Emotions and Aesthetic Judgements Elicited by the Nature Experience

Positive evaluations of the nature experience and explicit descriptions of the beauty of nature were associated with the secondary school students' perception of personal relevance. More specifically, students perceived their nature experiences as more personally relevant (p = .011) when they evaluated their nature experience explicitly positively and/or mentioned the beauty of nature (N = 49; M 3.33 SD .65) compared to students who neither evaluated their nature experiences, nor mentioned the beauty of nature (N = 17; M 2.53 SD 1.17). The effect size is medium (r = .31). There were too few students who evaluated their nature experiences explicitly negatively (N = 3; M 2.67 SD 1.52) to include them in the tests of statistical significance.

19.5.3   Findings for RQ 2: To What Extent Are the Perceived Personal Relevance of the Nature Experience and Its Frequency Associated?

Secondary school students rated the frequency of their nature experience (N = 68; M 2.68; SD .92) slightly above the theoretical mean of the scale (M 2.5).

We observed statistically significant age differences (p = .011) between students in grade 5, 10–11 years old (N = 22; M 3.09 SD .81), grade 8, 13–15 years old (N = 26; M 2.65 SD .93) and grade 11, 16–18 years old (N = 29; M 2.25 SD .85), $(F_{(2,65)} = 4.88; p = .011)$. The statistically significant difference was between grade 5 and grade 11 (p = .008). The effect size is medium (d = .46). Furthermore, in the whole sample, female students (N = 39; M 2.85 SD .87) rated the frequency of their nature experience higher than male students (N = 29; M 2.45 SD .94), but the difference was not statistically significant (p = .075). In grade 8, however, females (N = 18; M 3.00 SD .76) rated the frequency of their nature experiences significantly higher (p = .003) than males (N = 8; M 1.88 SD .83). The effect size is large (r = .55). In grade 5, differences between females (N = 10; M 3.30 SD .65) and males (N = 12; M 2.92 SD .90) were not significant (p = .091). Likewise, in grade

11, differences between females (N = 11; M 2.18 SD .87) and males (N = 9; M 2.33 SD .89) were not significant (p = .769) either.

Note that we asked the students to rate the frequency of the nature experience they had described and not of their nature experiences in general. The specificity of the information allows us to test the hypothesis that students who find a specific nature experience personally relevant seek to re-engage in it. Correlational analyses revealed a low positive (r_s = .42) and statistically significant relationship (p = .001) between the secondary school students' perceived relevance of their nature experience (all three types combined) and the frequency of such nature experiences. We calculated correlation coefficients for each of the three types of interactions (exploratory, recreational and other). For recreational nature experiences, the relationship was moderate (r_s = .51) and statistically significant (p = .006). For exploratory nature experiences (r_s = .29; p = .102) and for other nature experiences (aesthetic, instrumental and social combined) (r_s = .49; p = .237), in contrast, we found no statistically significant associations between personal relevance and the frequency of the nature experience.

Almost all secondary school students (98%) indicated that the nature experience described did not occur in the context of biology education. The only student, whose moment occurred in the context of biology education, described the activity of taking nature photos for a biology class. The majority (N = 59, 82%) advocated that biology instruction provide more frequent opportunities for the kinds of nature experiences the students had described, whereas 11 students (15%) did not. Advocating and non-advocating students did not differ in their assessment of perceived personal relevance of the nature experience (p = .105) and in the frequency of such nature experiences (p = .118). The main reason the advocating students gave was that nature experiences provide opportunities for learning and can enrich biology instruction, which they described as often too far removed from direct contact with nature (N = 31, 43%). For example, one student argued that biology instruction should acknowledge personally relevant nature experience "because in biology classes we are rarely in nature".

19.6 Discussion

Studying students' nature experiences is a well-established research strand in biology education research (Kellert, 2002). This study focused on variables associated with the perception of personal relevance of a nature experience (RQ 1), and the relationship between the perceived personal relevance of a nature experience and the frequency of experiences of the same kind (RQ 2). The contribution of this study lies in studying students' nature experiences from their own authentic portrayals – and not by using retrospective designs or pre-formulated items to investigate the different dimensions of high school students' anticipated nature experiences. This methodological approach allows for high internal validity because the students portrayed a moment in nature they remembered very well and rated the personal relevance of the same moment. Using a quantitative approach, we were then able to identify inner and outer aspects of the students' nature experiences associated with the perceived personal relevance of the same nature experience, which we found positively correlated with the frequency of nature experiences of the same kind. From this, we derive the recommendation that outdoor educators make nature experiences personally relevant, and thus motivating for secondary students to re-engage in them to counteract the extinction of experience.

As main findings for RQ 1, age, gender, the setting of the nature experience and positive evaluations and/or descriptions of the beauty of nature were associated with the secondary school students' perceived relevance of their nature experience (Table 19.1). In particular, younger secondary school students (10–11 years old) perceived their nature experiences more personally relevant than the two age groups of older secondary students (13–15 years old and 16–18 years old). This finding is consistent with that of Kaplan and Kaplan (2002), who argue that children attach

Table 19.1 Summary of variables associated with secondary school students' perceived relevance of their nature experience

Associated variable	Description	Effect size /strength of correlation
Age	Younger students (10–11 years old) perceive higher personal relevance than older students (13–15 years old and 16–18 years old)	Medium $r = .46$ for the comparison between students (10–11 years old) and students (13–15 years old); $r = .45$ for the comparison between students (10–11 years old) and students (16–18 years old)
Gender	Female students (10–11 years old) perceive higher personal relevance than male students (10–11 years old)	Large $r = .50$
Setting of the nature experience	Students (10–18 years old) perceive higher personal relevance in human-made settings than in natural settings	Small $r = .27$
Positive evaluation and/ or perception of beauty	Students (10–18 years old) who evaluate their nature experience explicitly positively and/or comment on the beauty of nature perceive their nature experience as more personally relevant than students who do not explicitly evaluate their nature experience positively and/or do not comment on its beauty	Medium $r = .31$
Frequency of the nature experience	For students (10–18 years old) who engage in recreational nature experiences, correlational analyses revealed a moderate positive and statistically significant relationship between the perceived relevance and the frequency of such nature experiences	Moderate $r_s = .51$

more importance to nature than adolescents (for an extended discussion of this finding see also Price et al., 2022). Furthermore, this finding is consistent with research in school-aged children's nature connectedness, which included students of the same age groups and found a dip in nature connectedness from age 10–12 to age 13–15, as well as a similar levels of nature connectedness between age 13–15 and age 16–18 (Richardson et al., 2019). These parallel findings are also interesting from a methodological perspective: The six-item scale for measuring nature connectedness used by Richardson et al. (2019) contained the item "Spending time in nature is very important to me", which is similar to the item used in this study to assess the personal relevance of the nature experience portrayed by the student ("How personally relevant are such moments/situations in nature for you?"). Although the rating scales differed and the item used by Richardson et al. (2019) is formulated in a more general way than the item used in this study, further efforts need to be made to disentangle the theoretically different variables.

Furthermore, we observed significant gender differences among children (10–11 years old), with girls reporting higher personal relevance of their nature experience than boys. This is also consistent with prior findings about gender differences (Müller et al., 2009). Contrary to expectations, however, this study did not find a significant gender difference for adolescents, perhaps because the sample size was small and the data were not normally distributed. The type of interaction with nature was not associated with students' perceived personal relevance of the nature experience. In particular, secondary school students who described recreational activities rated their perceived personal relevance at the same level as secondary school students who described exploratory nature experiences. This is a somewhat surprising finding because Kaplan and Kaplan (2002) argue that adolescents attach less importance to nature herself and take a "time out" from nature by seeking activities with peers which are set in nature, but not directed at nature herself. In the present study, in contrast, age differences between students describing exploratory and recreational activities were not statistically significant (although the former were younger than the latter), and the relevance ratings were almost identical. In conclusion, because the personal relevance of a nature experience seems individually constructed and a function of individual and biographical factors (Gebhard, 2020), it is obviously insignificant for the perception of personal relevance as to whether nature is an essential feature of the experience – as in exploratory nature experiences – or nature sets the context for recreational nature experiences.

Concerning RQ 2, correlational analyses supported the hypothesis that secondary school students who perceive their own nature experience as personally meaningful, also report to engage in it more frequently (and vice versa) than students who do not perceive it as personally meaningful. This finding is crucial because the motivation to engage (and re-engage) in nature experiences is positively correlated with pro-environmental attitudes and behavior (Wells & Lekies, 2006; De Ville et al., 2021). Interestingly, we found a significant positive correlative relationship between personal relevance and frequency only for recreational nature activities, and not for exploratory nature activities. The reason for this may be related to a limitation of the research method we used because we asked the students to provide information about the frequency of a single experience. An unusual encounter with an animal (which we attributed to the exploratory dimension), therefore, can be rated as personally relevant, but its frequency can be rated as rare or very rare because of the scarcity of opportunities for the encounter.

Regarding educational implications, outdoor educators are recommended to seek activities that make students – in particular older secondary students – aware of the fact that nature experiences can be personally relevant and thus motivating to re-engage in them. Generally, interventions that aim at increasing the personal relevance of nature experiences need to acknowledge the diversity of personal preferences and individual experiences because the perceived personal relevance of nature experiences is subjective and probably formed through individual prior experiences in nature. For example, encouraging secondary students to self-reflect on personally relevant nature experiences – as an educational activity – may make secondary school students aware of their preferences and habits, and increase the likelihood that they re-engage with nature experiences of the same kind. Such interventions still need to be developed and evaluated, but for nature connectedness, a related construct, reflective self-attention was associated with an increase in connection to nature (Richardson & Sheffield, 2015).

Another type of intervention aiming at increasing the awareness of the personal relevance of nature experiences can involve the simple activity of noticing aspects of one's own nature experience that make it personally relevant. Again, such aspects are subject to personal preferences and individual experiences, although this study suggests that positive evaluations (feelings, emotions), as well as the perception of beauty, are positively correlated with the perceived personal relevance of the nature experience across the different age groups of the secondary school students studied. Similar approaches have been developed in the related field of nature connectedness, where they proved effective in empirical evaluations. One of these interventions involved the simple activity of noticing the good things in nature (McEwan et al., 2019). Another intervention focused on pathways to nature connectedness, and one of the pathways – the beauty pathway – encouraged people to pay attention to beautiful surroundings and think about how what they saw made them feel when they were walking in the countryside (Lumber et al., 2017). To counteract the extinction of experience, such approaches are very valuable and congruent with the findings of this study.

References

Angeli, C., Alexander, J., & Hunt, J. A. (2015). 'Draw, write and tell': A literature review and methodological development on the 'draw and write' research method. *Journal of Early Childhood Research, 13*(1), 17–28.

Baird, J., Dale, G., Holzer, J. M., Hutson, G., Ives, C. D., & Plummer, R. (2022). The role of nature-based program in fostering multiple connections to nature. *Sustainability Science, 17*, 1899–1910.

Bögeholz, S. (1999). *Qualitäten primärer Naturerfahrung und ihr Zusammenhang mit Umweltwissen und Umwelthandeln* [The qualities of primary nature experiences and their relationship with environmental knowledge and action]. Leske+Budrich.

Bögeholz, S. (2006). Nature experience and its importance for environmental knowledge, values and action: Recent German empirical contributions. *Environmental Education Research, 12*(1), 65–84.

Chawla, L. (1999). Life paths into effective environmental action. *The Journal of Environmental Education, 31*(1), 15–26.

Cheng, J. C.-H., & Monroe, M. C. (2012). Connection to nature: Children's affective attitude toward nature. *Environment and Behavior, 44*(1), 31–49.

Clayton, S. (2003). *Identity and the natural environment: The psychological significance of nature.* MIT Press.

De Ville, N. V., Tomasso, L. P., Stoddard, O. P., Wilt, G. E., Horton, T. H., Wolf, K. L., Brymer, E., Kahn, P. H., & James, P. (2021). Time spent in nature is associated with increased pro-environmental attitudes and behaviors. *International Journal of Environmental Research and Public Health, 18*, 7498.

Gebhard, U. (2020). *Kind und Natur: Die Bedeutung der Natur für die psychische Entwicklung* [Child and nature: The significance of nature for the mental development]. Springer.

Hidi, S., & Renninger, K. A. (2006). The four-phase model of interest development. *Educational Psychologist, 41*, 111–122.

Hughes, J., Richardson, M., & Lumber, R. (2018). Evaluating connection to nature and the relationship with conservation behavior in children. *Journal of Nature Conservation, 45*, 11–19.

Kaplan, R., & Kaplan, S. (2002). Adolescents and the natural environment: A time out? In P. H. Kahn & S. R. Kellert (Eds.), *Children and nature: Psychological, sociocultural and evolutionary investigations* (pp. 227–257). The MIT Press.

Kellert, S. R. (2002). Experiencing nature: Affective, cognitive, and evaluative development in children. In P. H. Kahn & S. R. Kellert (Eds.), *Children and nature: Psychological, sociocultural and evolutionary investigations* (pp. 117–151). The MIT Press.

Landesamt für Statistik Niedersachsen. (2007). *Bezirk Weser-Ems* [District Weser-Ems]. Retrieved from https://www.statistik.niedersachsen.de/download749189

Landis, J. R., & Koch, G. G. (1977). The measurement of observer agreement for categorial data. *Biometrics, 33*(1), 159–174.

Lude, A. (2001). *Naturerfahrung und Naturschutzbewusstsein: eine empirische Studie* [Nature experience and environmental awareness: An empirical study]. Studienverlag.

Lumber, K., Richardson, M., & Sheffield, D. (2017). Beyond knowing nature: Contact, emotion, compassion, meaning and beauty are pathways to nature connection. *PLoS One, 12*(5), e0177186.

McEwan, K., Richardson, M., Sheffield, D., Ferguson, F. J., & Brindley, P. (2019). A smart-phone app for improving mental health through connecting with urban nature. *International Journal of Environmental Research and Public Health, 16*(18), 3373.

Müller, M. M., Kals, E., & Pansa, R. (2009). Adolescents' emotional affinity toward nature: A cross-societal study. *The Journal of Developmental Processes, 4*(1), 59–69.

Oh, R. R. Y., Fielding, K. S., Nghiem, L. T. P., Chang, C. C., Carrasco, L. R., & Fuller, R. A. (2021). Connection to nature is predicted by family values, social norms and personal experiences of nature. *Global Ecology and Conservation, 28*, e01632.

Price, E., Maguire, S., Firth, C., Lumber, R., Richardson, M., & Young, R. (2022). Factors associated with nature connectedness in school-aged children. *Current Research in Ecological and Social Psychology, 3*, 100037.

Priniski, S. J., Hecht, C. A., & Harackiewicz, J. M. (2018). Making learning personally meaningful: A new framework for relevance research. *The Journal of Experimental Education, 86*(1), 11–29.

Pyle, R. M. (2003). Nature matrix: Reconnecting people and nature. *Oryx, 37*(2), 206–214.

Richardson, M., & Sheffield, D. (2015). Reflective self-attention: A more stable predictor of connection to nature than mindful attention. *Ecopsychology, 7*(3), 166–175.

Richardson, M., Hunt, A., Hinds, J., Bragg, R., Fido, D., Petronzi, D., Barbett, L., Clitherow, T., & White, M. (2019). A measure of nature connectedness for children and adults: Validation, performance, and insights. *Sustainability, 11*, 3250.

Soga, M., & Gaston, K. J. (2016). Extinction of experience: The loss of human-nature interactions. *Frontiers in Ecology and the Environment, 14*(2), 94–101.

Tal, R. T., & Morag, O. (2009). Reflective practice as a means for preparing to teach outdoors in an ecological garden. *Journal of Science Teacher Education, 20*(3), 245–262.

Wells, N. M., & Lekies, K. S. (2006). Nature and the life course: Pathways from childhood nature experiences to adult environmentalism. *Children, Youth and Environments, 16*(1), 1–24.

Zylstra, M. (2014). *Exploring meaningful nature experience, connectedness with nature and the revitalization of transformative education for sustainability* (Unpublished doctoral dissertation). Stellenbosch University. https://ciret-transdisciplinarity.org/biblio/biblio_pdf/Matthew_Zylstra.pdf

Chapter 20
Challenges Emerged During an Action Research Approach Applied in a Schoolgarden Project: Reflections and Revisions

Anthi Christodoulou and Konstantinos Korfiatis (iD)

20.1 Introduction

In this paper, we describe the challenges that emerged during a school project, where elementary school students participated in the cultivation of a school kitchen-garden. We also explain how the educational team transformed its research and teaching strategies to overcome these challenges. The aim of the project was to develop, through gardening, the self-determined environmental motivation of a group of elementary students from a disadvantaged background. As it has been shown in many studies, school gardens in their various forms (botanic gardens, kitchen gardens; recreational gardens, etc.) enhance students' motivation to partici-pate not only in the garden experience, but also in the school and community life (Blair, 2009; Darner, 2012; Murakami et al., 2018; Christodoulou & Korfiatis, 2019). In all the cases reported in the literature, the key to a successful school gar-den project is the degree of active participation of the children in the management and the various activities of the project (Korfiatis & Petrou, 2021).

Furthermore, an Action Research (AR) approach was adopted as it was consid-ered to be the best way to achieve the goal for which the program was designed. According to the classic definition by Elliot (1991), an AR is the study of a social situation with a view to improving the quality of action within it. The aim of AR is to improve work practices through collaborative inquiry following a spiral of plan-ning, acting, observing, and reflecting (Zuber-Skerritt, 2018): The *planning* stage includes the design of the project and the clarification of the target. The *acting* stage includes the implementation. The *observing* stage includes data collection. Finally, the *reflecting* stage includes conversations and critical thinking between the research team members about the collected data, the challenges that emerged during the

A. Christodoulou (✉) · K. Korfiatis (✉)
Department of Education, University of Cyprus, Nicosia, Cyprus
e-mail: achris14@ucy.ac.cy; korfiati@ucy.ac.cy

K. Korfiatis et al. (eds.), *Shaping the Future of Biological Education Research,*
Contributions from Biology Education Research,
https://doi.org/10.1007/978-3-031-44792-1_20

277

implementation and the changes they should consider applying to the next research cycle. AR is a cyclical iterative process of action and reflection on and in action. Theory and practice are not separated; however, they are integrated as well as research and development. In that way AR allowed the continuous revision and improvement of the project, so as to better meet the needs of the participating students whose environmental motivation was about to be identified after the implementation.

AR has been used in a variety of environmental education studies to help practitioners examine and transform their practice and improve education programmes (Doyle & Krasny, 2003; Can et al., 2017; Do Carmo Galiazzi et al., 2018; Servant-Miklos, 2022). In the research presented here, AR had a dual role: to facilitate the evaluation of project implementation and to support the participating educators in examining and transforming their teaching practice.

Our aim was to understand how this project worked, what challenges the action research team faced and how they managed to overcome them. We hope that the outcomes of this research offer insights to educators aiming to motivate and engage vulnerable youth in environmental and sustainability actions.

Therefore, this paper aims to answer the following question: How were the researchers' strategies transformed to overcome the challenges that emerged during the action research project?

20.2 Research Design and Method

20.2.1 Participants

Thirteen, six- to twelve-year-old students from a rural primary school participated in the project. Participation was voluntary and anonymous, and all the ethical considerations had been taken into account, including written agreement from the participating children and their guardians, as well as written permission from the ethical committee of the Ministry of Education. All participants were members of a group following afterschool activities organised by the school they were studying, after the end of the formal curriculum programme. Four of them were boys and nine were girls. Most of the students came from families with low socio-economic level that find it difficult to obtain their basic survival needs (ex. daily food). They also expressed low motivation, low interest in participating in school life, high obesity rates and detachment from nature, despite the fact that they were living in a rural environment.

The AR team consisted of the Teacher/Researcher (T/R), the critical friend, the facilitators (i.e. the TR's academic advisor and the school's principal) and three more teachers in the same school, who were the research partners.

20.2.2 Roles

According to Reason and Bradbury (2008), participation is the defining characteristic of action research. Participation is central to AR methods, which talks of research 'with' people rather than 'on' or 'for' them, involving all participants as co-inquirers (Heron & Reason, 2006). The research team may include as few as two persons, or it may include several teachers and administrators working with university staff or any other external agency (Caro-Bruce, 2000). During the AR process of the present study, all participants had a specific role, as it is described in other AR settings (Can et al., 2017; Hawkins, 2015). However, until the end of the intervention, some responsibilities needed to be explained or changed. The flexibility of the AR team roles has been stated by other researchers (Avgitidou, 2009).

The roles assumed by the participants in this action research project are outlined as follows:

20.2.2.1 The Teacher/Researcher

The T/R, who is the first author of this paper, had multiple responsibilities during the project, as it is often the case in educational AR projects (Kember, 2000). As a *"Coordinator"*, she explained the main goals of the project to the members of the team, outlined their roles and responsibilities, facilitated their cooperation and organised the meetings. As a *"Teacher"*, she implemented the activities with the students in rotation with the critical friend. Finally, as a *"Researcher"*, she was observing and taking notes on students' participation during the project.

20.2.2.2 Critical Friend

In the AR literature, a critical friend is defined as a trusted person who is invited to join an AR project based on the qualities of knowledge, experience, and skills (Campbell et al., 2004). Generally, the role of a critical friend is to ask questions, provide data to be examined through another lens, and offer a critique of a person's work as a friend (Noor & Shafee, 2021). The critical friend was chosen by the T/R from her colleagues (teachers in the same school), because of her previous teaching experience with the same group of students, as well as her previous research experience in similar research settings.

The critical friend worked as *"the Partner"*, giving advice and professional feedback to the T/R, as an *"Evaluation Advisor"*, controlling and evaluating the applied theoretical practices of the project, and finally as a *"Teaching Assistant & Consultant"*, providing students with the necessary tools or any other materials they needed for the activities and encouraging students to participate, cooperate and express their thinking with their team members.

20.2.2.3   Outside Researchers

Students facilitated as 'outside researchers', a characterization adopted from Doyle and Krasny (2003). They provided the T/R with comments, thoughts, initiatives, and ideas through their reflections. They communicated the facts they observed with the T/R and they were active participants by making recommendations and taking the final decisions about the actions they wanted to get involved in during the project.

20.2.2.4   Facilitators

The T/R's academic advisor and the school's principal were the facilitators of the AR process. T/R's academic advisor played mostly a consulting role by providing scientific information about the methodology and the theoretical approaches of the project. He also inspired, mobilised and encouraged the T/R.

The school principal was responsible for designing the teaching programme during the working hours of the elementary school and making sure that it was being applied effectively. Also, the school principal provided the T/R with useful information about the participating students (academic performance, social conditions, etc).

20.2.2.5   Research Partners

Three teachers in the same school served as research partners by keeping notes about participating students' collaboration, as well as students' attitudes or expression of ideas or emotions during their involvement with gardening.

20.2.3   The School Kitchen-Garden Project

A school kitchen-garden is distinguished from other types of school gardens (e.g. a botanical garden, or a recreational garden), participation in a kitchen-garden project also includes participation in the preparation and cooking of meals using products harvested from the garden (Gibbs et al., 2013). After convincing the school principal of its benefits, the authors of this article initiated the project. The activities of the project were designed by the T/R based on students' thoughts and decisions. Some examples of students' involvement in the project include the selection of the vegetables they wanted to plant and the location of the garden in the school yard, decorating the garden, deciding how to utilize their harvest, how to take care of the garden and deal with its 'enemies' (i.e. bad weather, parasites and weeds). Students also cooked food using their own harvest. Overall, students had the most active role, while the adult participants in the project (i.e. the T/R, other teachers, the principal, etc.) functioned as collaborators, facilitators, and partners. The project included weekly routines concerning the cultivation of vegetable plants, as well as seventeen 40-min meetings between the children and the T/R.

20.2.4   The Action Research Process

During the present study, meetings were held during the planning stage of the first AR cycle to ensure that the adult participants in the AR acknowledged the main goals of the project, the research and teaching strategies they needed to apply, as well as their specific role.

Furthermore, meetings were held regularly following the circular stages of the AR process. All meetings were scheduled by the T/R. However, sometimes unscheduled meetings needed to take place to deal with a challenging situation that had emerged. The unscheduled meetings were mostly held between the T/R and the critical friend to deal with challenges relating to students' cooperation.

During the meetings, the team members discussed their observations and the students' reflective notes. Also, they evaluated the research tools and the teaching strategies they used. Finally, during the reflection stage of the AR cycles, the team discussed possible changes and revisions needed for the kitchen-garden project.

20.2.5   Data Collection

Observation Notes During the project, the T/R, the critical friend, and the research partners recorded any kind of observation they made on students' participation during gardening, as well as concerning student comments, or expressions of feelings or ideas. Additionally, the T/R kept notes concerning research-partners' pedagogical approaches and implementation of research tasks.

Students Scheduled & Unscheduled Reflecting Notes The scheduled reflective notes were open-ended forms with more specific questions about students' feelings and experiences that were completed by the students in the middle and at the end of the project. The unscheduled reflective notes were open-ended forms that students filled in whenever they wanted to express their personal reactions, feelings, concerns, ideas, initiatives, or decisions concerning gardening. These were completed independently by each student during the project.

20.2.6 Data Analysis

All observations and reflective notes were studied by the T/R as the project was running, in order to identify if the research or teaching strategies which had been applied in each cycle were effective or not towards the aim of the project, or if revisions were needed. For example, a note during the second cycle of the project was as follows: "Student 4 expressed non-participatory behavior. Specifically, he refused to help his team members water the crops." After this observation, a new approach to help this student was applied by the T/R, and changes in his attitude and behavior were recorded during the third cycle.

Simultaneously, the T/R identified all the challenges that emerged during the cycles of the AR process and came up with four broader categories of those challenges by using the method of content analysis (Cohen et al., 2018). Some of the challenges emerged before the beginning of the project, during the planning stage of the first AR cycle. Some challenges emerged during the implementation of the project, and some emerged towards the end of the project. The first category included challenges regarding *"The coordination of the AR team"* (Table 20.1). The second category included challenges about *"The AR team pedagogical skills"* (Table 20.2). The third category included challenges concerning *"The AR team research skills"* (Table 20.3). Finally, in the fourth category the T/R included challenges about *"Students' participation"* (Table 20.4). For each challenge, a specific strategy was recommended to overcome the situation (see Tables 20.1, 20.2, 20.3, and 20.4).

20.3 Findings

Four research cycles were carried out from the beginning until the end of the project. The T/R and the rest of the adult participants in the project concluded that no further revision of the implemented practices was necessary after the fourth cycle.

Table 20.1 Coordination of the AR team

Challenges	Solution/new strategy
AR team members: Need to participate voluntarily in the AR process. Also, they should feel confident about participating in the AR process and develop cooperative relations among themselves.	The T/R organised meetings before the implementation while the AR team members were informed analytically about the main goals of the project and, the problems which lead to the necessity to implement the project. The AR team members were convinced about the necessity and value of implementing this AR.
Format students' groups	Students' gender, age, academic level and social relationships were considered by the AR team in order to format their students' groups.

Table 20.2 AR team members' pedagogical skills

Challenges	Solution/new strategy
The AR team members lack pedagogical skills, which lead to over-structured and teacher-centered activities instead of open inquiry ones	The T/R decided to implement (by herself) most of the activities with the students and to record more systematically the role of the critical friend who was providing feedback on students' thoughts and initiatives.

Table 20.3 AR team research skills

Challenges	Solution/new strategy
During the beginning of the project, limited data was collected about students' behaviour during project activities. This was due to a lack of researcher skills.	The T/R gave specific guidelines to the AR team about "what" to observe. She changed the methodological tools by: (a) Collecting students' reflecting notes (b) Introducing an observation rubric for use by the members of the AR team.
The reporting of students' initiatives relating to the project.	The T/R developed purposeful conversations with students trying to elicit their thoughts and ideas.
Data triangulation and validity.	AR team observations were cross-checked with students' personal reflections and some data derived from the post interviews between the T/R and the students.

Table 20.4 Students' participation (interest & cooperation)

Challenges	Solution/new strategy
Attract students' attention at the beginning of the project.	The T/R introduced the participants to a real issue concerning their lunch preparation in their school.
The T/R identified students' concerns about satisfying their basic survival needs (e.g., daily food). Due to that concern, they were unable to focus on other activities such as environmental activities (low environmental motivation).	The T/R took advantage of their concern and managed to make students understand the usefulness of participating in the specific environmental activities of the project for their lives (e.g., find a way to produce their own food).
Non-participatory behaviour of some students led to collaboration problems between students in the same group. Also, there was reduced interest from some students during the last sessions of the project.	The T/R approached students to understand the reasons for their non-participatory behaviour. She tried to convince them about the meaning of cooperating with their group members to achieve their common goals (e.g., giving them examples of similar situations).
Non-participatory behaviour of some students during specific activities.	The T/R applied activities promoting students' creativity, mobility and autonomy. The T/R decided to introduce external motives (e.g., diplomas) to attract those students' interest.

20.3.1   Action Research Cycles: Cycle 1

20.3.1.1   Plan: The Preparations

The educational meetings which were organised by the T/R aimed to inform the AR team about the main goals of the project and to understand its value and the need for its implementation. Also, semi-structured interviews were held between the T/R and the students at this stage. At the same time, the T/R was informed by the school's principal and other teachers about participating students' school performance and social condition.

20.3.1.2   Action: The Beginning of Implementation

Students were introduced to a realistic situation (our food and how it is produced) and took the decision to create a kitchen-garden in their school. They started working in mixed groups (concerning age, gender, school performance) of 3–4 members that were formed by the T/R. The students initiated conversations between their group members and the AR team. They also expressed their first thoughts about the cultivation activities (e.g., where, what, when to plant) and ideas about how to take care of their kitchen-garden (e.g., watering their crops, observing and keeping the planting area clean from litter and weeds). The T/R and the critical friend coordinated the activities of the project.

20.3.1.3   Observations & Reflections

The AR members collected limited data about students' reflections and reactions towards the activities of the project and about students' psychological needs of autonomy and competence. However, they managed to identify evidence of students participating effectively during their conversations. The T/R decided to transform the methodological tools to assist the AR members in collecting data.

20.3.2   Action Research Cycles: Cycle 2

20.3.2.1   (Re)Plan: The First Cycle Changes Applied to the Second Cycle of the Implementation

The T/R applied two additional methodological tools: The scheduled and unscheduled reflecting notes that needed to be filled out by the students and the use of an observation rubric by the AR team for a better systematical recording of information about students' actions and behaviour. Also, the T/R drew the other AR team members' attention to the fact that their pedagogical practices needed to be more student-centered than teacher-guided. Finally, the T/R designed a diary of students' responsibilities for their kitchen-garden.

20.3.2.2 Action: Implementing the First Cycle Changes and Students' Decisions into the Second Research Cycle

Students continued working and collaborating in groups. They created their kitchen-garden and started following a diary of responsibilities. They also decorated their garden and invited members of the community (municipality officials, parents) to visit it.

20.3.2.3 Observations & Reflections

The AR team members noticed that most, but not all, of the students' groups cooperated effectively. Some students felt disappointed and worried about their crops due to rainy days and because they noticed some marks on their crop's leaves. Also, not all students followed the diary of responsibilities and some students expressed non-participatory behaviour. The AR members noted this and wrote down their comments. Furthermore, the AR members mentioned and discussed specific moments of students' motivation. These moments were either observed by the AR team or were written by the students in their reflective notes. Finally, the T/R evaluated the use of the two additional methodological tools and the AR members teaching approaches.

20.3.3 Action Research Cycles: Cycle 3

20.3.3.1 (Re)Plan: The Second Cycle Changes Applied to the Third Cycle of the Implementation

The AR team decided not to change students' groups. However, they decided to provide more help, support, and encouragement to students to help them overcome the challenges that had upset them. For example, they encouraged students to search for different ways of dealing with their garden "enemies". Also, the AR team applied a "personal approach" toward students expressing non-participatory behavior in order to gather information about their behaviour. The T/R decided to be the only member responsible for conducting activities with the students in order to apply more student-centered teaching approaches. The diary of responsibilities was cancelled, and the students were free to choose when and who would be involved in taking care of their garden.

20.3.3.2   Action: Implementing the Second Cycle Changes and Students' Decisions

The students collaborated with their group members and organised a plan to deal with their crop's "enemies" by applying organic solutions (e.g., they placed ashes in the soil to prevent snails approaching their crops). Each student decided independently how many times per week he/she would deal with the gardening. Finally, students started discovering the advantages of creating a kitchen-garden (physical activity; enjoyment; watching plants grow; food production).

20.3.3.3   Observations & Reflections

The AR team noticed that some students continued to reveal non-participatory behaviour and tried to find the reasons behind this behaviour. They decided to apply an external motive to attract those students' attention. At the same time, they noted instances where students expressed relatedness, autonomy and competence. Also, they noticed that fewer students asked for help and expressed concerns about the kitchen-garden.

20.3.4   Action Research Cycles: Cycle 4

20.3.4.1   (Re)Plan: The Third Cycle Changes Applied to the Fourth Cycle of the Implementation

The AR team members continued to encourage and support students. They also continued to apply the "personal approach" and engaged in more discussions with the students about their feelings and their intentions regarding the kitchen-garden. Additionally, they decided on the introduction of the external motive of giving diplomas to all students who were seen to care about their garden.

20.3.4.2   Action: Implementing the Second Cycle Changes and Students' Decisions

Students continued to work collectively, being encouraged, and rewarded (with diplomas) by the AR team when they got involved in their kitchen-garden care.

20.3.4.3   Observations & Reflections

The number of students who worked collectively significantly increased. At the same time, students who revealed a high level of interest from the beginning of the project increased the frequency of visiting and working in their kitchen-garden.

Also, most of the students made sure that their participation did not go unnoticed by the T/R, who would provide them with the diplomas. It seems that the external motive of diplomas worked effectively in students' participation in the project.

20.3.5   Challenges Emerged During the Action Research Cycles

Tables 20.1, 20.2, 20.3, and 20.4 present all the categorised challenges that emerged during the above-mentioned AR cycles and the new strategies that were applied as the solution to each challenge.

20.4   Discussion

This AR study highlights core issues that emerged through the implementation of a kitchen-garden programme with a group of vulnerable elementary school children.

During an AR that takes place in a school, problematic issues can appear concerning either the research team members, or the students participating in the procedure (Zhou, 2012; Can et al., 2017). This was apparent in the present study as all the challenges concerned either the AR team or students' participation in gardening. The issues referred to four different aspects: The coordination of the AR team, the AR team's pedagogical skills, the AR team's research skills and the students' participation.

The process of an AR requires much effort, concentration and systematicity. The AR team members, having a participatory role, are responsible for identifying not only the challenges that emerged during the AR process but also the strategies that lead to the solution of those challenges. The reflective character of an AR is a huge advantage in this process (Palmer, 2017; Pine, 2009; Reason & Bradbury, 2008). According to Katsenou et al. (2015), a key and fundamental characteristic of AR is its participatory and critical-reflective nature.

In the present study, most of the challenges that emerged were about students' participation. Challenges concerned their interests and their collaboration in the activities of the kitchen garden project. The T/R applied specific strategies to attract students' attention and to support their collaboration. Strategies (such us engaging in a real scenario deriving from students) and applying games (such as role playing or debates) provided students with the opportunity to choose and apply their own decisions concerning the activities of the project. These strategies were applied successfully in the present project, but they are also recommended by other researchers (Liao & Wang, 2008).

Another important outcome was that during the specific AR no challenges emerged concerning the research team's relationship and collaboration. This was probably due to the educational meetings that were held between the AR team members before the implementation. During those educational meetings, the AR team members were convinced about the need to implement the project. Therefore, the AR team members expressed their personal interest and volition in participating in the AR process. According to Barber et al. (2021), collaboration between action researchers is the key to AR.

Another outcome of the specific study that comes into agreement with previous findings is that the teachers participating in an AR need professional development with respect to methods and techniques for doing AR (Zhou, 2012; Kalaitzidaki & Filippaki, 2021). Even though the T/R organised educational meetings before the implementation during which she tried to explain the observatory role of each AR team member, the student-centered pedagogical approaches they needed to apply and the AR methods, the outcome didn't meet the preferred standards as this strategy did not work very effectively. It seems that more than educating meetings is required to develop the research skills of the teachers. More frequent experiences of participating in different research settings would be helpful to develop the appropriate research skills.

Finally, an important outcome of the specific study was that the success of implementing the project was due to the effective collaboration between the AR team members and the students. The relationship between the co-researchers and the participants is very important during an AR process (Jacobs, 2016).

Nobody can be sure in advance about the process and the outcomes of an AR project. The specific research highlights core issues (challenges about the AR team coordination, the AR team members pedagogical and research skills) which can emerge through the implementation of a kitchen-garden project with vulnerable elementary school children. Nevertheless, it would be very useful to know the possible challenges that may arise during an AR process as well as some specific recommendations to deal with those challenges.

During this project, a research group committed to dialogue, shared knowledge and action, managed to create a participatory and effective project with participating vulnerable children. Specific strategies such us organising regular meetings between the research team members, developing a friendly relationship between teachers and participating students, having a personal approach towards students who participate in the applied project (e.g., private conversations), identifying students' personal worries or social difficulties and applying activities promoting students' creativity, mobility and autonomy, could be helpful to other environmental educational projects, as well.

References

Avgitidou, S. (2009). Participation, roles and processes in a collaborative action research project: A reflexive account of the facilitator. *Educational Action Research, 17*(4), 585–600.

Barber, C. R., Palasota, J. A., Steiger, M. A., Bagnall, R. A., Reina, J. C., Wagle, J., & Bai, Y. (2021). Enhancing STEM equity programs with action research. *Action Research, 19*(4), 614–631.

Blair, D. (2009). The child in the garden: An evaluative review of the benefits of school gardening. *The Journal of Environmental Education, 40*(2), 15–38.

Campbell, A., McNamara, O., & Gilroy, P. (2004). Critical friendship, critical community and collaboration. In *Practitioner research and professional development in education* (pp. 106–124).

Can, Ö. K., Lane, J. F., & Ateşkan, A. (2017). Facilitating place-based environmental education through bird studies: An action research investigation. *Environmental Education Research, 23*(5), 733–747.

Caro-Bruce, C. (2000). *Action research facilitator's handbook.* National Staff Development Council.

Christodoulou, A., & Korfiatis, K. (2019). Children's interest in school garden projects, environmental motivation and intention to act: A case study from a primary school of Cyprus. *Applied Environmental Education & Communication, 18*(1), 2–12.

Cohen, L., Manion, L., & Morrison, K. (2018). *Research methods in education* (8th ed.). Routledge.

Darner, R. (2012). An empirical test of self-determination theory as a guide to fostering environmental motivation. *Environmental Education Research, 18*(4), 463–472.

Do Carmo Galiazzi, M., Paula Salomão de Freitas, D., Aguiar de Lima, C., da Silva Cousin, C., Langoni de Souza, M., & Launikas Cupelli, R. (2018). Narratives of learning communities in environmental education. *Environmental Education Research, 24*(10), 1501–1513.

Doyle, R., & Krasny, M. (2003). Participatory rural appraisal as an approach to environmental education in urban community gardens. *Environmental Education Research, 9*(1), 91–115.

Elliot, J. (1991). *Action research for educational change.* McGraw-Hill Education (UK).

Gibbs, L., Staiger, P. K., Townsend, M., Macfarlane, S., Gold, L., Block, K., et al. (2013). Methodology for the evaluation of the Stephanie Alexander Kitchen Garden program. *Health Promotion Journal of Australia, 24*(1), 32–43.

Hawkins, K. A. (2015). The complexities of participatory action research and the problems of power, identity and influence. *Educational Action Research, 23*(4), 464–478.

Heron, J., & Reason, P. (2006). The practice of co-operative inquiry: Research "with" rather than "on" people. In J. Heron & P. Reason (Eds.), *Handbook of action research* (pp. 144–154). Sage.

Jacobs, S. (2016). The use of participatory action research within education--Benefits to stakeholders. *World Journal of Education, 6*(3), 48–55.

Kalaitzidaki, M., & Filippaki, A. (2021). Action research in environmental education in Greece. *PEEKPE, 22*(67).

Katsenou, C., Flogaitis, E., & Liarakou, G. (2015). Action research to encourage pupils' active participation in the sustainable school. *Applied Environmental Education & Communication, 14*(1), 14–22.

Kember, D. (Ed.). (2000). *Action learning, action research: Improving the quality of teaching and learning.* Routledge.

Korfiatis, K., & Petrou, S. (2021). Participation and why it matters: Children's perspectives and expressions of ownership, motivation, collective efficacy and self-efficacy and locus of control. *Environmental Education Research, 27*(12), 1700–1722.

Liao, H. C., & Wang, Y. H. (2008). Applying the ARCS motivation model in technological and vocational education. *Contemporary Issues in Education Research (CIER), 1*(2), 53–58.

Murakami, C. D., Su-Russell, C., & Manfra, L. (2018). Analyzing teacher narratives in early childhood garden-based education. *The Journal of Environmental Education, 49*(1), 18–29.

Noor, M. S. A. M., & Shafee, A. (2021). The role of critical friends in action research: A framework for design and implementation. *Practitioner Research, 3*, 1–33.

Palmer, P. J. (2017). *The courage to teach: Exploring the inner landscape of a teacher's life* (10th anniv. ed.). Wiley.

Pine, G. J. (2009). *Teacher action research: Building knowledge democracies.* Sage.

Reason, P., & Bradbury, H. (Eds.). (2008). *The SAGE handbook of action research: Participative inquiry and practice* (2nd ed.). Sage.

Servant-Miklos, V. (2022). Environmental education and socio-ecological resilience in the COVID-19 pandemic: Lessons from educational action research. *Environmental Education Research, 28*(1), 18–39.

Zhou, J. (2012). Problems teachers face when doing action research and finding possible solutions: Three cases. *Chinese Education & Society, 45*(4), 68–80.

Zuber-Skerritt, O. (2018). An educational framework for participatory action learning and action research (PALAR). *Educational Action Research, 26*(4), 513–515.

Part IV
Biology Teachers' Professional Development

Chapter 21
Assessing the Quality of Mentoring: Evidence-Based Development of a Training Programme for Biology Mentors and Pre-service Biology Teachers for Scientific Issues

Emanuel Nestler, Carolin Retzlaff-Fürst, and Jorge Groß

21.1 Introduction

The first lessons which pre-service biology teachers give in school are mentored by experienced biology teachers – called 'biology mentor teachers' (BMTs) or 'mentors.' These placements are called 'practical experience settings', 'year of practice', or, in our case, 'practical exercises in school'. This is the first time where pre-service biology teachers need to apply their professional knowledge in the complex teaching situation. Biology mentor teachers help pre-service biology teachers develop into reflective practitioners (Schön, 1983).

What we know is that a good teacher does not necessarily have to be a good biology mentor teacher. As experts, mentors must make tacit knowledge explicit again (Bromme & Jucks, 2014), bring this knowledge profitably into conversations (Ellis et al., 2020; Kreis, 2012), and develop an appropriate mentor role (Crasborn et al., 2011; Hennissen et al., 2008). A serious weakness of biology teacher training is that only a few projects and even fewer studies (Elster, 2008; Nestler et al., 2022) exist that focus on the task to train experienced biology teachers to become mentors.

In 2015, a SWOT analysis of teacher training in Mecklenburg-Western Pomerania revealed this deficient lack of BMT training. Afterwards, a BMT training with one focus on subject-specific mentoring was created and introduced. One major focus of the mentor training was to support pre-service biology teachers in teaching Scientific Inquiry.

E. Nestler (✉) · C. Retzlaff-Fürst
University of Rostock, Rostock, Germany
e-mail: emanuel.nestler@uni-rostock.de; carolin.retzlaff-fuerst@uni-rostock.de

J. Groß
Leibniz University of Hannover, Hannover, Germany
e-mail: gross@idn.uni-hannover.de

© The Author(s) 2024
K. Korfiatis et al. (eds.), *Shaping the Future of Biological Education Research*,
Contributions from Biology Education Research,
https://doi.org/10.1007/978-3-031-44792-1_21

This paper is part of a larger design-based research study (e.g. Reinmann, 2022; van den Akker, 1999) accompanying this process and opening the field of research on BMT training. In the first design-cycle, we aimed to connect and compare the designed mentor training to previous research on mentoring and mentor training. The impact on teaching quality is reported in Nestler et al. (2022). We conducted this present study to evaluate the mentoring quality and content of the mentoring dialogues before and after BMT training.

The first aim of this paper is to describe our theoretical advancements to connect the generic view on mentoring quality with biology-related mentoring. Therefore, in the theoretical background, we report on previous studies regarding BMT training, describe how we adapted the tetrahedron model (based on Prediger et al., 2017), and define the core concepts, considering mentoring quality and the content of mentoring dialogues.

The second aim of this study is to report and discuss the results of evaluating mentoring quality and the content of mentoring dialogues.

21.2   Theoretical Background and Research Question

Barnett and Friedrichsen (2015) point out that mentoring can improve the professional knowledge of pre-service biology teachers. They focus on pedagogical content knowledge (PCK) as part of professional knowledge (Baumert & Kunter, 2006; Shulman, 1986). Furthermore, there is a growing number of quantitative and qualitative studies describing the impact of mentoring on the professional development of pre-service teachers (e.g. Hobson et al., 2009; Kindall et al., 2017; Nguyen & Parr, 2018). However, as noted by Shulman (1986), we do not have specific domains of biology mentors' knowledge for teachers' professional knowledge.

Additionally, only a few researchers have addressed biology-related mentoring (Barnett & Friedrichsen, 2015; Elster, 2008; Wischmann, 2015; Nestler et al., 2022). Barnett and Friedrichsen (2015) report on a case study of educative mentoring to support pre-service biology teachers, yet they do not mention training for mentors at all. Elster (2008) describes subject-related mentor training for biology mentors, while focussing on mentoring for innovative gender-proofed practices in biology teacher training. Wischmann (2015) then did research on mentoring dialogues with six pre-service biology teachers. In summary, apart from these, there is no broad scientific field of BMT research with common methods and instruments.

The range of concepts associated with this area of research thus makes it a very complex field.

21.2.1 The Tetrahedron Model for a BMT Training

To arrange the concepts needed, we adapted the tetrahedron model for a BMT training (Nestler & Retzlaff-Fürst, 2020; Nestler et al., 2022) from the three-tetrahedron model for content-related professional development research (Prediger et al., 2017; Prediger, et al., 2019; Roesken-Winter et al., 2021).

Prediger et al. (2017) highlight one of many advantages of their three-tetrahedron model: Researchers can focus on single points, the connections between points of one tetrahedron and the connections between different tetrahedrons. We recreated this for our specific case, deducing tetrahedrons for biology teacher placements, BMT training, and the connected topic of teaching Scientific Inquiry (Fig. 21.1).

The three tetrahedrons of teaching biology, biology teacher training, and BMT training are interconnected because every lower tetrahedron is the subject of the tetrahedron above. Having a closer look, biology-related mentoring shows itself to be the connection between biology mentor teachers, biology pre-service teachers and the biology-related subject of the mentoring. To keep our focus on developing biology-specific BMT training, we therefore include 'biology' in every one of these three points of the tetrahedron. In the studied units, our biology mentors should support pre-service biology teachers in teaching lessons with experiments and living animals.

Therefore, in our case, we can adapt the model for the subject of Scientific Inquiry (Fig. 21.1).

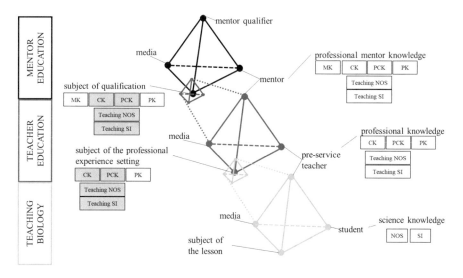

Fig. 21.1 The tetrahedron model of biology education

Teaching biology requires professional knowledge of biology teachers (Coe et al., 2014; Förtsch et al., 2016). This includes content knowledge (CK), pedagogical content knowledge (PCK) and pedagogical knowledge (PK) (Baumert & Kunter, 2006; Shulman, 1986).

The PCK of biology teachers, in turn, includes two of many more parts: teaching Nature of Science (NOS; Dittmer & Zabel, 2019; Paul et al., 2016) and teaching Scientific Inquiry (SI; Nerdel, 2017).

However, there is no given set of biology mentor knowledge (MK) or skills to support pre-service biology teachers in teaching scientific inquiry. For example, one review-study about biology mentor teachers' skills was published in 2020 (Ellis et al., 2020) – but it only contains generic MK. Additionally, this study was published 3 years after our study was conducted, but it lacks a focus on subject-specific aspects of mentoring.

Furthermore, as we know from Kirkpatrick (1959) and in relation to it from Lipowsky (2010), learning is only the second level of success of training programmes. However, an impact of BMT training on biology-related mentoring quality in the biology teacher placement would be the third and thus higher level of influence.

The lack of research and suitable instruments to test biology-related MK and the higher level of evaluation of mentor training leads us to biology-related mentoring quality.

21.2.2 Mentoring Quality

We define mentoring quality as the connection between the biology mentor teacher and pre-service teacher (Fig. 21.1). Most studies exclude the subject of the mentoring, learning goals and the specific learning of the mentees (e.g. Kreis, 2012). The biology-specific study of mentoring (Wischmann, 2015) describes the mentoring and the connection to a specific content. However, since there is no established field of biology-related mentoring quality with empirical evidence, we can only focus on mentoring quality as a generic concept. With this qualitative approach, we follow the theoretical framework of Michelsen et al. (2022) by defining concepts as mental constructs and their importance for teaching and learning. This includes mainly the relationship between mentor and mentee, as well as their dialogues. In the tetrahedron model (Fig. 21.1), this is illustrated by the connection between the points of the mentor and the pre-service teacher. The subject of the professional experience setting is discussed in Sect. 21.2.3 as the content of mentoring dialogues.

These insights serve as the basis of our two-part research question:

What effect does BMT training have on mentoring quality and the content of mentoring dialogues?

Since there are many generic studies, not only on the (generic) quality of mentoring, but also on the training of mentors, we can focus on studies that combine these.

Hennissen et al. (2011, p. 1049) trained mentors and reported 'shifts in their frequency of use of distinct skills'. Previously they published their work on mentors' roles and skills for mentoring dialogues (Crasborn et al., 2008) based on video analyses. Similarly, Kreis & Staub (2011) conducted extensive research on content-focussed coaching. Following this, Kreis (2012) trained mentors to use specific coaching techniques associated with better learning of pre-service teachers. This list of mentor training may be extended with the addition of the work of Langdon & Ward (2015), Cooke (2018) and Beutel & Spooner-Lane (2009).

We follow Crasborn's, Hennissen's, Staub & Kreis's understanding of mentoring quality because of their extensive work. Mentoring quality in this case includes two dimensions: The biology mentor teachers create a situation of co-constructive development of the lessons (Kreis, 2012), while the pre-service biology teachers have the chance to submit their own ideas to the discussions (Crasborn et al., 2008; Hennissen et al., 2011).

Based on the research on mentoring quality (Crasborn et al., 2008; Hennissen et al., 2011; Kreis, 2012), our first hypothesis is: *The quality of mentoring increases after the BMT training.* This hypothesis describes the main connection between the BMT training and the mentoring as such.

Biology mentor teachers and pre-service biology teachers may vary in their assessment of mentoring quality. Mentor teachers often underestimate their share in the mentoring dialogues (Hennissen et al., 2011). Additionally, the BMT training may have an impact on the assessment of mentoring quality by biology mentor teachers because they are sensitised to mentoring quality. Therefore, our second hypothesis is: *Observers vary in their assessment of mentoring quality.*

Additionally, in doing so, we avoid the often-described problems of self-assessment (Carter & Dunning, 2008).

21.2.3   Content of Mentoring Dialogues and Professional Knowledge of Biology Teachers

If we want to ensure that BMT training has a biology-related impact, focussing on the content of mentor dialogues may be one way to achieve this. The subject of mentoring dialogues is a challenging field of research, since it is attached to several conditions: First, the pre-service biology teachers give their first lessons. Mentoring needs to be adapted to the concrete situation and challenges of the pre-service teacher in their individual development. For example, a lesson on ecology can result in mentoring dialogues about classroom management, training scientific inquiry or the connections between plants and fungi. Mentoring has to be adapted to the specific situation. Therefore, standardisation of the content of mentoring dialogues is limited. Second, dialogues connect different aspects of teaching, which makes it hard to differentiate between pedagogical and didactics aspects. Third, the complex connection between universities and schools results in a limited scope for action.

For example, focussing on teaching scientific inquiry in the field of genetics is nearly impossible from a practical perspective.

The studies that focus on content knowledge are often studies using recordings of mentoring dialogues. Strong & Baron (2004), for instance, stated that only 2% of the topics of mentoring dialogues were part of the content knowledge (CK). Crasborn et al. (2011) analysed mentoring discussions and discovered that only 7% of the topics discussed represented the subject matter. We are very uncertain whether these low percentages also apply to the mentoring dialogues in our specific case.

As mentioned above, increasing professional knowledge of pre-service teachers should be one aim of mentoring (Barnett & Friedrichsen, 2015). With the conditional limitations in mind, we followed Wischmann (2015) and asked for the dimensions of professional knowledge of teachers as content of mentoring dialogues which should include content from biology (CK) and teaching biology (PCK). Our third hypothesis addresses this instant and refers to the biology-related content of mentoring dialogue: *Subject-specific parts of mentoring dialogues increase after the BMT training.*

Moreover, we need to focus on the perspectives of biology mentor teachers and pre-service teachers. Distinguishing between their respective views, we developed our fourth hypothesis: *Observers vary in their assessment of the content of mentoring dialogues.*

21.3   Research Design and Method

Design-based research (e.g. Reinmann, 2022) recognises the circumstance that research in the educational field is very much shaped by the concrete system, making it sometimes difficult to generalise the findings. As pictured in the tetrahedron model (Fig. 21.1), BMT training connects three levels of training: teaching biology, teacher education, and mentor training. Our research design is based on the concrete BMT training and the limitations of our educational system affecting this mentor training.

We follow the design-based research methodology and address these limitations with a clear description of the training and situations in Sects. 21.3.1 and 21.3.2, so that subsequent teacher trainers can adapt their training. In the following subsections (Sects. 21.3.3 and 21.3.4), the method and data analysis are provided.

21.3.1   Teacher Training of Pre-service Biology Teachers and Practical Exercises in School

In the year of the study, in Mecklenburg-Western Pomerania, about 30 pre-service biology teachers were supported each semester in placements organised by the head of biology teacher education. This pre-service biology teacher placement is called 'practical exercises in school'. Five pre-service biology teachers in their fifth semester took turns in teaching biology. They were supported over one semester by one out of seven biology mentors. The lessons given at this stage were the very first biology lessons of pre-service teachers in practical exercises in school. Mentors responsibly supported pre-service teachers in preliminary discussions. The given lesson was discussed with all pre-service teachers during debriefing. Although these lessons were the first ones for pre-service biology teachers, there was a focus on biology-related content in mentoring dialogues with reference to the second hypothesis.

21.3.2 Biology Mentor Teacher Training (BMT Training)

In this federal state, BMT training is designed for every active biology mentor teacher in this part of biology teacher education. This study is an interventional study without a control group because there are no more (comparable) pre-service biology teachers or biology mentors who are in the same educational system.

The BMT training (Fig. 21.2) took place between the winter semester and the summer semester as described by Nestler et al. (2022, 132): 'This qualification involved an overall workload of 90 h (45 h general, 45 h biology-related) for biology mentors ($N = 7$, age: $M = 42.3$).' Design principles of the general mentor training (45 h) are described by Malmberg et al. (2020). During the 2017/18 winter semester, the mentors supported 25 pre-service teachers (11 women, 10 men, four without information; age $M = 25.6$) in practical exercises in school. The BMT

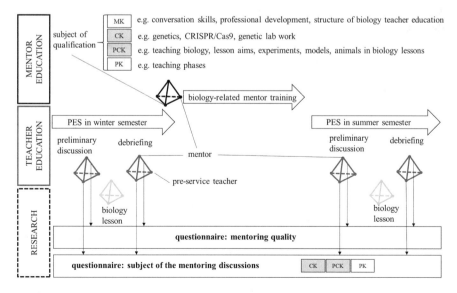

Fig. 21.2 Mentor training and research design

training was implemented during the semester break. During the summer semester 2018, the mentors supported 28 pre-service teachers (16 women, 10 men, two without information, age $M = 23.4$).

The BMT training (Fig. 21.2) is designed to support mentoring for scientific inquiry: After a brief introduction to scientific inquiry as part of the BMT training, the participants observed the milk snake *Lampropeltis triangulum*. They acted out a preliminary discussion of one biology lesson by taking the different roles of biology mentor teachers, pre-service biology teachers and observer. In these discussions, they used conversational techniques as mentioned by Kreis (2012) and Hennissen et al. (2011). Afterwards, they reflected on their application of these techniques. On another day, they carried out experiments together with a professor of genetics, as mentioned in Nestler et al. (2022).

21.3.3   Method

The main objective of this research is to evaluate the quality of BMT training and connect it to previous research on mentoring quality and the content of mentoring dialogues. Only when connecting these research fields are we able to derive design principles that constitute the field.

Therefore, in the first design cycle, the BMT training was conducted between practical exercises in school in the winter and summer semester (Fig. 21.2: Mentor Training and Teacher Training).

To answer the research question, the quality of mentoring and content of mentoring dialogues was evaluated during the winter semester prior to mentor training, as well as during the summer semester right after the BMT training (Fig. 21.2 Research).

Mentoring Quality was evaluated with a self-created questionnaire as a reaction to a lack of suitable research instruments. Within the questionnaire, we designed a six-item scale of mentoring quality based on the findings of Crasborn et al. (2008), Hennissen et al. (2011) and Kreis (2012) (e.g. '*The atmosphere in this meeting was constructive*' or '*Pre-service teacher had the opportunity to address their questions and concerns*'). This scale has a high internal consistency ($\alpha = .848$) and was intentionally kept short to avoid survey fatigue.

Regarding the second part of our concept of biology-related mentoring quality, the content of mentoring dialogues was assessed with the help of three single items, i.e. '*Content knowledge was discussed in the conversation*' (for content knowledge), '*Pedagogical content knowledge was discussed in the conversation*' (for pedagogical content knowledge), and '*Pedagogical knowledge was discussed in the conversation*' (for pedagogical knowledge). All items were rated on a 6-level scale, ranging from 1 (do not agree) to 6 (agree).

21.3.4 Data Analysis

60 participants filled out 633 questionnaires to assess mentoring quality and the content of the mentoring dialogues. The participants were the seven biology mentor teachers and their 53 pre-service biology teachers, who are mentioned in Sect. 21.3.2. The programme SPSS was used to analyse the data. To validate the hypotheses, we used Mann-Whitney U tests and Kruskal-Wallis tests.

21.4   Findings

We discovered that the biology mentor teachers were curious during the BMT training and gave good feedback, which is level one of successful training (Kirkpatrick, 1959; Lipowsky, 2010).

21.4.1   Mentoring Quality

In general, mentors and pre-service teachers reported that mentoring quality was higher than the expected mean value of the scale (Fig. 21.3).

To assess the impact of our BMT training, our first hypothesis was as follows: *The quality of mentoring increases after the BMT training.* The mean values for the quality of mentoring increased after mentoring training (Fig. 21.3). To test for significant differences, a Mann-Whitney U test was run. There was a significant

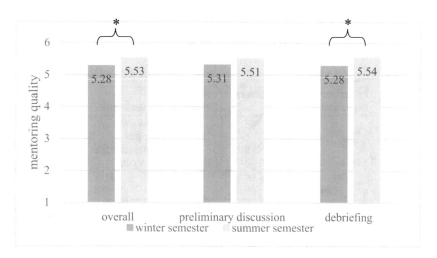

Fig. 21.3 Mean mentoring quality for preliminary discussion and debriefing before and after the mentor training. *$p < .001$

difference for the overall measurement ($z = -4.030, p < .001, r = .16$) and debriefing ($z = -4.253, p < .001, r = .19$). The analysis did not indicate a significant result for preliminary discussions ($z = -.705, p = .481$). Overall, these results support our hypothesis but have a small effect size.

21.4.2   Views on Mentoring Quality by Different Observers

These results call for a closer look at our second hypothesis: *Observers vary in their assessment of mentoring quality.* Table 21.1 shows the descriptive statistics for the quality of mentoring. All groups of participants rated the mean mentoring quality higher after BMT training (summer semester 2018) than before. Interestingly, mentors generally perceived mentoring quality lower than pre-service biology teachers. Applying Mann-Whitney U tests for preliminary discussions indicates significant differences between observers during the winter semester ($z = -2.917, p = .004, r = .32$) and the summer semester ($z = -3.827, p < .001, r = 0.46$). However, our results for debriefing paint a different picture: For the winter semester, the Kruskal-Wallis test reveals no significant differences for observers. Surprisingly, while mentors do not rate mentoring quality much higher (+ 0.06), pre-service teachers (+0.29 & +0.31) in fact do. Overall, the Kruskal-Wallis test evidenced significant differences between the mentors and the observing pre-service teachers ($z = -5.395, p < .001, r = .54$), with a strong effect. On the contrary, between the mentors and the pre-service biology teachers responsible for the lesson ($z = -5.026, p < .001, r = .39$), only a medium effect could be detected. These differences are not found between both pre-service biology teachers groups. It is interesting to note that BMT training may have an impact on mentoring quality, although mentors do not have to necessarily see this effect. Overall, our results support our second hypothesis.

Table 21.1 Sample size (n), means (M) and standard deviation (SD) for mentoring quality

| | Preliminary discussion before the lesson | | | | Debriefing after the lesson | | | | | |
| | Mentor | | Pre-service teachers responsible for the lesson | | Mentor | | Pre-service teachers responsible for the lesson | | Pre-service teacher observing | |
	n	M (SD)	n	M (SD)	n	M (SD)	n	M (SD)	n	M (SD)
Winter semester 2017/18	32	5,07 (0.96)	52	5.46 (0.79)	48	5.19 (0.77)	59	5.35 (0.79)	144	5.28 (0.75)
Summer semester 2018	22	5.22 (0.42)	46	5.64 (0.34)	40	5.25 (0.33)	59	5.64 (0.32)	127	5.59 (0.52)

21.4.3   Content of Mentoring Dialogues

A Mann-Whitney U test was calculated to examine the third hypothesis: *The subject-specific parts of mentoring dialogues increase after BMT training* (Fig. 21.4). The mean values for all content of conversation between mentors and mentees increased after the mentoring training, except for the PCK in the preliminary discussion. These increases were statistically significant, but had small effect sizes for all parts of conversations for the overall measure (CK: $z = -4.015, p < .001$, $r = .16$; PCK: $z = -2.527, p = .011, r = .10$; PK: $z = -4.271, p < .001, r = .17$) and debriefing (CK: $z = -4.251, p < .001, r = .20$; PCK: $z = -3.323, p = .001, r = .15$; PK: $z = -4.557, p < .001, r = .21$). Again, no significant differences between the estimations of the observers were found for the preliminary discussions (CK: $z = -.640, p = .522$; PCK: $z = -.713, p = .476$; PK: $z = -.722, p = .470$). Our results show slight indications of support for our third hypothesis. However, we must note that after the BMT training, general content of conversation also grew.

21.4.4   Views on the Content of Mentoring Dialogues by Different Observers

Further analyses for our fourth hypothesis: *Observers vary in their assessment of the content of mentoring dialogues* are illustrated in Table 21.2. In preliminary discussions, significant differences in estimating topics in content knowledge (CK) were found in the winter semester ($z = -2.106, p = .035, r = .23$), as well as a larger effect in the summer semester ($z = -3.821, p < .001, r = .46$). As we can see in Table 21.2, the assessment continues to diverge during the summer semester. Regarding our second hypothesis, there is not only no significant increase of topics

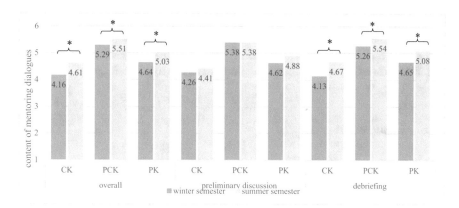

Fig. 21.4 Mean ratings of content knowledge (CK), pedagogical content knowledge (PCK), and pedagogical knowledge (PK) *$p < .001$

Table 21.2 Sample size (n), means (M) and standard deviations (SD) for specific parts of conversations

| | | Preliminary discussion before the lesson | | | | Debriefing after the lesson | | | | | |
| | | Mentor | | Pre-service teacher in responsibility | | Mentor | | Pre-service teacher in responsibility | | Pre-service teacher observing | |
		n	M (SD)	n	M (SD)	n	M (SD)	n	M (SD)	n	M (SD)
Winter semester 2017/18	CK	32	3.84 (1.51)	52	4.52 (1.39)	46	3.85 (1.61)	58	4.34 (1.38)	141	4.13 (1.38)
	PCK	32	5.31 (1.31)	52	5.42 (0.89)	45	5.42 (0.99)	59	5.39 (0.91)	141	5.15 (0.99)
	PK	32	4.78 (1.60)	50	4.52 (1.33)	47	4.77 (1.15)	57	4.67 (1.19)	141	4.59 (1.07)
Summer semester 2018	CK	22	3.45 (1.47)	46	4.87 (1.09)	40	3.68 (1.40)	59	4.83 (1.26)	127	4.91 (1.07)
	PCK	22	5.41 (0.80)	46	5.37 (0.90)	40	5.38 (0.59)	58	5.60 (0.59)	127	5.57 (0.73)
	PK	22	4.77 (1.07)	46	4.93 (1.14)	40	4.87 (0.88)	58	5.14 (1.02)	127	5.11 (0.99)

in content knowledge in preliminary discussions, but also a wider gap in the estimation after the BMT training in this field of professional knowledge.

The Kruskal-Wallis test for debriefing during the winter semester did not reveal significant differences in the content of the conversations ($p = .101$). In the summer semester, these differences were found for content knowledge (CK) with a medium effect between mentors and pre-service teachers responsible for the lesson ($z = -4.225, p < .001, r = .42$), and between mentors and observing pre-service biology teachers ($z = -4.819, p < .001, r = .37$). Furthermore, in the summer semester we could find a weak effect in the estimation of the content of mentoring dialogues in pedagogical content knowledge between mentors and observing pre-service teachers ($z = -2.422, p = .046, r = .24$). Therefore, referring back to our third hypothesis, the significant differences in the debriefing are driven by the estimation of the pre-service teachers. Consequently, our second hypothesis needs to be re-evaluated. Nevertheless, our results support our fourth hypothesis.

21.5   Discussion

All the biology mentors for these specific pre-service biology teacher placements called practical exercises in school were trained to an up-to-date level by the department didactics of biology at the University of Rostock. Regarding the few studies on biology mentors training (Barnett & Friedrichsen, 2015; Elster, 2008; Nestler et al., 2022), this is a first success.

21.5.1   The Tetrahedron Model of Biology Education

Our first aim was to describe our theoretical advancements to connect the generic view on mentoring quality with biology-related mentoring. In Chap. 2, Theoretical Background, we pointed out the difficulties in the field of BMT training. The adaption of the tetrahedron model (Prediger et al., 2017) shows the interdependencies between the three levels mentor training, biology teacher training, and teaching biology. Thus, the adapted model shines a light on the underdeveloped field of subject-related mentoring. While mentoring quality is generally well researched (Kreis, 2012; Hennissen et al., 2008), content knowledge, for example, is a marginalised topic of mentoring dialogues (Crasborn et al., 2011; Strong & Baron, 2004). We therefore require the existence of a coherent concept of biology-specific mentoring quality. Initially, we have to focus on generic mentoring quality and CK, PCK and PK as content of mentoring dialogues in order to approach such a concept.

Future studies can use the tetrahedron model to describe new BMT training. Additionally, this model can be specified for different areas of the professional development of pre-service biology teachers creating a coherent concept of biology teacher training. For every subject and field of research in didactics of biology education the question remains: How can we train the biology mentor teachers to support the pre-service biology teachers in this specific area of professional development?

21.5.2   Mentoring Quality

In our case, we were interested in the generic quality of mentoring. Overall, the high mean values of mentoring quality show a satisfactory support for our pre-service biology teachers. We saw that the pre-service biology teacher rated the mentoring quality even higher than the biology mentor teachers. Our biology mentor teachers were good biology mentor teachers before the BMT training and even better thereafter.

Therefore, mentoring quality increased after performing the BMT training. This supports our first hypothesis. Additionally, these results are consistent with previous research (Kreis, 2012). BMT training seems to be a satisfactory way of improving biology teacher training.

Surprisingly, the biology mentor teachers are more careful than the observing pre-service teachers. Therefore, our mentor training could have an impact on practical experience settings without the mentors realising it. The different mean values support the hypothesis of Kreis (2012) that preliminary discussions and debriefings have different effects. Further research on these differences is needed. The questionnaire of mentoring quality is characterised by consistency and is therefore appropriate for this study, in which different concepts such as teaching quality (Nestler et al.,

2022) and mentoring quality are connected. Prospectively, a comparison with data of videotaped mentoring dialogues could improve the quality of this questionnaire.

In the future, more research is needed on different aspects of mentoring quality related to biology-specific mentoring. The views of different observers should be part of these studies.

21.5.3 Content of Mentoring Dialogues

The tetrahedron model of biology education leads to the different challenges of subject mentoring – which, in our case, is biology. One of the challenges lies in linking the three levels of mentor training, teacher training and biology teaching, all of which are difficult to research.

After our BMT training, we saw an increase in every area of professional knowledge. However, biology mentor teachers are considerably more cautious in assessing it, and the increase is driven by the pre-service biology teachers. At first glance, our findings do not support the low shares of content knowledge in mentoring dialogues (Crasborn et al., 2011; Strong & Baron, 2004). A second glance, however, reveals that this might be a methodical difference between videotaped sessions and the assessment performed by means of questionnaires.

In summary, our BMT training is compatible with previous research on mentoring quality and – by focussing on biology – progressively enhances this field of research. We know that mentoring dialogues are important for the professional development of pre-service teachers (Kreis, 2012). If we want to ensure that mentoring dialogues have an impact on PCK and CK, we need to train biology mentor teachers to focus on these areas of professional knowledge. This is a major desideratum in biology teacher training.

21.5.4 Limitations

Following the argumentation of design-based research (Reinmann, 2022), this study provides small to no evidence for a generic view on a BMT training. The differences in the structure of teaching biology, practical experience settings, and possible BMT training may have a huge impact on the results. Absence of a control group, different pre-service teachers in the winter and summer semester, and the brevity of questionnaires add further factors to these limitations.

Additionally, we cannot redo this study in Mecklenburg-Western Pomerania because our group of biology mentor teachers for the practical exercises in school is now trained. A larger study including more states or countries would change the initial conditions on all levels: teaching biology, teacher training, and mentor training. While this is challenging, we support Prediger et al. (2017) in their request to describe this potential chain of impacts (or effects), and assess them with a quantitative or mixed-method approach. This complements the previous qualitative work of Wischmann (2015), Barnett and Friedrichsen (2015) and many other researchers.

In summary, our quantitative data on the possible impact of the BMT training shows the need for more theoretical, empirical, and practice-oriented developments in this core field of biology teacher professionalisation.

21.5.5   Conclusion

The current study is one of the first studies in BMT training. In combination with our previous study on BMT training (Nestler et al., 2022) and its impact on teaching quality, we have obtained initial indications of level 3 behavioral change (mentoring quality) and level 4 changing results (teaching quality) (e.g. Kirkpatrick, 1959; Lipowsky, 2010). In the course of the study, we acquired empirical support for the given feedback of one mentor trainee: *'The training ensured a better structure and debriefing sessions and an improvement in the way we talk to students, as well as a better understanding of the students' initial situation.'*

Our research emphasizes the need for more BMT training. The pre-service teacher placements are one of the few curricular opportunities for biology teacher training to apply content and pedagogical content knowledge, and to improve biology-related teaching skills. This paper sheds light upon this and possible chains of effects.

Acknowledgments   Many thanks to Daniel Rühlow for his expertise and support in language editing.

References

Barnett, E., & Friedrichsen, P. J. (2015). Educative mentoring: How a mentor supported a pre-service biology teacher's pedagogical content knowledge development. *Journal of Science Teacher Training, 26*, 647–668.

Baumert, J., & Kunter, M. (2006). Stichwort: Professionelle Kompetenz von Lehrkräften. *Zeitschrift für Erziehungswissenschaft, 9*, 469–520.

Beutel, D., & Spooner-Lane, R. (2009). Building mentoring capacities in experienced teachers. *International Journal of Learning, 16*, 1–10.

Bromme, R., & Jucks, R. (2014). Fragen Sie ihren Arzt oder Apotheker: Die Psychologie der Experten-Laien-Kommunikation. In M. Blanz, A. Florack, & U. Piontkowski (Eds.), *Kommunikation. Eine interdisziplinäre Einführung* (pp. 237–246). Kohlhammer.

Carter, T. J., & Dunning, D. (2008). Faulty self-assessment: Why evaluating one's own competence is an intrinsically difficult task. *Social and Personality Psychology Compass, 2*(1), 346–360. https://doi.org/10.1111/j.1751-9004.2007.00031.x

Coe, R., Aloisi, C., Higgins, S., & Major, L. E. (2014). *What makes great teaching? Review of the underpinning research, project report.* Sutton Trust.

Cooke, D. M. (2018). *The influence of professional development, in educative mentoring, on mentors' learning and mentoring practices.* Faculty of Culture and Society, Auckland University of Technology. (Dissertation).

Crasborn, F., Hennissen, P., Brouwer, N., Korthagen, F., & Bergen, T. (2008). Promoting versatility in mentors' use of supervisory skills. *Teaching and teacher training, 24,* 499–514.

Crasborn, F., Hennissen, P., Brouwer, N., Korthagen, F., & Bergen, T. (2011). Exploring a two-dimensional model of mentor roles in mentoring dialogues. *Teaching and Teacher training, 27,* 320–331.

Dittmer, A., & Zabel, J. (2019). Das Wesen der Biologie verstehen: Impulse für den wissenschaftspropädeutischen Unterricht. In J. Groß, M. Hammann, P. Schmiemann, & J. Zabel (Eds.), *Biologiedidaktische Forschung: Erträge für die Praxis* (pp. 93–110). Springer Spektrum.

Ellis, N. J., Alonzo, D., & Nguyen, H. T. M. (2020). Elements of a quality preservice teacher mentor: A literature review. *Teaching and Teacher training, 92,* 103072. https://doi.org/10.1016/j.tate.2020.103072

Elster, D. (2008). *Subject-related mentoring in biology teacher education.* Paper presented at 'Impact of science education research on public policy", NARST annual international conference (national association for research in science teaching) 29th March-2nd April 2008, Garden Grove. https://doi.org/10.13140/2.1.3341.1366

Förtsch, C., Werner, S., von Kotzebue, L., & Neuhaus, B. J. (2016). Effects of biology teachers' professional knowledge and cognitive activation on student achievement. *International Journal of Science Education, 38,* 2642–2666.

Hennissen, P., Crasborn, F., Brouwer, N., Korthagen, F., & Bergen, T. (2008). Mapping mentors' roles in mentoring dialogues. *Educational Research Review, 3,* 168–186. https://doi.org/10.1016/j.edurev.2008.01.001

Hennissen, P., Crasborn, F., Brouwer, N., Korthagen, F., & Bergen, T. (2011). Clarifying pre-service teacher perceptions of mentors' developing use of mentoring skills. *Teaching and Teacher training, 27,* 1049–1058.

Hobson, A. J., Ashby, P., Malderez, A., & Tomlinson, P. D. (2009). Mentoring beginning teachers: What we know and what we don't. *Teaching and teacher training, 25,* 207–216.

Kindall, H. D., Crowe, T., & Elsass, A. (2017). Mentoring pre-service educators in the development of professional disposition. *International Journal of Mentoring and Coaching in Education, 6,* 196–209. https://doi.org/10.1108/IJMCE-03-2017-0022

Kirkpatrick, D. L. (1959). Techniques for evaluating training programs. *Journal of the American Society of Training Directors, 13*(11), 21–26.

Kreis, A. (2012). *Produktive Unterrichtsbesprechungen: Lernen im Dialog zwischen Mentoren und angehenden Lehrpersonen.* Haupt.

Kreis, A., & Staub, F. C. (2011). Fachspezifisches Unterrichtscoaching im Praktikum. Eine quasiexperimentelle Interventionsstudie. *Zeitschrift für Erziehungswissenschaft, 14,* 61–83.

Langdon, F., & Ward, L. (2015). Educative mentoring: A way forward. *International Journal of Mentoring and Coaching in Education, 4,* 240–254.

Lipowsky, F. (2010). Lernen im Beruf – Empirische Befunde zur Wirksamkeit von Lehrerfortbildung. In F. Müller, A. Eichenberger, M. Lüders, & J. Mayr (Eds.), *Lehrerinnen und Lehrer lernen – Konzepte und Befunde zur Lehrerfortbildung* (pp. 51–70). Waxmann.

Malmberg, I., Nestler, E., & Retzlaff-Fürst, C. (2020). Qualitäten der Mentor*innenqualifizierung M-V. Eine Design Based Research Studie zu einem Lernbegleitungsprogramm an der Schnittstelle zwischen Schule und Hochschule. In F. Hesse & W. Lütgert (Eds.), *Auf die Lernbegleitung kommt es an! Konzepte und Befunde zu Praxisphasen in der Lehrerbildung.* Klinkhardt.

Michelsen, M., Groß, J., Paul, J., & Messig, D. (2022). Elaboration of practical diagnostic competence in context of students' conceptions on plant nutrition. *Science Education International (SEI), 33*(2).

Nerdel, C. (2017). *Grundlagen der Naturwissenschaftsdidaktik. Kompetenzorientiert und aufgabenbasiert für Schule und Hochschule.* Springer.

Nestler, E., & Retzlaff-Fürst, C. (2020). Die Mentor*innenqualifizierung im Fach Biologie zur Unterstützung der Reflexion von fachwissenschaftlichen und fachdidaktischen Konzepten Studierender. In Y. Völschow & K. Kunze (Hrsg.), *Reflexion und Beratung in der Lehrerinnen- und Lehrerausbildung. Beiträge zur Professionalisierung von Lehrkräften* (S. 365–380).

Nestler, E., Retzlaff-Fürst, C., & Groß, J. (2022). Train the trainer in the jigsaw puzzle of biology education: Effects of mentor training on teaching quality. In K. Korfiatis & M. Grace (Eds.), *Contributions from biology education research. Current research in biology education: Selected papers from the ERIDOB community* (pp. 127–140). Springer. https://doi.org/10.1007/978-3-030-89480-1_10

Nguyen, M. H., & Parr, G. (2018). Mentoring practices and relationships during the EAL practicum in Australia: Contrasting narratives. In A. Fitzgerald, G. Parr, & J. Williams (Eds.), *Reimagining professional experience in initial teacher training: Narratives of learning* (pp. 87–105). Springer.

Paul, J., Lederman, N. G., & Groß, J. (2016). Learning experimentation through science fairs. *International Journal of Science Education, IJSE, 38*(15), 2367–2387.

Prediger, S., Leuders, T., & Rösken-Winter, B. (2017). Drei-Tetraeder-Modell der gegenstandsbezogenen Professionalisierungsforschung: Fachspezifische Verknüpfung von Design und Forschung. *Jahrbuch für allgemeine Didaktik, 2017*, 159–177.

Prediger, S., Roesken-Winter, B., & Leuders, T. (2019). Which research can support PD facilitators? Research strategies in the three-tetrahedron model for content-related PD research. *Journal for Mathematics Teacher Education, 22*(4), 407–425. https://doi.org/10.1007/s10857-019-09434-3

Reinmann, G. (2022). Was macht Design-BasedResearch zu Forschung? Die Debatte um Standards und die vernachlässigte Rolle des Designs. *EDeR – Educational Design Research, 6*(2), 1–22.

Roesken-Winter, B., Stahnke, R., Prediger, S., et al. (2021). Towards a research base for implementation strategies addressing mathematics teachers and facilitators. *ZDM Mathematics Education, 53*, 1007–1019. https://doi.org/10.1007/s11858-021-01220-x

Schön, D. A. (1983). *The reflective practitioner: How professionals think in action*. Basic Books.

Shulman, L. S. (1986). Those who understand: Knowledge growth in teaching. *Educational Researcher, 15*(2), 4–14.

Strong, M., & Baron, W. (2004). An analysis of mentoring conversations with beginning teachers: Suggestions and responses. *Teaching and Teacher Education, 20*(1), 47–57. https://doi.org/10.1016/j.tate.2003.09.005

Van den Akker, J. (1999). Principles and methods of development research. In J. v. Akker, R. M. Branch, K. Gustafson, N. Nieveen, & T. Plomp (Eds.), *Design approaches and tools in education and training* (pp. 1–14). Kluwer.

Wischmann, F. (2015). *Mentoring in fachbezogenen Schulpraktikum: Analyse von Reflexionsgesprächen*. (Ph. D), Bremen.

Chapter 22
Exploring a Theory-Practice Gap: An Investigation of Pre-service Biology Teachers' Enacted TPACK

Alexander Aumann and Holger Weitzel

22.1 Introduction

Digital technology and media (DTM) have the potential to expand the field of representation (Hsu et al., 2015) and (inter)active learning (Chi & Wylie, 2014). In science teaching this can promote a deeper understanding of abstract and complex phenomena (Hsu et al., 2015). However, it is important to note that simply implementing DTM in class does not ensure higher learning outcomes as its effectiveness depends on the way it is implemented (Chien et al., 2016). Whether the potential of DTM is realized in biology classrooms is significantly determined by the teacher's preparedness (Drossel et al., 2019). To achieve a high level of preparation, teacher education programs incorporate digitization-related competences into the curriculum (Ning et al., 2022). The Technological Pedagogical and Content Knowledge (TPACK) framework (Mishra & Koehler, 2006) is frequently applied for modeling this complex body of knowledge (Njiku et al., 2020). However, there remains a gap between what pre-service teachers are taught at university and their actual use of DTM in practice (Tondeur et al., 2012). One reason why there is only little knowledge about the practical application of TPACK (Mouza, 2016; Ning et al., 2022) is the scarcity of studies analyzing the enactment of TPACK in the field (Willermark, 2018) and thereby taking contextual conditions into account (Rosenberg & Koehler, 2015).

According to the Transformation Model of Lesson Planning, teachers transfer professional knowledge into practice through reflective planning, implementation, and subsequent reflection of lessons (Stender et al., 2017). In order to contribute to research on the enactment of TPACK, this study aims to obtain insights into the

A. Aumann (✉) · H. Weitzel (✉)
University of Education Weingarten, Weingarten, Germany
e-mail: alexander.aumann@ph-weingarten.de; weitzel@ph-weingarten.de

© The Author(s) 2024
K. Korfiatis et al. (eds.), *Shaping the Future of Biological Education Research*,
Contributions from Biology Education Research,
https://doi.org/10.1007/978-3-031-44792-1_22

process of TPACK transfer via the planning, implementation, and reflection of biology lessons by collecting and analyzing data on these three steps of secondary school pre-service biology teachers (PSBTs).

22.2   Research Design and Method

Since the development of TPACK is considered a highly individual process (Niess, 2015; Tondeur et al., 2021), a case study approach was conducted to gather a comprehensive insight including various contextual data. This research approach has been applied in previous studies examining (pre-service) teachers' ability to link theory and practice (Upmeier zu Belzen & Merkel, 2014) and the TPACK development of in-service science teachers (Jaipal-Jamani & Figg, 2015). In the sense of an explorative case study, the role of this study is to generate initial hypotheses and thus to provide requirements for quantitative investigations (Yin, 2012).

22.2.1   Sample

A cohort of 16 PSBTs was accompanied in the classroom. All PSBTs were in their master's studies. According to the module handbook of the study program, the PSBTs attended subject didactic courses to the extent of nine credit points as part of their bachelor's studies. Media didactic topics were also covered, but these were not related to the use of DTM in biology teaching. Using a purposeful sampling method, three heterogeneous cases were selected from the cohort in terms of their TPACK (see Table 22.2). Thus, the cases represent (1) a low, (2) an average, and (3) a high self-reported TPACK level.

22.2.2   Context

To gain insights into the field, the PSBTs were accompanied during their internship semester of 14 weeks at school at the end of their master's studies. The TPACK mandatory for a specific use of DTM (student-generated explainer videos in biology) was conveyed via a workshop at the university (Aumann et al., 2023). Following the workshop, the PSBTs were asked to plan, implement, and reflect on a biology lesson in which their students create explainer videos (see Fig. 22.1). The PSBTs were free to choose hardware and software to use in their lessons. To compensate for the different conditions at the various schools, they had the opportunity to borrow equipment from the university. These precautions address the extrinsic first-order barriers regarding the implementation environment according to Ertmer (1999), as the PSBTs had access to technical equipment, support, and training.

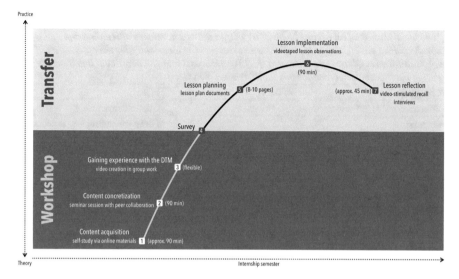

Fig. 22.1 Course of the workshop and data collection throughout the internship semester in terms of theory and practice

22.2.3 *Instruments*

22.2.3.1 The Survey

The survey was compiled from established scales measuring TPACK (Bernhard & Grassinger, 2022). By including these variables, the PSBTs intrinsic second-order barriers regarding teachers' professional beliefs are considered (Ertmer, 1999). Items were measured on a 6-point Likert scale. One item was added to record the number of lessons in which the PSBTs had previously integrated DTM. Survey data was analyzed using IBM SPSS version 28.0.

22.2.3.2 The EnTPACK Rubric

The EnTPACK rubric depicts instructional criteria for biology lessons in which students create explainer videos. The rubric was developed based on a systematic literature review as an instrument for measuring TPACK by comparing lesson plans, lesson observations, and lesson reflections containing the categories Pedagogical Content Knowledge (PCK), Technological Content Knowledge (TPK), and TPACK (Aumann et al., 2023) The rubric divides these categories into further subcategories, which in turn have been assigned several criteria. The expression of these criteria is assessed via observable indicators (see example Table 22.1).

Table 22.1 Sample of the EnTPACK rubric

"The identification of all organs involved as well as the two hormones adrenaline and noradrenaline, which trigger biological stress reactions, is elementary for the application of special strategies of stress management. [...] The HPA axis and glucocorticoid functions are not part of the required expertise for this topic" (lesson plan C3, p. 6–8).

Category	PCK
Subcategory	Content structuring
Criteria	Focus on central key aspects
Indicators	1. Central aspects of the content are highlighted/2. Central aspects are distinguished from non-essential aspects.

Examples from the data sets as well as exemplary criteria were translated from German into English

22.2.4   Data Collection and Analysis

Besides a survey conducted between workshop and transfer, data were collected regarding planning, implementation, and reflection of the lesson. Lesson planning was collected via lesson plan documents (1. Subject analysis, 2. instructional and 3. methodological planning, 4. lesson outlines) and lesson materials. Lesson implementation was videographed. PSBT's lesson reflection was collected via semi-structured video-stimulated recall interviews. Data analysis was conducted using evaluative qualitative content analysis (Kuckartz & Rädiker, 2022) via VERBI MAXQDA version 2020. First, the data sources were reviewed, commented on, and particularly noteworthy sections were marked. Subsequently, the materials were coded deductively on the basis of the EnTPACK rubric (Aumann et al., 2023) and inductively based on the data sets. Afterwards, all data sources were evaluated again, including the inductively derived categories. The coded data sets were then analyzed qualitatively and interpretatively. Contentious issues were validated communicatively between the two authors.

22.3   Results

Table 22.2 presents the survey results of the three cases in comparison with the total sample. In this regard, case one shows a below-average, case two an average, and case three an above-average self-reported TPACK compared to the sample.

The following description focuses on those results that show relevant aspects for the specific objectives of the use of DTM and their implementation.

Table 22.2 Survey data of the three cases

Scale	Case one	Case two	Case three
TPACK	M = 2.60 (↓)	M = 3.80 (↔)	M = 4.40 (↑)
Self-efficacy	M = 2.29 (↓)	M = 3.43 (↔)	M = 1.43 (↓)
Motivational orientation	M = 2.83 (↓)	M = 3.33 (↔)	M = 1.83 (↓)
Attitudes	M = 3.63 (↓)	M = 5.13 (↑)	M = 5.38 (↑)
Perceived usefulness	M = 4.00 (↓)	M = 6.00 (↑)	M = 5.75 (↑)
Intentions	M = 3.00 (↓)	M = 4.80 (↑)	M = 4.40 (↔)
Number of lessons taught	1–5 (↓)	6–10 (↑)	1–5 (↓)

All scales measure the constructs in relation to the use of DTM in class
↓ = *below average*   ↔ = *average*   ↑ = *above average of the total sample*

22.3.1 Case One

Case one is a female PSBT with the lowest self-reported TPACK of the sample. Although considering the use of DTM in class as potentially beneficial, her motivation and intention to use them is limited. This could be related to her low self-reported ability and self-efficacy to use DTM in class, as well as to her lack of experience in this regard. Her lesson utilized the creation of explainer videos to introduce insects' physique to a sixth grade class.

Content

The *Lesson planning* is characterized by a focus shift in the transition from instructional to methodological planning. Whereas instructional planning is content-centered, methodological planning is DTM-centered. The methodological planning focuses exclusively on the question of how to create an explainer video. As a result, DTM-related phases in the lesson outline significantly exceed those of content-related phases (lesson plan C1, p. 9ff).

The *Lesson implementation* is congruent with the methodological planning. The PSBT discusses a variety of design-related criteria for the creation of explainer videos, which are irrelevant to the biological subject. Those criteria build the guideline for both the development and the feedback phase (lesson implementation C1, 2:38–8:55).

Lesson reflection: Case one notices that the students' lost sight of the content in the lesson: "I don't think they paid much attention to the content, so the event was more about creating the video actually" (lesson reflection C1, l. 120ff). However, she ultimately attributes this problem to external factors such as the students' fascination with creating a video and a lack of time in class.

TPACK

The *Lesson planning* does not consider reasons for using explainer videos to teach the subject content and to reach the learning objectives. Case one rather seems to interpret the creation of an explainer video and the subject content as separate lesson contents: "The lesson focuses on the physique of the different insects and on the creation of an explainer/learning video" (lesson plan C1, p. 7).

Lesson implementation: Congruent with the lesson plan, no systematic relationship between subject content, learning objectives, and the DTM use is evident. For instance, the exemplary explainer video she selected to provide orientation for the students deals exclusively with cutout animations and does not provide any reference to biological subject matter (lesson implementation C1, 9:47–12:22).

Lesson reflection: Case one reports insecurity in the use of DTM: "I found it kind of stressful, because sometimes you don't really have it under your control" (lesson reflection C1, l. 502ff) and identifies technological challenges as her "main concern" (lesson reflection C1, l. 511). Accordingly, she justifies the choice of the production procedure (One Take) with preparation effort, rather than with instructional considerations: "it was actually the simplest [procedure] that I can explain the quickest" (lesson reflection C1, l. 410ff).

PCK

Lesson planning: While case one writes down subject-related instructional considerations in subject analysis and instructional planning, these are not considered again in the further course of the lesson plan. In the instructional planning, she puts emphasis on the fact that the students use "precise biological terms or technical terms" only sporadically (lesson plan C1, p. 7). Nevertheless, she uses a large number of technical terms in her lesson materials, partly irrelevant in regard to lesson content (e.g. polarized light). The lesson objective is clearly outlined in terms of content, but she misses the opportunity to prepare the students for the content. Despite her own problems with technology, she expects students to simultaneously learn the subject matter by themselves while transferring it into an explainer video.

Lesson implementation: Congruent with the lesson outlines, instructional considerations are not incorporated into teaching and students are left alone with content acquisition while creating the videos.

Lesson reflection: Case one comments on the cognitive overload of the students in content acquisition: "I think if you actually take one hour before roughly introducing insects' physique, it is clearer to them what you are aiming at later on and they are not that overwhelmed, maybe" (lesson reflection C1, l. 107ff). However, she justifies her approach with time constraints.

22.3.2   Case Two

Case two is a male PSBT with an average self-reported TPACK within the sample. He perceives DTM as helpful to enrich learning, and is motivated and confident to use it himself. Furthermore, he reports the highest amount of own experience in teaching with DTM. His lesson served to consolidate knowledge about hormone regulation in an eighth grade class.

Content

Lesson planning: Case two emphasizes the consolidation of the subject content as a lesson objective in his entire lesson plan. This focus is reflected in the lesson materials, which are intended to "clarify the content again" (lesson plan C2, p.12) and to support students in "visualizing the subject content" (lesson plan C2, p.13).

The *Lesson implementation* is congruent with his lesson plan. The ratio content (10 min) to DTM-related input (3 min) shows a clear bias towards subject content. Referring to an example video on a biological content, case two highlights the consolidation of the subject content and classifies design aspects as secondary: "This is now a rather simply produced explainer video. I think today we can't produce it in a more elaborated way" (lesson implementation C2, 6:37–6:45).

The focus on content is explained in the *lesson reflection* more deeply: "It is actually supposed to be a biology lesson utilizing this very method" (lesson reflection C2, l. 259f). In addition, he justifies the design of his lesson materials with a focus on content: "I have provided an outline of what is expected in terms of content in the video, just to clarify the task again and to direct the focus on the subject content which you finally want to have as a result" (lesson reflection C2, l. 339ff).

TPACK

Lesson planning: Case two selects the subject content deliberately based on the instructional potential offered by the DTM: "Methodologically, the use of models is particularly suitable for this topic in order to make the complex process of hormone regulation visually tangible for the students. One way to visualize such processes and procedures by means of models is the use of digital media, particularly the use of explainer videos" (lesson plan C2, p. 8). In the following the DTM is characterized as a tool promoting students' motivation and encouraging "exploratory behavior regarding the subject content" (lesson plan C2, p. 12).

Lesson implementation: In accordance with the lesson plan, DTM is used as a tool to help students visualize the process of the hormonal regulatory cycle. The introduction to the software focuses on basic functions.

Lesson reflection: Case two expresses confidence in DTM usage due to his prior experience: "I have already worked with some digital media from time to time, and that makes it immensely easier, of course" (lesson reflection C2, l. 533ff). He justifies the selection of a specialized video production software (stop-motion studio) with reference to the cognitive load theory, through the "structuring and outlining" (lesson reflection C2, l. 290), which the software enables in comparison to One Take: "You call it cognitive load, that is somehow overlaid too much by the method and the subject content doesn't get enough space" (lesson reflection C2, l. 276ff). Thus, he "also limits the introduction" to the software to basic functions, which are necessary but "not entirely self-explanatory" (lesson reflection C2, l. 196f).

PCK

Lesson planning: Instructional considerations identified in the instructional planning continue to be addressed and are clearly emphasized throughout the remaining lesson plan. For example, case two deliberately refers to a "phenomenon known to

the students (sweating, freezing)" (lesson plan C2, p.8) as an exemplary starting point.

The *lesson implementation* is consistent with his lesson plan in this regard.

In the *lesson reflection*, case two repeatedly refers to instructional aspects of his
lesson implementation and aligns them with his considerations in the lesson plan.
For example, he criticizes a lack of structure in the feedback phase, which he would
optimize by means of an "evaluation catalog" (lesson reflection C2, l. 73f) if the
lesson was repeated.

22.3.3   Case Three

Case three is a female PSBT with one of the highest self-reported TPACK levels of
the sample. She is convinced of the usefulness of DTM and reports a high attitude
towards it. Although she has been able to gain a lot of vicarious experience, she
seems to have little experience in using DTM for teaching and an accordingly low
level of motivation and self-efficacy in this regard. Her lesson served to present
individual strategies for dealing with stress in a ninth grade class.

Content
Lesson planning: Case three describes the biological content in the subject analysis
and prepares the content instructionally in the transition to instructional planning.
However, in the remaining lesson plan these considerations are not further taken up.
The lesson outlines and materials address exclusively the creation and design of an
'explainer video.

Case three's *lesson implementation* accordingly concentrates on creating an
explainer video. This is illustrated by a low proportion of content-related phases and
her response to students' content-related questions, where she gives the advice to
disregard the biological content (lesson implementation C3, 34:47–34:55).

The PBST articulates a product-oriented focus in her *lesson reflection*, neglect-
ing content-related criteria. She states as the central concern of the elaboration
phase: "that something comes out. What that is, I did not define. Only a product
should result" (lesson reflection C3, l. 432f).

TPACK
Lesson planning: Case three does not connect subject content and the use of DTM
in her lesson plan. She seems to regard DTM as an additional learning content,
identifying technical terms ("tablet, cutout animation, hard- & software" (lesson
plan C3, p.8)) next to subject terms. Furthermore, she analyzes the topic explainer
videos in detail next to the subject content stress (lesson plan C3, p. 5ff).

Lesson implementation: As in the lesson plan, there is no connection between the
subject content and the DTM.

During *lesson reflection*, it becomes apparent that her considerations were dominated by her uncertainty in dealing with the technology: "So the technology

preparation, it was the biggest insecurity. [...] and then I decided for that [procedure] I used the first time, and which I thought was the easiest option [...] because the introduction to another program might have required another hour that I didn't have available" (lesson reflection C3, l. 483ff).

PCK

Lesson planning: Case three verbalizes fuzzy content-related lesson objectives and criteria for the creation of the explainer videos. Although she mentions the "classification of the biological subject content" (lesson plan C3, p. 6) and the "naming of all organs involved, as well as the two hormones adrenaline and noradrenaline" (lesson plan C3, p. 6) as central requirements in her instructional planning, these considerations are not taken into account in the remaining lesson plan.

Congruent with her lesson outlines and materials, instructional considerations are not included in the *lesson implementation*. Likewise, the vague criteria for the video design are not specified. Thus, they do not provide any orientation for the students during video production and during the feedback phase (lesson implementation C3, 1:29:10–1:34:04).

Lesson reflection: Case three justifies her problems in lesson implementation with external factors like time constraints or the students' insufficient level of cognitive development: "Well, this is actually a problem on the students' side, that they are not able to give feedback. [...] I would just attribute that to their cognitive developmental level probably" (lesson reflection C3, l. 152ff).

22.4   Discussion

Based on the cases, the following discussion identifies differences in the enactment of TPACK leading to the definition of two types of DTM usage in a biology lesson.

The salient feature of **Type one** is the separation between the methodological approach and thus the DTM usage and content considerations. Although the latter are addressed in the first chapters of the planning documents, they are not pursued further thereafter. Instead, the focus is on designing the DTM deployment. The methodological planning is then implemented without further disruptions and the reflection is also largely limited to the methodological implementation. If problems occur during the implementation, they are explained with external circumstances or with insufficient skills of the students. Moreover, the reflection hardly refers to the considerations made in lesson planning. Type one can be divided into two subtypes with regard to the handling of subject content:

Subtype 1 tends to cognitively overload students. The cognitive overload can be illustrated, for example, by the fact that the students have to work out an unfamiliar content themselves and at the same time transfer it into a form of representation that is also unfamiliar to them (Case one).

Subtype 2 differs from subtype one in that neither clear content-related objectives nor criteria for the design of the explainer videos are defined either during the planning or the implementation of the lesson. This results in an insufficient clarity of content-related lesson objectives (Case three).

Type two systematically relates subject content and DTM. The creation of the explainer videos is planned, implemented, and reflected as a useful method for consolidating the biological subject content. In contrast to type one, type two already focuses on the content-related lesson objectives during planning and consequently includes content-related instructional considerations in the lesson outlines and the lesson implementation. In this respect, type two reflects problems in lesson implementation with regard to his lesson plan (Case two).

22.4.1 TPACK and DTM Usage

The cases with both the highest and the lowest self-reported TPACK levels of the sample are classified as type one, while the case designated as type two has an average self-reported TPACK. Consequently, at least with regard to the selected cases, it is not possible to draw conclusions from the self-reported TPACK to the enacted TPACK.

The two cases assigned to type one report little experience in using DTM for teaching compared to type two. In addition, both cases of type one show a low digital media self-efficacy and report insecurities in the use of DTM in class. As a result, both cases choose the technologically less complex video production procedure (One Take), which does not require specific software. In contrast, type two shows an average digital media self-efficacy and reports confidence in using DTM for teaching, attributing this to his experience. Unlike type one, he chooses a specific software under instructional considerations. Consistent with existing studies, teaching experience with DTM seems to represent a central prerequisite for the PSBTs digital media self-efficacy and TPACK (Valtonen et al., 2015), with self-efficacy playing a mediating role (Wang & Zhao, 2021).

One explanation for the discrepancy between content-related instructional objectives and the DTM usage in type one could be that, due to a lack of teaching experience and a low digital media self-efficacy, the PSBTs concentrate on the methodological use of DTM and thus have no resources for processing the content-related information. This is in line with Ling Koh et al. (2014), who proposed that the more teachers focus on external factors during lesson planning, the less they consider deeper pedagogical aspects and the less the lesson is aligned with TPACK. The fact that case three rates her own TPACK as particularly high despite her low enacted TPACK might be linked to a low ability of self-assessment. According to Max et al. (2022), this phenomenon is particularly evident in PSBTs with low TPACK, especially for increasing task complexity.

22.5 Limitations and Future Directions

Due to the objectives and associated methodology of the present study, the following limitations must be considered: (1) The results of this study are limited to a specific DTM usage in a selected subject domain. (2) The study exclusively analyzes the teaching offered by the PSBTs. The student level is not included in this regard. (3) Since the present study is a qualitative case analysis with a correspondingly small sample (n = 3), the results do not allow representative statements about the population. (4) Since data was collected in the field, it was not possible to create identical conditions.

Nevertheless, based on the sample, initial connections can be established by including a large number of contextual data. In addition, the data suggest that it may be appropriate not to have all PSBTs plan full lessons right away, from which they may leave frustrated (Type one). Instead, there could be intermediate steps already at the university (e.g. micro-teaching), which could help to reduce difficulties with lesson planning and implementation in the school.

The study also indicates that the consideration of contextual factors and the comprehensive comparison of TPACK at the levels of lesson planning, implementation, and reflection enable the identification and explanation of fractures in PSBTs knowledge enactment. Accordingly, future studies should further concentrate on examining enacted TPACK from a more holistic perspective, considering the unique contextual conditions.

Acknowledgement This project is funded by the German Federal Ministry of Education and Research (BMBF) (ref.01JA2036). Responsibility for the content published in this article, including any opinions expressed therein, rest exclusively with the authors.

References

Aumann, A., Schnebel, S., & Weitzel, H. (2023). The EnTPACK rubric: Development, validation, and reliability of an instrument for measuring pre-service science teachers' enacted TPACK. *Frontiers in Education, 8.* https://www.frontiersin.org/articles/10.3389/feduc.2023.1190152/abstract

Bernhard, G., & Grassinger, R. (2022). Förderung der Motivation zum Einsatz digitaler Medien im Unterricht während des Lehramtsstudiums. In S. Dippelhofer & T. Döppers (Eds.), *"Qualität im Hochschulsystem": Eine Rundumschau im Posterformat: Die Beiträge zur 16. Jahrestagung der Gesellschaft für Hochschulforschung (GfHf)* (pp. 19–22). Universitätsbibliothek Gießen. https://doi.org/10.22029/JLUPUB-656

Chi, M. T. H., & Wylie, R. (2014). The ICAP framework: Linking cognitive engagement to active learning outcomes. *Educational Psychologist, 49*(4), 219–243. https://doi.org/10.1080/00461520.2014.965823

Chien, Y.-T., Chang, Y.-H., & Chang, C.-Y. (2016). Do we click in the right way? A meta-analytic review of clicker-integrated instruction. *Educational Research Review, 17*, 1–18. https://doi.org/10.1016/j.edurev.2015.10.003

Drossel, K., Eickelmann, B., Schaumburg, H., & Labusch, A. (2019). Nutzung digitaler Medien und Prädiktoren aus der Perspektive der Lehrerinnen und Lehrer im internationalen Vergleich. In B. Eickelmann, W. Bos, J. Gerick, F. Goldhammer, H. Schaumburg, K. Schwippert, M. Senkbeil, & J. Vahrenhold (Eds.), *ICILS 2018 #Deutschland. Computer- und informationsbezogene Kompetenzen von Schülerinnen und Schülern im zweiten internationalen Vergleich und Kompetenzen im Bereich Computational Thinking* (pp. 205–240). Waxmann.

Ertmer, P. A. (1999). Addressing first- and second-order barriers to change: Strategies for technology integration. *ETRD, 47*(4), 47–61. https://doi.org/10.1007/BF02299597

Hsu, Y.-S., Yeh, Y.-F., & Wu, H.-K. (2015). The TPACK-P framework for science teachers in a practical teaching context. In Y.-S. Hsu (Ed.), *Development of science teachers' TPACK* (pp. 17–32). Springer. https://doi.org/10.1007/978-981-287-441-2_2

Jaipal-Jamani, K., & Figg, C. (2015). A case study of a TPACK-based approach to teacher professional development: Teaching science with blogs. *Contemporary Issues in Technology and Teacher Education, 15*(2), 161–200.

Kuckartz, U., & Rädiker, S. (2022). *Qualitative Inhaltsanalyse: Methoden, Praxis, Computerunterstützung: Grundlagentexte Methoden* (5th ed.). Beltz Juventa.

Ling Koh, J. H., Chai, C. S., & Tay, L. Y. (2014). TPACK-in-action: Unpacking the contextual influences of teachers' construction of technological pedagogical content knowledge (TPACK). *Computers & Education, 78*, 20–29. https://doi.org/10.1016/j.compedu.2014.04.022

Max, A., Lukas, S., & Weitzel, H. (2022). The relationship between self-assessment and performance in learning TPACK: Are self-assessments a good way to support preservice teachers' learning? *Journal of Computer Assisted Learning, 38*(4), 1160–1172. https://doi.org/10.1111/jcal.12674

Mishra, P., & Koehler, M. J. (2006). Technological pedagogical content knowledge: A framework for teacher knowledge. *Teachers College Record, 108*(6), 1017–1054. https://doi.org/10.1111/j.1467-9620.2006.00684.x

Mouza, C. (2016). Developing and assessing TPACK among pre-service teachers. A synthesis of research. In M. C. Herring, M. J. Koehler, & P. Mishra (Eds.), *Handbook of technological pedagogical content knowledge (TPCK) for educators* (2nd ed., pp. 169–190). Routledge.

Niess, M. L. (2015). Transforming teachers' knowledge: Learning trajectories for advancing teacher education for teaching with technology. In C. Angeli & N. Valanides (Eds.), *Technological pedagogical content knowledge* (pp. 19–37). Springer. https://doi.org/10.1007/978-1-4899-8080-9_2

Ning, Y., Zhou, Y., Wijaya, T. T., & Chen, J. (2022). Teacher education interventions on teacher TPACK: A meta-analysis study. *Sustainability, 14*(18), 11791. https://doi.org/10.3390/su141811791

Njiku, J., Mutarutinya, V., & Maniraho, J. F. (2020). Developing technological pedagogical content knowledge survey items: A review of literature. *Journal of Digital Learning in Teacher Education, 36*(3), 150–165. https://doi.org/10.1080/21532974.2020.1724840

Rosenberg, J. M., & Koehler, M. J. (2015). Context and technological pedagogical content knowledge (TPACK): A systematic review. *Journal of Research Technology & Engineering, 47*(3), 186–210. https://doi.org/10.1080/15391523.2015.1052663

Stender, A., Brückemann, M., & Neumann, K. (2017). Transformation of topic-specific professional knowledge into personal pedagogical knowledge through lesson planning. *International Journal of Science Education, 39*(12), 1690–1714. https://doi.org/10.1080/09500693.2017.1351645

Tondeur, J., van Braak, J., Sang, G., Voogt, J., Fisser, P., & Ottenbreit-Leftwich, A. (2012). Preparing pre-service teachers to integrate technology in education: A synthesis of qualitative evidence. *Computers & Education, 59*(1), 134–144. https://doi.org/10.1016/j.compedu.2011.10.009

Tondeur, J., Howard, S. K., & Yang, J. (2021). One-size does not fit all: Towards an adaptive model to develop preservice teachers' digital competencies. *Computers in Human Behavior, 116*, 106659. https://doi.org/10.1016/j.chb.2020.106659

Upmeier zu Belzen, A., & Merkel, R. (2014). Einsatz von Fällen in der Lehr- und Lernforschung. In D. Krüger, I. Parchmann, & H. Schecker (Eds.), *Methoden in der naturwissenschaftsdidaktischen Forschung* (pp. 203–212). Springer. https://doi.org/10.1007/978-3-642-37827-0_17

Valtonen, T., Kukkonen, J., Kontkanen, S., Sormunen, K., Dillon, P., & Sointu, E. (2015). The impact of authentic learning experiences with ICT on pre-service teachers' intentions to use ICT for teaching and learning. *Computers & Education, 81*, 49–58. https://doi.org/10.1016/j.compedu.2014.09.008

Wang, Q., & Zhao, G. (2021). ICT self-efficacy mediates most effects of university ICT support on preservice teachers' TPACK: Evidence from three normal universities in China. *British Journal of Educational Technology, 52*(6), 2319–2339. https://doi.org/10.1111/bjet.13141

Willermark, S. (2018). Technological pedagogical and content knowledge: A review of empirical studies published from 2011 to 2016. *Journal of Electronic Commerce Research, 56*(3), 315–343. https://doi.org/10.1177/0735633117713114

Yin, R. K. (2012). *Applications of case study research* (3rd ed.). Sage.

Chapter 23
Promoting Digitally Supported Inquiry Learning in Diverse Classrooms Through Teacher Training

Patrizia Weidenhiller, Susanne Miesera, and Claudia Nerdel

23.1 Introduction

The heterogeneity of students involves diverse learner needs. This is why inclusive teaching concepts are needed to enable all students to participate, especially in science education and its specific processes and procedures, like inquiry learning. Inquiry-based learning is an active method of learning. It begins with posing questions, problems, or scenarios, and can involve scientific methods, such as conducting experiments. The process of generating one's own hypotheses and problem-solving approaches is important. In this complex process, many barriers can arise, such as the handling of materials and instructions (Stinken-Rösner & Abels, 2021) or methodological difficulties, e.g. hypothesizing, reflecting on measurement inaccuracies (Baur, 2018). However, there are many possibilities, for example, to prepare the scientific work methods in a differentiated and student-oriented way. This could be through their graduated complexity, the level of abstraction and the observation level (Bruckermann et al., 2017). Another approach can be the targeted use of digital media, which may enable access and avoid barriers through multimedia design (Kerres, 2018). In order to effectively use digital media to promote inquiry learning and reduce barriers, teachers need professional knowledge and competencies in inclusion and digitization and how to link them.

P. Weidenhiller (✉)
Institute for Biology Education, LMU Munich, Munich, Germany
e-mail: Patrizia.weidenhiller@bio.lmu.de

S. Miesera (✉) · C. Nerdel (✉)
Associate Professorship of Life Sciences Education, Technical University of Munich, Munich, Germany
e-mail: Susanne.miesera@tum.de; Claudia.nerdel@tum.de

© The Author(s) 2024
K. Korfiatis et al. (eds.), *Shaping the Future of Biological Education Research*,
Contributions from Biology Education Research,
https://doi.org/10.1007/978-3-031-44792-1_23

23.2   Theoretical Background

Following the Index for Inclusion, inclusion is about minimizing all barriers in education and learning for all students (Boban & Hinz, 2003, p. 11). This broad understanding of inclusion encompasses all dimensions of heterogeneity that may present barriers, such as gender, ability, special education needs, religion, and more. In terms of teachers' professional knowledge, this shows a need in pedagogical knowledge, which encompasses the pedagogical aspects of teaching and the individual needs of students. Pedagogical knowledge (PK), together with content knowledge (CK) and technological knowledge (TK), forms the basis for the TPACK model, which describes teachers' professional knowledge as the interplay of the different knowledge domains (Mishra & Koehler, 2006). CK refers to the subject matter being taught and goes beyond the scope of the school subject matter. PK includes the pedagogical aspects of teaching and the individual needs of students. TK includes the handling and knowledge of digital media and their application. In addition, the intersections of the individual disciplines are considered (Fig. 23.1). The intersection of pedagogical content knowledge (PCK) describes the didactic contexts of instructional design. This is specifically about the didactic preparation of subject matter content for the needs of the students. Technological content knowledge (TCK), on the other hand, is the knowledge of the technical and digital applications used in the subject matter. The intersection of technological pedagogical knowledge (TPK) realizes the use of digital media and technical possibilities tailored to the needs of the students. The eponymous intersection of the three areas of technological pedagogical content knowledge – TPACK – unites them to form the professional knowledge of teachers. TPACK thus describes teachers' knowledge of how technologies can be used for a specific content, considering the needs of the students. In addition to professional knowledge, there are other factors that influence teaching actions. These include among others attitudes, motivation and self-regulation (Baumert & Kunter, 2006). If we take a closer look at attitudes, we can

Fig. 23.1 Teachers' professional knowledge: TPACK Modell. Representation according to (Mishra & Koehler, 2006)

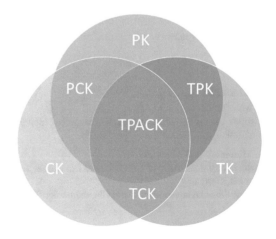

see that positive attitudes together with high self-efficacy expectations are an important predictor of planned behavior (Fishbein & Ajzen, 2010).

Based on the theory of planned behavior, it can be seen that these factors are multiple and interact with each other (Fishbein & Ajzen, 2010). Thus, attitudes are just as decisive in achieving a desired behavior as self-efficacy expectations and subjectively perceived norms. These in turn are influenced by beliefs, which are based on background factors. The background factors are divided into three categories: personal factors relate to, among other things, general attitude, intelligence or emotion; social factors include, for example, age, gender or education; information-related factors, on the other hand, refer to the experience and knowledge of the respective person on a particular issue (Fishbein & Ajzen, 2010). The interaction of these entire factors ultimately leads to behavior that can be observed. By measuring individual factors, more or less reliable statements can be made about future behavior. In the field of inclusive education, correlations have already been proven in studies. For example, teachers with positive attitudes towards inclusion show more effective teaching for all students (Jordan et al., 2009). Attitude and self-efficacy are predictors of planned behavior in inclusive teaching (Sharma & Jacobs, 2016). Another finding is that attending courses on inclusion has a positive impact on attitudes toward inclusion (Sharma, 2012; Miesera & Gebhardt, 2018; Miesera & Will, 2017). Similarly, teachers' attitudes towards digital media are decisive for their implementation in the classroom (Eickelmann & Vennemann, 2017). All of these findings indicate that attitudes and self-efficacy expectations are important factors in the implementation of specific instructional practices. For both fields, inclusion and digitization, there are already study results and valid assumptions on which to build. However, in order to apply these general statements about inclusive teaching and the use of media to subject-specific problems, such as inquiry learning, the topics must be examined in conjunction with one another.

23.3 Aim of the Study and Research Question

The aim of the study is to prepare teachers specifically for the use of digital media in heterogeneous classes in biology lessons. In particular, the heterogeneity dimensions of ability and special needs will be focused on in order to investigate the use of digital media in scientific work methods. On the one hand, the question arises whether the targeted knowledge transfer in an intervention has an effect on the self-assessment in the TPACK domains. On the other hand, the teachers' attitudes toward the topics of inclusion and digitization will be investigated, which will lead to the question of what connections exist between the self-assessment in the TPACK domains and the attitudes toward inclusion and digitalization.

23.4 Methods

23.4.1 Design and Intervention

The study design consists of a teacher training with pre and post survey. Biology teachers from Bavarian secondary schools took part in a one-day digital teacher training session dealing with "digitally supported inquiry learning for all students". The study was verified and approved by the Bavarian State Ministry of Education and Cultural Affairs. Participation in both the study and the training was voluntary and unpaid. The training aimed at increasing teachers' self-efficacy assumptions according to the TPACK model as shown in Fig. 23.2 (Mishra & Koehler, 2006). Therefore, teachers planned and performed an experiment on the enzymatic browning of apples, which was digitally supported in all phases: planning, implementation, evaluation (Weidenhiller et al., 2022). In addition, they considered the needs of students and elaborated on possible barriers. The outcome of the training was a planned experiment supported by digital media to differentiate the phases of the inquiry process. The teachers focused on the heterogeneity of their own classes and

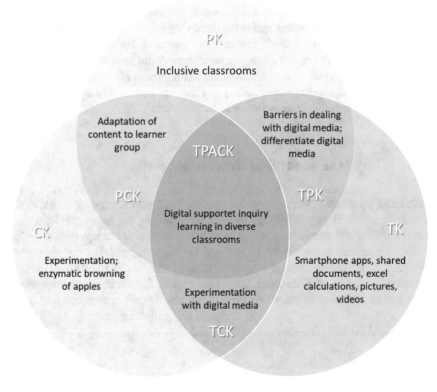

Fig. 23.2 TPACK content in the intervention

differentiated according to the needs of the students. Participants were randomly assigned to three experimental groups. The experimental groups differed in the instruction phases in advance of the work phase. The first group focused on the design of digital media, media didactics, and the use of digital media in science teaching (a.o. Hamilton et al., 2016, Kerres, 2018; Mayer, 2005; Mayer & Pilegard, 2014; Puentedura, 2006; Sweller, 2005). The second group focused on approaches to inclusive didactics, concepts for differentiation and their implementation in science lessons, as well as legal aspects of inclusion in Bavaria (a.o. Reich, 2013, 2014; Stinken-Rösner et al., 2020). The last group had an integrated format for the instructions. This included the mentioned aspects of digital media from the first experimental group, as well as the aspects of differentiation of the second group.

The pre-post survey contained scales about teachers' attitudes toward digitalization and inclusion. The scale attitudes toward digitalization addressed different aspects of learning with digital media, such as anchoring in the curriculum, influence on the teaching level and on the student's activity (Vogelsang et al., 2019). The scale attitudes toward inclusion contained two main constructs: 'schooling and support' and 'social inclusion' (Kunz et al., 2010). Furthermore, the survey covered self-efficacy assumptions regarding inclusion and digitalization in accordance with the TPACK model (modified according to Graham et al., 2009; Weidenhiller, in preparation). For this purpose, the TPACK scales of Graham et al. (2009) were modified and adapted to the content of the intervention. In total, five scales were used: TK, TPK, TCK, TPACK, DILAS (digitally supported inquiry learning for all students) (Weidenhiller, in preparation). The TK scale asks about the use and production of various media, such as presentations, videos, shared documents, or e-books unrelated to teaching. TPK is about the differentiated design of teaching materials or action sequences through the use of digital media. The TCK scale, on the other hand, contains items on the extent to which digital media help to make scientific phenomena easier to understand. The TPACK and DILAS scales now include questions about the instructional use of digital media in science. TPACK focuses on the general use of media in science lessons, whereas the DILAS scale deals specifically with science working methods in different class compositions.

23.4.2 Sample

The intervention took place from June 2021 to January 2022 as one-day teacher training sessions. A total of 141 Bavarian secondary school biology teachers (70% female) were trained in small groups of up to 10 people. The matched data sets of pretest and posttest correspond to approximately one-third of the total sample and are almost equally distributed across the three experimental groups (Table 23.1).

Table 23.1 Sample

	Digital media	Inclusion	Integrated	total
Participants	40	43	58	141
Posttest (matched with pretest)	22 (17)	25 (16)	24 (19)	72 (52)

Table 23.2 Rasch analysis of the scales

Scale (items)	Likert-Scale	Person separation (Reliability)	Item separation (Reliability)
Attitude inclusion: EZI (11)	0–5	2.22 (.83)	7.22 (.98)
Attitude digital Media (10)	0–3	1.98 (.80)	3.28 (.91)
TPACK (8)	0–4	2.67 (.88)	6.02 (.97)
TCK (7)	0–5	2.43 (.86)	2.49 (.86)
TPK (6)	0–5	2.34 (.85)	9.98 (.99)
TK (10)	0–5	2.63 (.87)	12.46 (.99)
DILAS (9)	0–5	2.38 (.85)	6.76 (.98)

23.5 Data Analysis and Results

A Rasch analysis was performed to determine the quality of the scales. The calculations were performed on the matching data sets with pretest and posttest and were carried out with the Winsteps software. A high person separation (>2) and reliability (>0.8) implies a sensitive instrument (with enough items), which distinguish between high and low performers (Boone et al., 2014). A high item separation (>4) and reliability (>0.9) implies that the sample is large enough to have a good confirmation of the item difficulty hierarchy (ibid.). This means a high construct validity. The real reliabilities and separation indices for all scales are quite satisfactory, except for the item separation and reliability of the TCK scale and the item separation of the attitudes to digital media scale, which are below the desired values (Table 23.2).

A major advantage of the Rasch analysis in contrast to classical test theory is that the response format is not assumed to be metric. The Rasch analysis calculates the difficulty of the items and the ability of the individuals and reports them on the same scale in units of logits, which is presented in a Wrightmap. As an example, the Wrightmap of the Attitude toward Inclusion scale is shown (Fig. 23.3). The structure of the Wrightmap is as follows: On the left side of the map we see the persons, represented as "X". These are plotted according to their person ability. The higher up a person is, the more capable they are. In the middle we see the "M", which represents the mean ability of the cohort. On the right side we see the items, which are plotted by difficulty. The higher up an item is, the more difficult, or harder to agree with, the item is. For example, the item EI10 stands out, because it seems to be very difficult. In the translated version, the item reads, "If learners with special educational needs spent most of their time in regular classes, they would also receive all the support there that they would otherwise have in a small class or special school".

Fig. 23.3 Wrightmap of
the Attitudes towards
Inclusion scale - Pretest

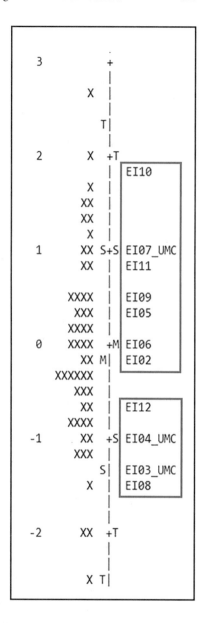

Thus, agreement with this item indicates that the same support is available at a regular school as at a special school. This question about support does not seem to be easily answered in the affirmative by the teachers. In comparison, the question about social inclusion, for example, can be affirmed more easily in this scale, as item EI12 shows. EI12 reads as follows: "The more time learners with special educational needs spend in a regular class, the more likely they are to be treated well by other classmates in their class". The original factor analysis of the attitudes towards inclusion scale (EZI) shows two components (Kunz et al., 2010). The first component "Schooling and Support" consists of seven items, the second component "Social Inclusion" consists of four items. The item structure of this factor analysis is also reflected in the Wrightmap in this study. The upper box in Fig. 23.3 represents component 1. The lower box represents component 2. From this arrangement in the Wrightmap, we can now assume that overall the items dealing with social aspects of inclusion are easier to agree with than direct support in the class. Further calculations without anchoring item difficulty in the posttest did not show a different arrangement. Thus, the difficulty of the constructs seems constant before and after the intervention. As a second example, the DILAS scale is considered. In terms of content, the scale deals with inquiry learning processes in science, such as experimentation and modeling with digital media in heterogeneous classes. Over the items TP11 – TP14, the degree of heterogeneity in the class is varied from homogeneous (TP11) to heterogeneous in performance (TP12) to behavioral problems (TP13) and special educational needs (TP14). The level of digitalization remains the same. In items TP15-TP19, the level of digitization varies from no digital media (TP15) to digital instruction (TP16), digital observation (TP17), digital measurement (TP18), and purely digital laboratories (TP19). The class is always described as heterogeneous without more specific information. As shown in Fig. 23.4, the difficulty of the items correspond to the increasing level of digitization and diversity as described above. The more digital media intervene in the doing science process or the more explicitly diversity is described, the more difficult the item is rated. In terms of the teachers' self-efficacy, this means that the higher the level of digitization and diversity, the less confident the teachers feel. To compare pretest and posttest results, the item difficulty of the pretest was anchored and then the person ability was calculated on the posttest. The mean of person ability in pre- and posttest is highlighted by circles in Fig. 23.4. The mean of the item difficulty is set to zero. The mean person ability on the pretest is below the mean item difficulty. In comparison, the posttest shows an improvement in mean person ability, which is now above the mean item difficulty. This means that an improvement in teachers' self-efficacy assumption occurred in the course of the intervention. This visual difference is also statistically detectable. The **one-way repeated measures MANOVA** is significant, as shown in Table 23.3. In order to calculate the differences of the scales in pretest and posttest, first the person abilities for the respective scale at the respective time of measurement were calculated. The item difficulties were anchored as described above. The person abilities were compared between the measurements using a **one-way repeated measures MANOVA**. There were significant improvements in teacher ratings after the intervention, $F(7, 45) = 6.791$, $p < .001$, partial $\eta^2 = .514$, Wilk's $\Lambda = .486$. These calculations were made across the three groups. The univariate testing shows an improvement of the following scales: Attitudes towards digital media, TPACK, TCK, TPK and DILAS. By this improvement, a positive impact of the intervention on teachers can be identified.

The pretest shows medium to strong correlations between the scales TPACK, TCK, TPK, TK, and DILAS (Table 23.4). These correlations show a close

Fig. 23.4 Wrightmap
DILAS scale in a pretest
(left) / posttest (right)
comparison

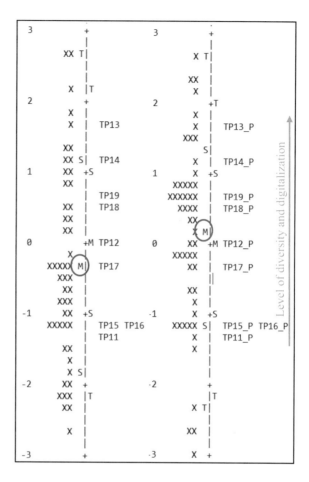

Table 23.3 univariate testing of the one-way repeated measure MANOVA (Greenhouse Geisser)

Scale	df	F	p	df(error)
Attitude inclusion	1.00	2.238	.141	51.0
Attitude digital media	1.00	4.385	.041	51.0
TPACK	1.00	33.056	<.001	51.0
TCK	1.00	28.901	<.001	51.0
TPK	1.00	4.510	.039	51.0
TK	1.00	1.153	.187	51.0
DILAS	1.00	11.214	.006	51.0

interconnectedness of the scales and suggest that the assumed individual components of the self-efficacy assumption in the TPACK domain influence each other.
There is also a weak correlation between attitudes towards digital media and the

Table 23.4 Pearson correlations of the TPACK model scales

	Pretest Pearson correlation N = 54				Posttest Pearson correlation N = 53			
	TPACK	DILAS	TPK	TCK	TPACK	DILAS	TPK	TCK
DILAS	.687**				.529**			
TPK	.563**	.413**			.568**	.637**		
TCK	.616**	.434**	.550**		.537**	.623**	.785**	
TK	.508**	.504**	.605**	.516**	.518**	.514**	.669**	.639**

**.Correlation is significant at the 0.01 level (2-tailed)

Table 23.5 Comparison of the three experimental groups

	V1: digital media (N = 17)	V2: inclusion (N = 16)	V3: integrated (N = 19)
Attitude inclusion	0.40 (SD = 0.86)	0.04 (SD = 1.07)	−0.03 (SD = 1.18)
Attitude digital media	1.91 (SD = 1.57)	1.96 (SD = 1.70)	1.31 (SD = 0.92)
TPACK	1.68 (SD = 1.63)	1.99 (SD = 2.52)	1.64 (SD = 1.86)
TPK	1.30 (SD = 2.02)	0.57 (SD = 2.36)	0.57 (SD = 1.96)
TCK	0.36 (SD = 1.22)	−0.72 (SD = 1.34)	−0.13 (SD = 1.41)
TK	1.06 (SD = 1.60)	0.82 (SD = 1.38)	0.66 (SD = 1.73)
DILAS	0.27 (SD = 1.26)	0.41 (SD = 1.48)	0.37 (SD = 1.24)

TPK self-efficacy assumption ($r = 0.293$, $p = 0.031$). The scale attitudes toward inclusion do not correlate with any other scale in the pretest.

In the posttest, the existing correlations remain (Table 23.4), but the scale attitudes toward digital media no longer correlate with TPK, but with TPACK ($r = 0.299$, $p = 0.03$). However, what is much more interesting is that there is a correlation between the scales attitudes toward inclusion and attitudes toward digital media with a medium effect ($r = 0.316$, $p = 0.021$), which was not there in the pretest. One interpretation could be that the two topics of inclusion and digital media are no longer understood in such a detached way as they were before the intervention and they are now thought of in a more linked way.

Comparing the three experimental groups, it is noticeable that there are no significant differences (Table 23.5). Again, the person abilities between the groups were tested for the single constructs. Afterwards, a group comparison between the three experimental groups – digital media, inclusion and integrated – was calculated. The one-way MANOVA found no statistically significant differences between the groups on the combined dependent variables, $F(14, 86) = 0.521$, $p = .914$, partial $\eta^2 = .078$, Wilk's $\Lambda = .850$. Despite supposed differences due to different instructional phases, no significance occurs, either because the groups are too small and thus the deviations are too large, which is reflected in a high standard deviation, or because no effects could be obtained. Table 23.5 shows the mean values and standard deviations of the three groups.

23.6 Discussion and Practical Implications

In summary, the following main findings can be described. Systematic variation in the instructional phase shows no difference in teachers' attitudes or self-efficacy assumptions. Attitudes toward inclusion are stable across the intervention, whereas attitudes toward digitization change. In the attitude towards inclusion, the questions about social inclusion are much easier to agree with than the questions about adequate support. The correlation between the two scales indicates a stronger link between the topics after the intervention. There are weak correlations, if any, between attitudes and self-efficacy assumptions, arguing for separate constructs.

The aim of the study was to promote teachers specifically for the use of digital media in heterogeneous classes in biology lessons. Following the theory of planned behavior (Fishbein & Ajzen, 2010), it can be assumed that the goal has been achieved because positive attitudes and high self-efficacy assumptions are strong predictors for a desired behavior, in this case, the targeted use of digital media for differentiation in doing science. The results show significant improvements in attitudes toward digital media and in self-efficacy expectations in the TPACK domains. As discussed earlier, teachers with positive attitudes and with high self-efficacy expectations are more likely to demonstrate a desired behavior in inclusive teaching compared to teachers with more negative attitudes and lower self-efficacy expectations (Sharma & Jacobs, 2016). The significant improvement now allows the assumption that the teachers feel more competent in using digital media for inquiry learning in heterogeneous classes. Consequently, it can be assumed that the use of digital media in inquiry learning may be more frequent in future lessons, adapted to the needs of the students. However, this remains an assumption derived from theory, which must be verified, for example, through follow-up interviews.

Looking at the second research question (What are the relationships between the TPACK domains and attitudes?), it is important to note that the attitude scales deal separately with the domains of digital media and inclusion, whereas the TPACK scales require an intertwined approach. The TPACK scales (and the DILAS scale) deal with the intertwining of digital media and subject content and student needs. Within these constructs, correlations among the scales occurred in both the pretest and posttest. This is as expected, as the domains are strongly interrelated due to the technological component.

The attitude scales, on the other hand, are less linked to the TPACK domains than expected. Although there are weak correlations between attitudes toward digital media and individual TPACK domains, these are neither strong nor consistent across pretest and posttest. A stronger linkage was assumed here because the TPACK scales used constructs of both inclusion and media use. The limited prevalence of this shows that attitudes are a rather independent construct. The theory of planned behavior assumes that both attitudes and self-efficacy assumptions are influenced by beliefs, which in turn are influenced by background factors (Fishbein

& Ajzen, 2010). Accordingly, only little cross-talk between attitudes and self-efficacy assumptions appeared here. The intervention seems to promote these two areas separately. It is all the more interesting that there is a correlation between attitudes toward inclusion and digitization in the posttest. These two constructs were still considered completely independently in the pretest. The linking of the two topics in the intervention therefore may has an effect here.

The Attitude Towards Inclusion scale shows that teachers find it easier to rate social aspects of inclusion positively than support aspects. This evaluation remains constant across the intervention. This is not in line with expectations, as the intervention focused strongly on the promotion of individual needs of students, but the social aspects were not in the foreground. One possible explanation is that attitudes themselves are a construct that is difficult to change. The underlying beliefs and factors have developed over long periods of time and require a lot of effort to change. The reportet changes of attitudes towards inclusion after an intervention in Miesera & Gebhardt (2018) and Miesera & Will (2017) belong to long term interventions. Accordingly, it can be assumed that a one-day intervention does not have the power to change them. Overall, multi day events in teacher training are considered more effective (Lipowsky & Rzejak, 2019, 2021). If we look at the DILAS scale, we see that teachers feel more confident the fewer digital media are used for scientific work and the more homogeneous the class is. This effect also remains constant, with the mean person ability increasing significantly after the training. In conclusion, this shows on the one hand the challenge of integrating innovations into the classroom and adapting them to heterogeneous needs, and on the other hand an improvement in self-efficacy.

In the post-test, we did not find any differences in the scales between the three intervention groups. The main change in the experimental design was the theoretical input in the instruction phase at the beginning. The experimental part and the discussion were basically the same in all three groups. Accordingly, we can say that the theoretical part of the training has too little influence to be measured with these scales. Rather, it can be assumed that application knowledge in the other phases of the training leads to the effect or overlaps the effect of the groups, or that the teachers already have a lot of knowledge, which makes a sharp separation of the groups impossible. This suggests, in accordance with the scientific discussion, that the attitude and self-efficacy assumptions do not only change based on theoretical discussion (Fishbein & Ajzen, 2010). Rather, other factors, such as experience, perceived control, and others, influence these personal characteristics as well (Fishbein & Ajzen, 2010). This effect is reflected in the significant improvements in almost all scales after the intervention. This suggests that the intervention itself is effective. Overall, it can be concluded that after the intervention, teachers feel more confident in using digital media during inquiry learning in heterogeneous classes and have more positive attitudes than before. According to the theory of planned behavior and the results of several studies as discussed (Eickelmann & Vennemann, 2017; Jordan et al., 2009; Sharma & Jacobs, 2016), we can conclude that the teachers may use more digital media for differentiation in future because of their attitudes and self-efficacy assumptions in the TPACK areas.

The following implications for teacher training can be derived from these results. Since there were no differences between the groups, none of the options can be emphasized. It is necessary to verify to what extent the instructional phase can be shortened or shifted in order to make it more effective. For example, the inclusion and digitalization of instruction could come into play after the barriers to the experiment identified. The practical phase should be placed more at the center of the training. Together with the subsequent discussion, it is a more effective component of the training than the instructional phase tested in this study. Overall, when designing teacher training, care should be taken to think about the increasing heterogeneity of students and incorporate it into the design. By linking the topics in the training, it is possible to provide teachers with differentiated options for action in the classroom through increased self-efficacy. However, there is the limiting factor that the training was offered as a webinar. In this respect, it is not possible to directly control what the teachers actually did in the instruction phase. Even with the camera switched on, it is not possible to guarantee that the teachers were not otherwise occupied outside of the work phases. The following additional limitations still need to be included in the results. There was only a small experimental group with 54 matching pre-posttests. Participation in the tests was voluntary and therefore the rate was only 50% and even lower for the matching data sets due to incorrect codes. Results are limited to secondary teachers. For future research, face-to-face sessions should be tested to examine instructional phases, while ensuring that teachers are not distracted. In addition, a better response rate would be expected. This would allow the possibility to verify the results of this study. To anchor the topic of inclusion, another practice phase could be integrated into the training, in which differentiation methods are run through by the teachers themselves, as in the previous practice phase. Another idea would be a training course of at least two sessions, which would allow for practical testing in the school, to see if this could improve teachers' attitudes.

References

Baumert, J., & Kunter, M. (2006). Stichwort: Professionelle Kompetenz von Lehrkräften. *Zeitschrift für Erziehungswissenschaft, 9*(4), 469–520. https://doi.org/10.1007/s11618-006-0165-2

Baur, A. (2018). Fehler, Fehlkonzepte und spezifische Vorgehensweisen von Schülerinnen und Schülern beim Experimentieren: Ergebnisse einer videogestützten Beobachtung. *Zeitschrift für Didaktik der Naturwissenschaften, 24*(1), 115–129.

Boban, I., & Hinz, A. (2003). *Index für Inklusion: Lernen und Teilhabe in der Schule der Vielfalt entwickeln*. Martin-Luther-Universität Halle-Wittenberg.

Boone, W., Staver, J., & Yale, M. (2014). *Rasch analysis in the human sciences*. Springer.

Bruckermann, T., Ferreira Gonzalez, L., Münchhalfen, K., & Schlueter, K. (2017). Inklusive Fachdidaktik Biologie. In K. Ziemen (Ed.), *Lexikon Inklusion* (pp. 109–110). Vandenhoeck & Ruprecht.

Eickelmann, B., & Vennemann, M. (2017). Teachers' attitudes and beliefs regarding ICT in teaching and learning in European countries. *European Educational Research Journal, 16*(6), 733–761.

Fishbein, M., & Ajzen, I. (2010). *Predicting and changing behavior. The reasoned action approach.* Psychology Press.

Graham, C. R., Burgoyne, N., Cantrell, P., Smith, L., St Clair, L., et al. (2009). TPACK development in science teaching: Measuring the TPACK confidence of inservice science teachers. *TechTrends, 53*(5), 70–79.

Hamilton, E. R., Rosenberg, J. M., & Akcaoglu, M. (2016). Examining the substitution augmentation modification redefinition (SAMR) model for technology integration. *Technology Trends, 60*(5), 433–441.

Jordan, A., Schwartz, E., & McGhie-Richmond, D. (2009). Preparing teachers for inclusive classrooms. *Teaching and Teacher Education, 25*(4), 535–542. https://doi.org/10.1016/j.tate.2009.02.010

Kerres, M. (2018). *Mediendidaktik. Konzeption und Entwicklung digitaler Lernangebote* (5th ed.). Walter de Gruyter GmbH.

Kunz, A., Luder, R., & Moretti, M. (2010). Die Messung von Einstellungen zur Integration (E-ZI). *Empirische Sonderpädagogik, 2*, 83–94.

Lipowsky, F., & Rzejak, D. (2019). Was macht Fortbildungenfür Lehrkräfte erfolgreich? – Ein Update. In B. Groot-Wilken & R. Koerber (Hrsg.), Nachhaltige Professionalisierung für Lehrerinnen und Lehrer. Ideen, Entwicklungen, Konzepte (S. 15–56). Bielefeld.

Lipowsky, F., & Rzejak, D. (2021). Fortbildungen für Lehrpersonen wirksam gestalten. Hg. v. Bertelsmann Stiftung. Gütersloh.

Mayer, R. E. (2005). Cognitive theory of multimedia learning. In R. E. Mayer (Ed.), *The Cambridge handbook of multimedia learning* (pp. 31–48). Cambridge University Press. https://doi.org/10.1017/CBO9780511816819.004

Mayer, R., & Pilegard, C. (2014). Principles for managing essential processing in multimedia learning: Segmenting, pre-training, and modality principles. In R. Mayer (Ed.), *The Cambridge handbook of multimedia learning* (Cambridge handbooks in psychology) (pp. 316–344). Cambridge University Press. https://doi.org/10.1017/CBO9781139547369.016

Miesera, S., & Gebhardt, M. (2018). Inklusive Didaktik in beruflichen Schulen – InkDibeS – ein Konzept für die Lehrerbildung: Videobasierte Fallkonstruktionen inklusiver Unterrichtssettings. In D. Buschfeld & M. Cleef (Eds.), *Vielfalt des Lernens im Rahmen berufsbezogener Standards* (QUA-LIS Schriftenreihe Beiträge zur Schulentwicklung). Waxmann.

Miesera, S., & Will, S. (2017). Inklusive Didaktik in der Lehrerbildung – Erstellung und Einsatz von Unterrichtsvideos. *Haushalt in Bildung und Forschung, 6*(3), 61–76. https://doi.org/10.3224/hibifo.v6i3.05

Mishra, P., & Koehler, M. J. (2006). Technological pedagogical content knowledge. A new framework for teacher knowledge. *Teachers College Record, 108*(6), 1017–1054.

Puentedura, R. (2006). *Transformation, technology, and education* [Blog post]. Retrieved from http://hippasus.com/resources/tte/

Reich, K. (2013). Inklusive Didaktik –Konstruktivistische Didaktik. In K. Zierer, D. Lamres, & C. v. Ossietzky (Eds.), Handbuch für Allgemeine Didaktik 2013 (pp.133–149). : Schneider Verlag.

Reich, K. (2014). *Inklusive Didaktik: Bausteine für eine inklusive Schule.* Beltz Verlag.

Sharma, U. (2012). Changing pre-service teachers' beliefs to teach in inclusive classrooms in Victoria, Australia. *Australian Journal of Teacher Education, 37*(10), 53–66. https://doi.org/10.14221/ajte.2012v37n10.6

Sharma, U., & Jacobs, D. K. (2016). Predicting in-service educators' intentions to teach in inclusive classrooms in India and Australia. *Teaching and Teacher Education, 55*, 13–23. https://doi.org/10.1016/j.tate.2015.12.004

Stinken-Rösner, L., & Abels, S. (2021). Digitale Medien als Mittler im Spannungsfeld zwischen naturwissenschaftlichem Unterricht und inklusiver Pädagogik. In S. Hundertmark, X. Sun, S. Abels, A. Nehring, R. Schildknecht, V. Seremet, & C. Lindmeier (Eds.), *Naturwissenschaften und Inklusion, 4. Beiheft Sonderpädagogische Förderung heute* (pp. 161–175). Beltz Juventa.

Stinken-Rösner, L., Rott, L., Hundertmark, S., Baumann, T., Menthe, J., Hoffmann, T., Nehring, A., & Abels, S. (2020). Thinking inclusive science education from two perspectives: Inclusive pedagogy and science education. *Research in Subject-matter Teaching and Learning (RISTAL), 3*, 30–45.

Sweller, J. (2005). Implications of cognitive load theory for multimedia learning. In R. E. Mayer (Ed.), *The Cambridge handbook of multimedia learning* (pp. 19–30). Cambridge University Press.

Vogelsang, C., Finger, A., Laumann, D., & Thyssen, C. (2019). Vorerfahrungen, Einstellungen und motivationale Orientierungen als mögliche Einflussfaktoren auf den Einsatz digitaler Werkzeuge im naturwissenschaftlichen Unterricht. *Zeitschrift für Didaktik der Naturwissenschaften, 25*(1), 115–129.

Weidenhiller, P. (in preparation). *Fachspezifische Arbeitsweisen in heterogenen Klassen digital unterstützen: Eine Interventionsstudie mit Biologielehrkräften an Gymnasien und FOS/BOS.*

Weidenhiller, P., Witzke, S., & Nerdel, C. (2022). Das Apfelexperiment – Enzymkinetik der Apfelbräunung mit digitalen Tools messen. In E. Watts, L. Stinken-Rösner, & M. Meier (Eds.), *digital unterrichten. Biologie* (6/2022). Friedrich Verlag.

Chapter 24
Implementation Processes: Sustainable Integration of Biotechnology Experiments into Schools

Sara Großbruchhaus, Patricia Schöppner, and Claudia Nerdel

24.1 Theoretical Background

Teachers are challenged to stay up-to-date with rapidly growing knowledge and technology (Borko, 2004). Molecular biology is a fast-expanding field with the ongoing development of new applications (Martin et al., 2021). The underlying basics, such as polymerase chain reaction (PCR) or gel electrophoresis (GE) are integrated in the German school curriculum (ISB, 2015). However, the teaching of these basics remains at a solely theoretical level. Some possible reasons for this circumstance are schools' lack of the necessary equipment and reagents, teachers' low confidence and content knowledge (Nerdel & Schöppner, 2021; Hanegan & Bigler, 2009; Borgerding et al., 2013). Especially in interdisciplinary domains like biotechnology, professional development (PD) can bridge these gaps by connecting teaching practices with innovations (King, 2014; Merchie et al., 2018). The general assumption is that PD improves teaching quality and, thereby, students' outcomes if it is just effective enough. However, PD and lesson teaching take place in a multidimensional structure and are influenced by factors on different systemic levels beyond the lesson or PD event itself.

Various models have been developed to represent this complex interplay and are based on the process product paradigm (Brühwiler et al., 2017; Lipowsky, 2010). Lipowsky claims PD can be effective on up to five levels (2020). First, teachers

S. Großbruchhaus (✉) · P. Schöppner
Technische Universität München, Munich, Germany
e-mail: sara.grossbruchhaus@tum.de; patricia.schoeppner@tum.de

C. Nerdel
Life Sciences, Technical University of Munich, Munich, Germany
e-mail: nerdel@tum.de

K. Korfiatis et al. (eds.), *Shaping the Future of Biological Education Research*, Contributions from Biology Education Research, https://doi.org/10.1007/978-3-031-44792-1_24

must be satisfied with the PD so that further engagement with the content takes place. Second, the PD content should enhance teachers' cognition and knowledge. Third, the quality of teaching should increase when teachers implement the PD content. Fourth, students' outcomes improve by the implementation. The fifth level is positioned beside the other four and concerns school development, which can be stimulated by PD (Lipowsky & Rzejak, 2020). These levels are generally seen as a causal chain, although their causality has not yet been empirically proven (Davis et al., 2017).

Based on Lipowsky's model, several studies assessed different characteristics that enhance PD effectiveness and should be considered while planning a new PD, for example, the duration or teachers' active role (Sims & Fletcher-Wood, 2021).

Particularly in subject-specific PD, teachers seem to benefit from a pedagogical double play in which they anticipate themselves with the learning process. Those PD showed effects on both teachers' knowledge and classroom behaviour, if they were confronted with similar challenges during the PD as their students during lessons (Lipowsky & Rzejak, 2021). This form of cognitive activation can thus be counted among the quality features of PD that facilitate implementation of PD content.

Implementation research itself postulates the influence of additional factors, for example, support by the school administration (Gräsel & Parchmann, 2004). For implementation, which in a broader sense represents the third level of Lipowsky's model, effectiveness is not defined uniformly, but rather two approaches emerged: Defined by the number of teachers implementing it or the quality of implementation (Gräsel, 2010; Gale et al., 2020). Assuming causality in the Lipowsky model, high-quality PD should be followed by implementation. In order to understand what teachers actually perceive helpful about PD for implementation, we need to know their approach to implementation.

A PD programme focussing on molecular biology contexts addresses both keeping teachers up to date with growing amounts of knowledge in biotechnology and enable them to implement experiments into their lessons (Schöppner et al., 2022). Following Lipowsky's model and recommendations from presented studies, we designed the PD accordingly with a high focus on cognitive activation as teachers can perform the biotechnology experiments themselves during the PD (Nerdel & Schöppner, 2021). The PD has been evaluated on the first of Lipowsky's levels and teachers are satisfied (Nerdel & Schöppner, 2021). After participating, teachers can borrow the equipment needed for implementation, such as thermocycler and reagents, free of charge, which addresses the stated problem of schools lacking these (Schöppner et al., 2022). Based on the theory, we expected participation in pairs and at schools directly to reduce implementation barriers regarding implementation quantity (Gräsel & Parchmann, 2004). This study aims at this implementation (Lipowsky's layer four). With respect to the PD's goals, two main questions arise: How did they implement the biotechnology experiments into their classroom? To what extent did the PD influence the implementation procedure? Both questions link the Lipowsky layers two and three in order to get a deeper understanding of their causality. What occurs following the participation and borrowing of equipment by teachers? We examined how teachers execute molecular biology experiments at their schools. Insights into the process and the different formats teachers used for implementation are presented.

24.2 Method

24.2.1 Professional Development

The starting point of this study is an evaluated PD that addresses experiments in DNA analysis and is aimed at biology teachers. It was extensively described previously (Nerdel & Schöppner, 2021; Schöppner et al., 2022). We embedded the DNA analysis into four different contexts (s. Fig. 24.1) firmly connected to the Bavarian biology curriculum (ISB, 2015). The contexts vary in thematical and methodical difficulty. Therefore, teachers may choose one that meets their needs ideally. Figure 24.1 shows an overview of the contexts starting with the thematically and methodically easiest: Crime scene.

Implementation refers to teachers who borrowed the equipment and carried out the practical molecular biology methods presented in Fig. 24.1 with their students at school in biology classes. Teachers are entirely free in their realisation regarding grade level, time spent or student numbers. Within the PD, we presented the modules in such a way that they can be implemented either in a regular 90-min biology lesson, or in two successive biology lessons with a break after PCR, which widens the possible implementation formats.

Analysing the various implementation formats that emerged in detail should give first insights into Lipowsky's third layer implementing PD contents (Lipowsky & Rzejak, 2020).

Module 1: *Crime Scene*		polymerase chain reaction		agarose gel electrophoresis
Module 2: *Circadian Rhythm*	DNA extraction	polymerase chain reaction		agarose gel electrophoresis
Module 3: *Bitter Taste Reception*	DNA extraction	polymerase chain reaction	restriction digest	agarose gel electrophoresis
Module 4*: *Lactose Intolerance*	DNA extraction	polymerase chain reaction	restriction digest	agarose gel electrophoresis

Fig. 24.1 Within the PD, teachers learn the theoretical background and perform the experiments of two modules. The modules grow in difficulty in both methodical scope and theoretical complexity. Module 1 is the easiest, as the reagents used are prefabricated DNA samples. Module 4 is advanced, requiring extensive genetic knowledge and clear practical procedure to ensure visible results

24.2.2 Sampling

Since 2017, a total of 289 teachers from 98 secondary schools have participated in the PD. Teachers from 38 schools borrowed the equipment and implemented the experiments at least once; 20 of those schools implemented the experiments more often. Implementation is described on a school level, as mostly one teacher carries out the borrowing process. However, more teachers are involved in the implementation process. Both participation in the PD and implementation occurred in groups of teachers.

In order to achieve the broadest possible coverage of participants, we followed theoretical sampling (Flick, 2006, p. 73) to include all known variables and their combination:

1. *Secondary school type:* gymnasium, upper secondary school, vocational high school
2. *PD participation mode*: alone, with colleagues
3. *Locations of PD*: school, university
4. *Implementation mode*: alone, with colleagues

We have to add that the PD mainly addresses teachers from German higher secondary schools (Gymnasium) and was specially designed for their curriculum (Nerdel & Schöppner, 2021). However, we wanted to include viewpoints from other school types as they show a significant interest in both the PD and implementing the biotechnology experiments. Therefore, we recruited teachers for the interviews via email, and they participated voluntarily. We interviewed a total of 20 teachers from 18 schools who implemented the biotechnology experiments. One interviewee presented the most complex implementation process in this study. For broader insights, we recruited a second interview partner from the same school.

24.2.3 Descriptive Statistics of the Sample

For the interviews, we recruited teachers for each variable and its expression. We even found a new combination of variables: Teachers who implemented the modules practically into their classes without attending the PD. Instead, they were trained internally by participating colleagues.

Regarding the *implementation mode*, the number of 20 interviews was exceeded as some teachers implemented more often. If they implemented several times alone **or** with colleagues, we counted them together, but if they implemented several times alone **and** with colleagues, we counted them separately. The following list shows the variable coverage of our interview partners. Notably, those variables do not correlate, e.g., the 13 teachers who participated with colleagues are not the same 13 teachers who implemented cooperatively.

Secondary school type:		Participation mode:
Grammar school: 16		Alone: 5
Upper vocational school: 3		With colleagues: 13
Comprehensive high school: 1		Not: 2
Location of PD:		Implementation mode:
University: 10		Alone: 9
School: 10		With colleagues: 13

We stopped recruiting due to information saturation and are therefore not covering all possible combinations of these variables. However, when looking at the variables alone, their different expressions seem not to impact implementation as it takes place in each.

24.2.4   Interviews

We conducted semi-structured interviews. The order and concrete wording of the open questions could vary (Krüger et al., 2018, p. 125). We asked our interview partners to describe the implementation process directly: 'Please describe how implementation was carried out in your classroom/school.' We started with face-to-face interviews (in February 2020) but switched to phone interviews due to the COVID-19 pandemic. The last interviews took place in May 2020. The duration of the interviews was M = 20 min (SD = 10). The interviews were transcribed following simple rules based on Dresing and Pehl (2020).

24.2.5   Examine Teachers' Implementation Formats

We summarised each interview as an individual case (Kuckartz et al., 2008) and mapped the individual implementation processes by the typecasting strategy of Mayring (Mayring & Fenzl, 2019). This four-step analysis generalises statements to identify types or categories: (1) paraphrasing (remove language that does not carry information), (2) generalising (abstract statement to a consistent level), (3) selection (select all abstracted statements that carry relevant information), (4) integration (summarise all statements that carry the same information). During mapping, we followed these steps. We focussed on actions undertaken by teachers and worked in pairs to verify decisions by continuous communicative validation.

In this paper, we present the *implementation strategies* that emerged from type-casting the 20 interviews and focus on a particular case: two interviews of teachers from the same upper vocational school. We call the two teachers Anna and Lisa, regardless of their true gender. In more detail, we extracted their implementation procedure to examine the interrelationships of factors predicting PD effectiveness and their implementation outcome as best practice examples corresponding to the research question.

24.3 Results

24.3.1 Implementation Strategies

Within the 20 interviews, we could identify three main implementation formats: *regular lessons* (N = 9), *block lessons* (N = 9) and *special event* (N = 6).

Teachers implemented the biotechnology experiments within the standard time frame of biology lessons during regular lessons. This strategy is presented within the PD (Sect. 24.2.1). It was supposed to reduce the organisational effort. Within *block lessons,* experiments were implemented with an expanded time frame. Teachers either exempted students from their following lessons or organised an afternoon session. In both cases, the participation for students was mandatory. Within a *special event,* teachers organised a whole project day or week. Thereby the participation of students was voluntary. Voluntary afternoon events also fall into this category.

Some interviewees implemented several times: If they implemented following the same strategy, we counted them together. If they implemented the following different strategies, we counted them separately. Therefore, the total number of implementations counted exceeded 20 interviews.

We were able to find every implementation strategy within every variable and its expressions which we defined in Sect. 24.2.3, with one exception: Teachers who did not participate in the PD only implemented cooperatively. Additionally, *block lesson* tends to be implemented cooperatively. However, further studies are needed to verify that tendency based on the sample size (N = 20) and the fact that some implemented block lessons alone. For the *participation mode* and *location of the PD*, this data shows no tendency towards an implementation strategy.

This was further supported by teachers who repeatedly implemented as they tended to choose the same implementation strategy again, regardless of the variables they fall into. We identified three scenarios:

Teachers who implement PD into one class either in *regular lessons* or *block lessons,* repeat this in the following year(s) if they teach a suitable class again.

Second, teachers who implemented the experiments collaboratively with the whole biology faculty for all biology students in the suiting grade level and repeat this annually.

Third, teachers who implemented it collaboratively with their colleagues in either *block lessons* or *special events,* and additionally implemented experiments with another class within *regular lessons.* Hence, teachers seem to only choose different implementation strategies when the implementation takes place in the same year.

From that, we drew two conclusions: Firstly, the variables defined in Sect. 24.2.3. have no influence on the chosen implementation strategy by teachers with the one stated exception. Secondly, a new characteristic to hint at quality emerges: annual repetition. In itself, repetition is a quantitative characteristic, but the practice that accompanies it could increase the quality of implementation in the long term. This gives a significant indication of PD quality because the PD content seems to be accepted and suitable for teachers to use it repeatedly. Teachers who implement repeatedly have the potential to adapt and develop their implementation process by gaining practice and routine. For PD addressing a specific content, this could define a first measure of success, as annual repetition automatically serves implementation quantity, not by the number of teachers implementing, but the coverage of students reached.

The only variable that could enhance this quality definition is cooperation: having many teachers repeatedly implementing together enhances the possibility of strengthening the implementation process through cooperation features, for example, feedback and team support. Additionally, if all teachers work together, they can address the whole grade level and create equal opportunities for all students. Subsequently, this scenario has great potential for sustainable integration of the PD content into the school curriculum. In the interviews with three teachers from two different schools, we found this described scenario. That proves that single PD events can initiate school curriculum development (fifth layer of Lipowsky's model) through collaborative work on innovative subject content for the classroom. In the following section, we closely focus on the implementation procedure and further development of that one school as we consider it a best practice example.

24.3.2 One School as a Role Model: Sustainable Integration of PD Content into Curriculum

Anna is involved in teaching pre-service teachers and continuously seeks for innovation as she wants young teachers to learn new teaching concepts to stay up-to-date. Consequently, she learned about PD early on. In a first step, she discussed the PD offer with the other biology teachers at her school. As a result, they decided democratically to attend the PD and implement the content collaboratively.

> Anna came up with the idea and democratically put [it] up for debate. We all thought it made a lot of sense because our students do not have any opportunity to practise biotechnologically. – Lisa

The teachers had two primary motivations for implementing the experiments: They highly value the chance for students to experience practical biological work, but suffer from the lack of equipment. In line with that, all students of suitable grades (12th and 13th) who selected biology as a subject should experience practical work. In their view, *block lessons* were more suitable. They divided things into different tasks to manage such an important occasion. Anna took over the superordinate organisation: contacting the school administration and all the teachers affected, finding a suitable week, etc.

Of course, the school organisation is another hype, [...] it is difficult to schedule [this event] correctly. – Anna

Lisa and the others shared tasks like preparing teaching material and students theoretically. In 1 week within defined time frames, all the biology classes went through the practical work one after another, while the teachers shared supervision and received support from student teachers. Thus, the students were released from their regular classes during their participation. Students evaluated the event regarding their satisfaction with sticky dots. Subsequently, the biology faculty reviewed the event, discussed the procedure and decided on adaptions. They went through this overall process over the following years, further defining their implementation procedure and establishing it as a fixed event at their school.

In summary, it can be stated that the faculty had both well-established cooperative processes into which the PD content was included and a change agent (Anna), continuously looking for improvement. This enables the faculty to define implementation success and reduce implementation barriers.

Figure 24.2 gives an overview of these findings.

The faculty decided on the following adaptions:

In **year two,** 13th graders should be able to implement module 3 (circadian rhythm), as they know module 1 from the previous year. The PD content directly enables this procedure of teaching over different grade levels. We assigned the different contexts to the factor of *adaptivity,* allowing teachers to implement it into various thematic fields of the national curriculum. This faculty used this as a learning opportunity, for the reason that students can focus on the higher thematic complexity of the content, as they know the handling of experiments, e.g. pipetting,

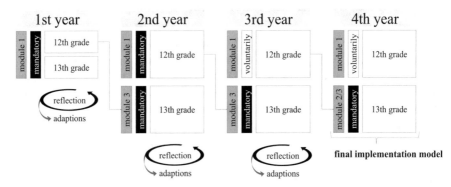

Fig. 24.2 Implementation strategy and its development over four years of an upper vocational school. First year: Whole biology faculty implemented the first module (crime scene), *blocked lessons* over one week, all biology students (12th and 13th grade). Second year: Same strategy, but 13th grade implemented module 3 (bitter taste perception). Third year: Same strategy, only 13th graders participated mandatorily, working groups reduced. Fourth year: 13th graders could choose between module 3 and 2 (circadian rhythm). This implementation module remained

from the previous year. In terms of the desired vertical connection of knowledge in school, this could directly influence teaching quality, the third layer of the Lipowsky model. During this implementation, no student teachers supported the biology faculty. They split the biology classes for implementation, so that one teacher supervises only half a class in the **third year**.

> The principal has now also approved having someone who does not teach a class still supervising students [at the event]. – Anna

This impressively showed how cooperation allows teachers to reduce implementation barriers by themselves and adapt their teaching flexibly. As the accessibility of student teachers could be an omnipresent barrier, this procedure is even more noticeable. Twelfth graders' participation was not mandatory anymore. This was to give the event a 'special flair', as it addresses only interested students, while assuring that everyone choosing biology in the 13th grade experiences biotechnology practically. With this change, their implementation procedure became a mixed version of *block lessons* and *special events* held in the same time frame. This combination of mandatory and voluntary participation could function as a factor for implementation quality by creating equal opportunity for everyone and simultaneously fostering interests. After the third implementation, the faculty was satisfied with the outcome. Despite many 12th graders joining the event, they had some remaining capacity and opened it for interested teachers with other subjects. They were rather surprised at how many of their colleagues participated and reflected on cross-connections of the experiments within the context of their own subject.

> In the meantime, [the organisational issues] have been well solved with a lot of talking and arguing within the school. – Lisa

In **year four**, 13th graders were able to choose the context they are most interested in, either circadian rhythm or bitter taste perception. The faculty's growing security with supervising the practical work made this flexibility possible. Within 4 years, this biology faculty established the experiments as an annually repeated event at their school and gained the support of other teachers, the principal and the whole school management. Consequently, they successfully integrated the PD content into their school curriculum. This allows us to draw three conclusions:

1. Regular repeated implementation (annually) is possible even if the PD is a one-time event. This is highly dependent on the teachers' school environment.
2. Teachers' scripts and beliefs (layer two) within a faculty could be understood as collective, affecting each other and contributing to the school environment. In this concrete case, this manifests in new teachers being obligated to participate in the PD and implement it cooperatively.
3. A faculty with established cooperative structures can reduce implementation barriers themselves.

24.4 Discussion

The PD underlying this study is in line with the constructivist view that learning with authentic contexts can increase students' scientific interest, in general, due to the close linkage of molecular biology topics with social issues, such as the COVID-19 pandemic (Nordqvist & Aronsson, 2019). With the option to borrow the equipment needed, teachers can allow students to analyse their own DNA and experience what it means to work in a molecular biology laboratory (Schöppner et al., 2022). Since this PD is a single event, it is affected by the critique of the current discussion on PD effectiveness, which generally questions the impact of such offers (Lipowsky & Rzejak, 2020). This study's goal was to evaluate teachers' implementation approaches to assess the PD's influence on implementation, which conforms to the overarching goal of the PD to counteract the missing equipment and enable teachers to implement molecular biology basics practically at school (Nerdel & Schöppner, 2021; Huang et al., 2018). As a starting point, we draw on Lipowsky's model of PD effectiveness (Lipowsky & Rzejak, 2020).

Our biotechnology content was presented in a regular lesson format. This should reduce implementation barriers due to connecting the content to classroom instruction and bringing innovation in line with existing teaching practice (Yurtseven Avci et al., 2020; Gräsel & Parchmann, 2004). Contrarily, we found more teachers that used other implementation strategies instead: *block lessons* or *special events*. Either the organisational effort in classroom management is not as crucial for our PD or overpowered by other factors influencing the implementation of the experiments, e.g. equipment borrowing. To make a conclusive statement about these linkages, we need to take a closer look at teachers' decision-making processes and reasoning. From our analysis, a new factor emerged: annual repetition. PD represents one of three main themes contributing to curriculum development (Langelotz & Olin, 2022). Annual repetition is the first step towards curriculum development, which, according to Lipowsky's model, is the fifth layer impacted by PD (2020). Curriculum development is a collaborative practice, and teachers remain the main agents in this process (Langelotz & Olin, 2022). In the past few decades, a conceptual change to merging top-down and bottom-up strategies shifted decision-making competencies and responsibilities to the individual school level (Maier-Röseler & Maulbetsch, 2022). Some researchers went a step further and expected teachers to not only implement innovation, but shape and influence the respective development themselves (Kneen et al., 2021). The presented case study impressively shows that teachers can shape their school development process with organisational communication on several school levels. We extracted several influencing factors, which are in line with current literature: change agent, highly cooperative structures, joint mission (a.o. Fussangel & Gräsel, 2009). Future studies should analyse the whole sample of teachers who implemented the biotechnology experiments in school for further evidence of correlations and their influence on implementation quantity. Nevertheless, those factors could influence the implementation strategies chosen by the teachers.

All these results must be confirmed by future studies in different PD settings and other subjects, as this study is limited by both sample size and the indications for teachers' self-statements. Nevertheless, our data first shed light on the implementation process's complexity. We could extract that *annual repetition* is a suitable predictor for PD effectiveness in the form of implementation that opens opportunities for school development via new topics and methods in the biology school curriculum. Thereby, we could demonstrate that a single PD addressing a certain topic can initiate school curriculum development if certain conditions are present, for instance, embodying a joint mission and cooperative structures at school. In the case presented, this could even lead to a feedback loop from school practice to PD because new teachers must contribute to the established PD content and are expected to participate in the PD. In terms of the Lipowsky model, this reveals a complex view on the postulated causality, because higher layers (e.g. implementation) influence lower layers (e.g. participation) in PD. Thus, at least for the present PD concept, it could be shown that one-time PDs are legitimate in the teacher training landscape.

References

Borgerding, L. A., Sadler, T. D., & Koroly, M. J. (2013). Teachers' concerns about biotechnology education. *Journal of Science Education and Technology, 22*(2), 133–147. https://doi.org/10.1007/s10956-012-9382-z

Borko, H. (2004). Professional development and teacher learning: Mapping the terrain. *Educational Researcher, 33*(8), 3–15. https://doi.org/10.3102/0013189X033008003

Brühwiler, C., Helmke, A., & Schrader, F.-W. (2017). Determinanten der Schulleistung. In M. K. W. Schweer (Ed.), *Lehrer-Schüler-Interaktion* (pp. 291–314). Springer Fachmedien Wiesbaden.

Davis, E. A., Palincsar, A. S., Smith, P. S., Arias, A. M., & Kademian, S. M. (2017). Educative curriculum materials: Uptake, impact, and implications for research and design. *Educational Researcher, 46*(6), 293–304.

Dresing, T., & Pehl, T. (2020). Transkription. In G. Mey & K. Mruck (Eds.), *Handbuch Qualitative Forschung in der Psychologie* (Band 2: Designs und Verfahren. 2., erw. u. überarb. Auflage 2020) (pp. 835–854). Wiesbaden.

Flick, U. (2006). *Qualitative Sozialforschung*. (4. Aufl). Rowohlt Taschenbuch Verlag.

Fussangel, K., & Gräsel, C. (2009). Die Kooperation in schulübergreifenden Lerngemeinschaften. Die Arbeit der Sets im Projekt „Chemie im Kontext ". *Kooperation und Netzwerkbildung. Strategien zur Qualitätsentwicklung in Schulen. Seelze,* 120–131.

Gale, J., Alemdar, M., Lingle, J., & Newton, S. (2020). Exploring critical components of an integrated STEM curriculum: An application of the innovation implementation framework. *IJ STEM Ed, 7*(1), 1–17. https://doi.org/10.1186/s40594-020-0204-1

Gräsel, C. (2010). Stichwort: Transfer und Transferforschung im Bildungsbereich. *Zeitschrift für Erziehungswissenschaft, 13*(1), 7–20. https://doi.org/10.1007/s11618-010-0109-8

Gräsel, C., & Parchmann, I. (2004). Implementationsforschung-oder: der steinige Weg, Unterricht zu verändern. *Unterrichtswissenschaft, 32*(3), 196–214.

Hanegan, N. L., & Bigler, A. (2009). Infusing authentic inquiry into biotechnology. *Journal of Science Education and Technology, 18*(5), 393–401. https://doi.org/10.1007/s10956-009-9155-5

Huang, A., Nguyen, P. Q., Stark, J. C., Takahashi, M. K., Donghia, N., Ferrante, T., et al. (2018). BioBits™ explorer: A modular synthetic biology education kit. *Science Advances, 4*(8), eaat5105. https://doi.org/10.1126/sciadv.aat5105

ISB (Ed.). (2015). LehrplanPLUS. Staatsinstitut für Schulqualität und Bildungsforschung 2015. München.

King, F. (2014). Evaluating the impact of teacher professional development: An evidence-based framework. *Professional Development in Education, 40*(1), 89–111. https://doi.org/10.108 0/19415257.2013.823099

Kneen, J., Breeze, T., Thayer, E., John, V., & Davies-Barnes, S. (2021). Pioneer teachers: How far can individual teachers achieve agency within curriculum development? *Journal of Educational Change.* https://doi.org/10.1007/s10833-021-09441-3

Krüger, D., Parchmann, I., & Schecker, H. (Eds.). (2018). *Theorien in der naturwissenschaftsdidaktischen Forschung.* Springer.

Kuckartz, U., Dresing, T., Rädiker, S., & Stefer, C. (2008). *Qualitative evaluation. Der Einstieg in die Praxis.* 2., aktualisierte Auflage. VS Verlag für Sozialwissenschaften.

Langelotz, L., & Olin, A. (2022). Action research and curriculum development with consideration of the Nordic context. In L. Langelotz & A. Olin (Eds.), *Oxford research encyclopedia of education.* Oxford University Press.

Lipowsky, F. (2010). Lernen im Beruf–Empirische Befunde zur Wirksamkeit von Lehrerfortbildung. *Lehrerinnen und Lehrer lernen. Konzepte und Befunde zur Lehrerfortbildung, 1,* 51–72.

Lipowsky, F., & Rzejak, D. (2020). Was macht Fortbildung für Lehkräfte erfolgreich? – Ein Update. In: B. Groot-Wilken & R. Koerber (Eds.), Nachhaltige Professionalisierung für Lehrerinnen und Lehrer. Ideen, Entwicklungen, Konzepte (pp. 15–56). wbv (Beiträge zur Schulentwicklung).

Lipowsky, F., & Rzejak, D. (2021). Welche Art von Fortbildung wirkt? *Was Lehrkräfte lernen müssen – Bedarfe der Lehrkräftefortbildung in Deutschland* (pp. 19–38). Netzwerk Bildung – Friedrich Ebert Stiftung.

Maier-Röseler, M., & Maulbetsch, C. (2022). Schulentwicklung and Schulentwicklungsforschung im Dialog – Meta-Reflexion als Transferstrategie. *Bildungsforschung: Gemeinsam mit Bildungspraxis? Wege, Dynamiken, Klärungen* (2), 1–14.

Martin, D. K., Vicente, O., Beccari, T., Kellermayer, M., Koller, M., Lal, R., et al. (2021). A brief overview of global biotechnology. *Biotechnology & Biotechnological Equipment, 35*(sup1), 5–14. https://doi.org/10.1080/13102818.2021.1878933

Mayring, P., & Fenzl, T. (2019). Qualitative Inhaltsanalyse. In N. Baur & J. Blasius (Eds.), *Handbuch Methoden der empirischen Sozialforschung* (pp. 633–648). Springer VS. https:// link.springer.com/chapter/10.1007/978-3-658-21308-4_42

Merchie, E., Tuytens, M., Devos, G., & Vanderlinde, R. (2018). Evaluating teachers' professional development initiatives: Towards an extended evaluative framework. *Research Papers in Education, 33*(2), 143–168. https://doi.org/10.1080/02671522.2016.1271003

Nerdel, C., & Schöppner, P. (2021). Evaluation einer Lehrerfortbildung zum praktischen Einsatz von biotechnologischen Methoden im Unterricht. In: S. Kapelari, A. Möller & P. Schmiemann (Eds.), "Naturwissenschaftliche Kompetenzen in der Gesellschaft von morgen". Internationale Jahrestagung der Fachsektion Didaktik der Biologie im VBIO und der Gesellschaft für Didaktik der Chemie und Physik, Wien 2019 (pp. 292–305). StudienVerlag (Lehr- und Lernforschung in der Biologiedidaktik, Band 9).

Nordqvist, O., & Aronsson, H. (2019). It is time for a new direction in biotechnology education research. *Biochemistry and Molecular Biology Education: A Bimonthly Publication of the International Union of Biochemistry and Molecular Biology, 47*(2), 189–200. https://doi. org/10.1002/bmb.21214

Schöppner, P., Großbruchhaus, S., & Nerdel, C. (2022). *Biotechnologie Praxisorientiert Unterrichten. Aktuelle Kontexte für Schule und Lehrerfortbildung.* Springer Spektrum.

Sims, S., & Fletcher-Wood, H. (2021). Identifying the characteristics of effective teacher professional development: A critical review. *School Effectiveness and School Improvement, 32*(1), 47–63. https://doi.org/10.1080/09243453.2020.1772841

Yurtseven Avci, Z., O'Dwyer, L. M., & Lawson, J. (2020). Designing effective professional development for technology integration in schools. *Journal of Computer Assisted Learning, 36*(2), 160–177. https://doi.org/10.1111/jcal.12394

Printed in the United States
by Baker & Taylor Publisher Services